The Professional Forecaster

The
Professional
Forecaster:

The Forecasting Process Through Data Analysis

James P. Cleary
Hans Levenbach

 LIFETIME LEARNING PUBLICATIONS
Belmont, California

A division of Wadsworth, Inc.

To my wife Lee and my children Beth and Kenneth.   JC

To my father and my children Jody and Amy.   HL

Designer: Richard Kharibian
Developmental Editor: Kirk Sargent

1 2 3 4 5 6 7 8 9 10———86 85 84 83 82

Library of Congress Cataloging in Publication Data

Cleary, James P.
  The professional forecaster.

  Bibliography: p.
  Includes index.
  1. Economic forecasting.   2. Business forecasting.
3. Forecasting.   I.   II. Title
HB3730.C546   338.5'442   81-13717
ISBN 0-534-97960-2   AACR2

Contents

vi   Contents

"Those who live by the crystal ball should learn to eat ground glass."
Ted Moskovitz, Economist

Preface

In recent years the need for improved statistical forecasting techniques in business and government has become increasingly clear, in part because of the uncertainty and frequency of change in the economic and financial worlds. At the same time, improved computer systems have been developed. It is now far easier to apply this technology to the planning and management of change than it was in the past; fortunately for the professional forecaster, computer-based techniques have greatly simplified access to data bases and quantitative techniques so that a wide variety of methods can be applied in a relatively short time at a reasonable cost. Still, the forecaster can easily be overwhelmed by the multitude of powerful (often newly discovered) forecasting techniques that are not readily classifiable into easy-to-understand methodologies. Moreover, even though access to these techniques is becoming easier for the professional forecaster, the manager or user of the forecasting process has until now been offered little guidance in how to make effective and appropriate applications of these techniques.

We have written *The Professional Forecaster* in order to help professional forecasters understand differences between many recently devised forecasting techniques and to provide managers or forecast users with guidance in applying them.

The Professional Forecaster is the second volume of a two-volume work. The first volume, entitled *The Beginning Forecaster,* describes a number of basic forecasting methods applicable to a wide variety of forecasting problems; the beginner will also find that the methods explained in that book provide initial models that can be used in conjunction with more complex models, to measure the improvements that can be achieved as a forecaster builds increasingly complex models.

The *Professional Forecaster* extends topics of the first volume that will be of interest to the *experienced* forecaster. It describes up-to-date statistical forecasting tools for forecasters experienced in modern data analysis, elements of robust/resistant methods, and the basics of regression analysis. However, we think that even experienced forecasters will find there is an advantage to beginning with Volume 1, since its development forms the basis for Volume 2.

In both volumes, we have emphasized the following:

- Establishment of a *process* for effective forecasting. Specific methods and techniques are presented within the context of the overall process.

- Selection of the forecasting and analytical techniques *most appropriate for any given problem*. The methods discussed, many representing the current state of the art, are the ones that have proved to be most useful and reliable to us as practicing forecasters.

- Refocusing the attention of practitioners *away from the mechanistic execution of computer programs* and towards a greater understanding of data and the processes generating data.

- Preliminary *analysis of data* before attempting to build models. Computer-generated graphic displays enable you to see in one picture what you might otherwise have to glean from a stack of computer printouts.

- Use of *robust/resistant methods* in addition to traditional methods to provide insurance that a few bad data values do not seriously distort the conclusions that are reached. Experience with a wide variety of practical applications has convinced us that data rarely behave well enough for the direct application of conventional modeling assumptions. The robust/resistant methods produce results that are less subject to the distortions caused by a few outlying data values. By comparing traditional and robust results, the practitioner is in a better position to decide which are most appropriate for the problem at hand.

- Performance of *residual analyses* to determine what "unexplained" variations might tell about the adequacy of the model. As in data analysis, the importance and usefulness of displaying data in residual analyses are emphasized throughout as essential in all phases of any effective model-building effort.

In addition, this volume shows how the results from the traditionally diverse fields of time series and econometric modeling can be combined into a decisive forecast, which can then be presented authoritatively and credibly.

A number of forecasting methods useful to analysts are not covered explicitly in either volume. The omitted methods are typically used when data are scarce or nonexistent. As an example, the whole area known as technological forecasting, which requires a grounding in probabilistic (in contrast to statistical) concepts, is not treated. Likewise, new-product forecasting, for which data are unavailable, also falls in this category. Since our volumes deal with exploratory data analysis along with confirmatory modeling, we have emphasized techniques for which a reasonable amount of data are available or can be collected.

Some practitioners may feel that we have given greater emphasis to data-analytic concepts than is necessary. However, many practicing forecasters and writers on forecasting methods tend to concentrate on making models more complex rather than keeping them simple: sources of forecast errors are difficult to analyze with increasingly complex models. Our experience suggests that, in practice, the undoing of many forecasting efforts begins with flaws in the quality and handling of data rather than in the lack of modeling sophistication. Thus an objective in both *The Beginning Forecaster* and *The Professional Forecaster* is to place greater emphasis on data-analytic methods (much of it intuitive and graphical) as a key to improved forecasting.

The *Professional Forecaster* is divided into five parts. The first part comprises two chapters that introduce the forecasting process. In Part 2, four chapters deal with regression methods for developing functional relationships between one or more variables—such methods are basic to all quantitative forecasting applications—and the uses and interrelationships of these methods: Chapter 3 treats some examples of short-term forecasting models as a motivation to Chapter 4, which deals with fitting and evaluating multiple linear regression models; Chapter 5 covers the key role of residual analysis, and its role in dealing with outliers; and Chapter 6 develops trend models for long-term forecasting applications.

Part 3 contains ten chapters dealing with demand analysis and econometrics. This part begins with an introduction to demand analysis (Chapter 7), then treats estimation of demand elasticities (Chapter 8), discusses demand analysis as a forecasting process (Chapter 9), and presents two case-study applications (Chapter 10). The econometric approach is treated in Chapters 11–15. This approach includes the use of dummy and lagged variables (Chapter 12), adjusting for serial correlation (Chapter 13), and other single-equation specification topics (Chapter 14). Specifying a system of econometric equations is introduced in Chapter 15. Chapter 16 serves as a bridge to Part 4 and deals with the pooling of cross-sectional and time series data.

Part 4 contains eight chapters on time series modeling that, to a large extent, use the Box-Jenkins methodology. The identification, estimation, and diagnostic checking of the ARIMA class of univariate time series models are covered extensively in Chapters 17–22. Chapter 23 deals with some useful multivariate techniques for forecasting time series data, and Chapter 24 presents two applications of these techniques.

The last part of the book summarizes management principles on which the forecasting process is based. Chapter 25 deals with measuring forecast performance, and Chapter 26 covers the management of the forecasting process itself.

Examples are used wherever possible throughout the book. The examples are predominantly drawn from the experience of the authors in the telecommunications business. While this may be distracting to some readers, it should be noted that *the characteristics of the data are what are important for the example,* not the fact that the data are or are not telephone-related. Other data sets from nontelecommunications sources have also been used where appropriate, to make certain points or illustrate a particular technique.

James P. Cleary
Hans Levenbach

Acknowledgments

The authors are indebted to a number of people in forecasting organizations throughout the Bell System whose involvement and contribution can only be indirectly recognized. Over the past decade several members of the technical staff at Bell Laboratories have made significant contributions to the methodological and computer software development of statistical forecasting in the Bell System. In particular, the efforts of Bill Brelsford, Dave Preston, Jim Inglis, and Pramila Agarwala stand out in this regard. Their software contributions are reflected in Bell System training courses and in the modeling work of many practicing forecasters in the Bell System.

The training courses provided a great deal of impetus to the introduction of improved quantitative forecasting techniques in the Bell System. This widespread acceptance of statistical forecasting by upper level management at AT&T and the Associated Telephone Companies has necessitated the introduction of management techniques in forecasting, to maximize forecast usefulness at minimum cost to the System. Some of the techniques discussed in this book are representative of this practice.

There are a number of individuals who have assisted the authors in their review of the manuscript. Bob Brousseau, Min-te Chao, Walt Paczkowski, Bapi Sen, and Monte Shultes of AT&T's Analytical Support Center provided detailed comments on at least one chapter apiece. Joe McCabe and Sam Thomas of New York Telephone, and Lorraine Denby, Jim Inglis, and Dave Preston of Bell Laboratories were also very helpful in reviewing specific chapters. Sherry McArdle Karas and Pramila Agarwala of the AT&T Analytical Support Center; and Liz Aquilinas and Josephine O'Connor of New York Telephone were instrumental in organizing the data files, running examples on the computer, and generating many of the visuals via computer graphics.

The authors would like to express their appreciation and indebtedness to the late Sir Ronald A. Fisher, F.R.S., Cambridge, and to Hafner Publishing Co., for permission to reprint Table IV from their book, *Statistical Methods For Research Workers;* to Professor E. S. Pearson and the *Biometrika* trustees for permission to reproduce the materials in Appendix A, Tables 1, 3, and 4; and to Professors J. Durbin and G. S. Watson for the values in Appendix A, Table 5.

The authors would also like to express their appreciation to the editorial and production staff of Lifetime Learning Publications for their courteous cooperation in the production of this book; to Lenore Pahler, Joan Mendez, and various members of the AT&T Word Processing Staff who typed the manuscript; and above all to Kirk Sargent for his numerous and valuable editorial suggestions for improving the text.

Lastly, the authors owe thanks to the management of the respective organizations in which they worked as forecasters and forecast managers. The Residence and Business Forecasting organizations, the Analytical Support Center, and the Demand Analysis group at AT&T have over the years provided leadership in the development of courses and in the implementation of quantitative techniques like those described in both volumes. However, any errors, obscurities, or omissions remain the sole responsibility of the authors. Any procedures described in this book should not necessarily be interpreted as representative of official forecasting practices of AT&T or its Associated Telephone Companies.

J. P. C.

H. L.

Phase 1: Design

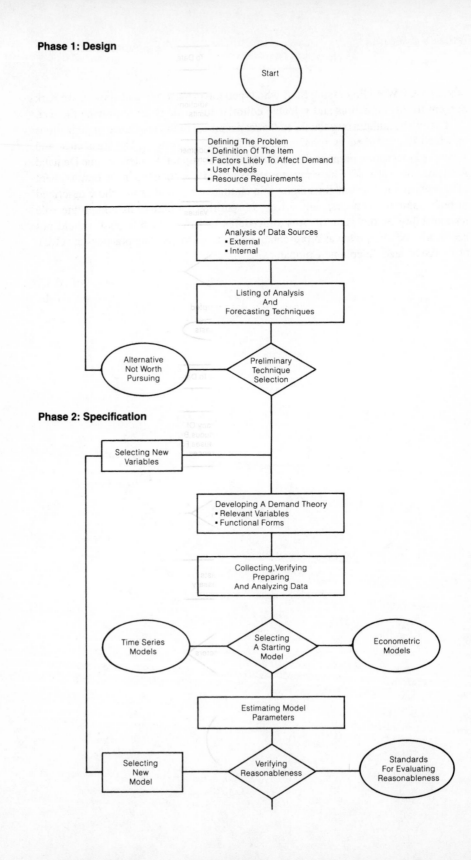

Phase 2: Specification

Phase 3: Evaluation

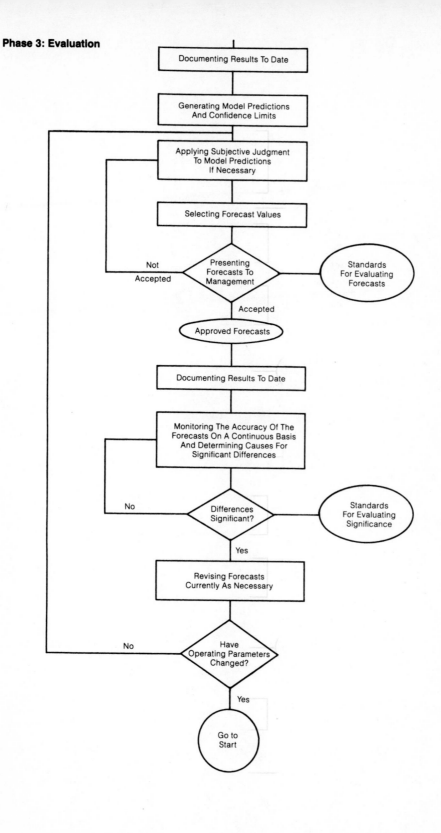

Part 1

The Forecasting Process

CHAPTER 1

Designing a Forecasting Process

A *forecasting process* can be described in terms of three phases:

- *Design phase;* the premodeling activities associated with some problem and the evaluating of cost-benefit tradeoffs that must be considered when building statistical models.

- *Specification phase;* the model-building activities.

- *Evaluation phase;* the forecasting and tracking activities that follow development of a model.

FORECASTING AS A PROCESS

Forecasting is a systematic process of decisions and actions performed in an effort to predict the future. More precisely, forecasting attempts to predict change. If future events represented only a readily quantifiable change from historical events, future events or conditions could be predicted through quantitative projections of historical trends into the future. Methodologies that are used to describe historical events with mathematical equations (or with a mathematical model) for the purpose of predicting future events are classified as *quantitative projection techniques*. However, there is much more to forecasting than projecting past trends.

A forecast is not an end product but rather an input to the decision-making process. A forecast is a prediction of what will happen—often, a prediction of future values of one or more variables—under an assumed set of circumstances; often, the assumed circumstances are that "business as usual" conditions will continue. A forecast that is a prediction of future values of one or more variables under "business as usual" conditions is often referred to as the *status quo* or "base" case. Forecasts are also required for a variety of "what if" situations and for the formulation of business plans to alter base-case projections that have proved unsatisfactory. The probability that these formulations will be successful can be greatly enhanced by

3

following a process that identifies all of the major activities of a business and their sequence, so that important considerations are not overlooked.

Typically, a forecasting problem is first formulated when a forecast user contributes information about the *environment* in which the user's business or institution operates and plans future operations, and about *probable variables* for consideration in projecting that planning into the future. The forecaster has familiarity with the forecasting process and recommends the specific forecasting techniques that are the most appropriate for the user's needs.

A structured method for attacking any business-forecasting problem generally leads to a better understanding of the factors that influence demand. These factors include demographic, economic, political, competitive, and pricing considerations, and many others. With a better understanding of these factors, a forecaster can become more expert in making or evaluating demand forecasts and in applying them to specific problems.

An important advantage in emphasizing the process by which forecasting is done is that it focuses attention on the right technique or techniques to be used for a given forecasting problem (Chambers, et al., 1974). Before you attempt to defend a forecast in fine detail, it is important to have ascertained that the approach you have used is appropriate for the problem. If a forecaster is using the right techniques, then he or she will arrive at the best forecast possible at the time a forecast is made. Of course, this also assumes that the forecaster *understands the methodologies* and can *exercise sound judgment* throughout the process.

The quantitative projection techniques presented in the present volume are applicable to forecasts of economic trends; market size and characteristics; business competition; finances and revenues; expenses and capital expenditures; marketing, and the demand for products and services; and production and inventory control. These techniques were introduced in *The Beginning Forecaster,* by Hans Levenbach and James P. Cleary, Lifetime Learning Publications, 1981.

The demand analysis techniques presented in the present volume allow for testing and confirmation of economic hypotheses. A frequent use of such techniques is to estimate the impact of changes in price on the demand for products and services. These techniques usually imply the existence of a sufficient amount of accurate data to fit models with an acceptable degree of precision.

THE DESIGN PHASE

We recommend a five-stage procedure for the design phase of a forecasting process, with each stage consisting of a series of activities, actions, and judgments. The five stages are:

- Defining the problem.
- Listing alternative forecasting techniques.

- Selecting among the alternatives for study.
- Evaluating the alternatives.
- Recommending the most appropriate technique(s).

Defining the Problem

Problem definition is a critical stage of any project. It is necessary in this stage to define what is to be done and to establish the criteria for successful completion of the project or forecast. Agreements are essential on the required outputs, time, and money to be devoted to solving the problem, the resources that will be made available, the time when a solution to the problem will be required, and the level of accuracy that may be achievable. The data analysis, model building, and forecasting steps follow, after these kinds of agreements are reached.

The specific operations that must be performed in these and other stages of the forecasting process are diagrammed in the Flowchart in the front of the book. It is clear from the Flowchart that the steps are sequential and iterative in nature. For any item that is to be forecast, there is a start point for the first forecast made of that item, but there is no end point until either the product no longer exists in its original form or there is no longer a need to forecast it.

The Flowchart highlights the activities involved in defining a problem and in deciding if a quantitative model-building approach is worthwhile. The purpose of this stage is to understand the real problem, as opposed to its symptoms, and to determine if cost-effective forecasts can be provided to assist in reaching a planning or operating decision.

Since all quantitative forecasting methods require data, the next step is to search through appropriate data sources. Corporate books will normally contain a reasonably rich history of *internal data* on revenues, expenses, capital expenditures, product sales, prices, marketing expenses, and so forth. *External data* are increasingly available from government publications and consulting firms in the areas of demographics (age, race, sex, households, etc.), employment, economic indicators, and related variables. Forecasts of these items for the United States have been available for some time, and now regional, state, and, in some cases, SMSA (Standard Metropolitan Statistical Areas) forecasts can also be obtained.

Listing Alternative Forecasting Techniques

The second stage of the design phase involves listing the specific analyses and forecasting techniques that may be of value in attacking a problem. The objective is to list all possible approaches. This is sometimes referred to as "green-lighting," since the goal is to develop a comprehensive list and not to make value judgments that would inhibit creativity. In subsequent stages, criteria are established and a screening process is implemented to limit the alternatives that have been studied to those that have the greatest likelihood of succeeding.

In Chapter 2 is shown a table that lists the forecasting techniques that are most commonly used and that may be appropriate for a given problem (Table 2.1). The techniques can be categorized as being either qualitative or quantitative in nature.

Qualitative projection techniques provide a framework within which quantitative techniques or other forms of quantitative analyses, such as decision trees and linear programming, are brought to bear on a particular problem. The objective of qualitative techniques is to bring together in a logical, unbiased, and systematic way all information and judgments that relate to the factors of interest. These techniques use human judgment and rating schemes to turn qualitative information into quantitative estimates. Qualitative techniques are most commonly used in forecasting things for which the amount, type, or quality of historical data is limited.

With the availability of sufficient and accurate data comes the possibility of using *quantitative projection techniques*. Such quantitative techniques can be classified into two major approaches. The *filtering* (time series) techniques focus entirely on patterns, pattern changes, and disturbances caused by random influences. The *causal* (econometric) techniques reflect the identification of explicit relationships between a variable that is to be forecast and other relevant, determining variables.

Time Series Analysis: A time series is a set of chronologically ordered points of raw data; an example would be revenue received, by month, for several years. An assumption often made in a time series analysis is that the factors that caused demand in the past will persist into the future (see, for example, Granger and Newbold, 1976).

Time series analysis also helps to identify and explain any regularly recurring or systematic variation due to *seasonality* in the data. In most cases, sales forecasters deal with monthly seasonality that occurs within a year. This is usually related to weather and human custom. Economic or business forecasters more often deal with quarterly time series.

Time series analysis also helps to identify trends in data and the growth rates of these trends. By *trend* is meant the basic tendency of quantities of some item to grow or decline over a long time period. For many consumer products, for example, the prime determinant of trend is household growth (change in number of households).

Finally, time series analysis can help to identify and explain *cyclical patterns* in the data, which repeat roughly every two, or three, or more years. A cycle is usually irregular in depth and duration, and tends to correspond to changes in economic expansions and contractions. Such a change is commonly referred to as the "business cycle."

Econometrics: The econometric approach may be viewed as a "causal" approach. Its purpose is to identify the factors responsible for demand. Econometric models of the U.S. economy, for example, are very sophisticated, and represent one extreme of econometric modeling. These models are built to depict the essential quantitative relationships that determine output, income, employment, and prices. The econometric examples presented in this book will be considerably smaller in scope than the large-scale models available in the literature (see, for example, Hickman, 1975).

It is general practice in econometric modeling to remove only the seasonal component of data prior to modeling. The trend and cyclical movements in the data should then be explainable by economic and demographic theories. Factors that are believed to influence demand are identified and forecast separately. Unlike what is done in the time series approach, in econometric modeling the assumption is generally not made that the factors that influenced demand in the past will persist into the future.

Different kinds of time series analyses: Within the listings of time series projection techniques, there are essentially two schools of thought regarding the analysis and prediction of time series. The first school of thought is best illustrated by the *time series decomposition technique*. The primary assumption on which this is based is that a time series can be decomposed into four components (trend, seasonal, cyclical, and irregular), and that the components can then be analyzed and projected into the future on an individual basis, one by one. The projection of the time series is then merely the combination of the projections for the components.

A second school of thought is associated with the *Box-Jenkins modeling approach*. This school of thought is purely statistical in nature (and based in control engineering), and does not explicitly assume that a time series is represented by the composition of separate components.

Selecting among the Alternatives for Study

The selection process begins with an understanding of the nature of the forecasting problem as it relates to the business function under consideration.

Executive managers, planners, and production managers, for example, have differing needs in terms of:

- The time horizon of a forecast.
- Accuracy requirements.
- The level of detail in a forecast.
- The quantity of forecasts that will be required.

Other considerations include the willingness of the user to accept the techniques that are recommended, and the compatibility of an approach or forecast with a user's planning processes.

The *time horizon* refers to the period of time into the future for which forecasts are required. The periods are generally short term (one to three months), medium term (three months to two years), and long term (more than two years). Wheelwright and Makridakis (1980) also refer to a time horizon equivalent to an immediate term (less than one month).

The *accuracy requirements* for a forecast are normally related to the cost of forecast error. In the case of an inventory control problem involving numerous relatively inexpensive parts that are readily available, the requirement for accuracy will be less restrictive than it is for an inventory control problem involving very expensive parts with long lead-times between order placement and delivery. In the latter case, the production line could be adversely affected, and sales could be lost to competitors, if the inventory were not adequate. Alternatively, an inventory level that is too great will cause a business firm to incur large carrying charges that could result in its need to raise prices to maintain profitability.

The particular situation will determine accuracy requirements for a forecast. It is not uncommon for a forecast user to expect accuracy levels that cannot be realistically achieved. It is up to the forecaster, as advisor, to state what can and cannot be achieved. The user may well have to establish contingency plans to deal with the potential imprecision in the forecast.

Level of detail is important because forecasting techniques that are appropriate for aggregate levels of demand may not provide satisfactory results for an individual product. For example, total automobile sales for all manufacturers may be related to certain aggregate economic variables, such as real income or Gross National Product (GNP). This relationship may not be an adequate way for an individual manufacturer to determine its share of the market, much less the sales for a given product.

Related to the level of detail is the *quantity of forecasts* that will be required. As a forecast problem moves from executive management to production line management, the quantity of all possible component-forecasts that could be supplied usually grows in a nonlinear manner. Resources are generally not available to devote the full range of modeling techniques to what could amount to thousands of required forecasts. Instead, simpler, more mechanical procedures must be applied. Then it is necessary for checks of reasonableness to be established to make sure that the sum of the piece parts is in reasonable agreement with more sophisticated forecasts of aggregate demand levels.

Evaluating the Alternatives

In evaluating alternatives, a forecaster will consider patterns or characteristics of data that will influence the final selection of the techniques for use. For example, which techniques are more accurate in predicting turning points or in predicting stable periods? Which techniques have the best overall accuracy when the forecast is tested? Do the models tend to overpredict or underpredict in given situations? Are the short-term predictions of one model better than another? How about long-term predictions? Do the coefficients of the model need to be evaluated in terms of sign or magnitude?

A forecaster should be careful to avoid using techniques based on assumptions that do not match data characteristics. Thus a basic principle is to utilize more than one projection technique. By observing this principle during the actual development of a forecast, a forecaster can be reasonably assured of avoiding biases that are inherent in a single projection technique or in the way in which that technique is used.

The purpose for using more than one technique is to ensure that the forecasting approach will remain as flexible as possible, and that the forecaster's judgment (which is so critical to the forecasting process) is not overly dependent upon one particular projection technique: it is not uncommon for a forecaster to develop a preference for one forecasting technique over another, and then to use that technique almost exclusively even in a new situation. Such a preference could be easily established because of the highly specialized nature of some of the techniques.

There is likewise a tendency for forecasters to use the most statistically sophisticated techniques that can be found. In many cases, this tendency can greatly reduce the effectiveness of the forecasting process because some sophisticated techniques are unresponsive to drastic pattern-changes in the time series. The degree of precision for a projection technique is not necessarily a direct function of the degree of its sophistication.

We recommend that *two or more* projection techniques be used to describe the historical behavior of data and to project this for the future. When a forecaster has a good working knowledge of the capabilities and limitations of the various projection techniques, the benefits of using more than one technique can be significant. In essence, you will be able to evaluate alternative views of the future. It is also possible to establish a probability level for each alternative. A comparison can then be made of the alternative views of the future; hence the chances that the selected forecast is reasonable will be enhanced.

SUMMARY

This chapter has introduced the design phase of the overall forecasting process that will be referred to throughout the book. The five steps are

- Defining the problem.
- Listing alternative forecasting techniques.
- Selecting among the alternatives for study.
- Evaluating the alternatives.
- Recommending the most appropriate technique(s).

In *The Beginning Forecaster,* simple linear regression and exponential smoothing techniques were presented as basic methods for time series decomposition and forecasting. In the present volume, these basics are extended to include (causal) econometric modeling and ARIMA time series modeling (Box-Jenkins approach). While details for specific forecasting and evaluation techniques may be different, the generic considerations that are made at each stage of the process are similar.

The Flowchart for the forecasting process helps to identify

- Where in the process you are at any time.
- The considerations to be made at that time.
- The next step to take along the way.

USEFUL READING

CHAMBERS, J. C., S. K. MULLICK, and D. D. SMITH (1974). *An Executive's Guide to Forecasting*. New York, NY: John Wiley and Sons.

GRANGER, C. W. J., and P. NEWBOLD (1976). *Forecasting Economic Time Series*. New York, NY: Academic Press.

HICKMAN, B. G. (1975). *Econometric Models of Cyclical Behavior*. New York, NY: Columbia University Press.

LEVENBACH, H., and J. P. CLEARY (1981). *The Beginning Forecaster*. Belmont, CA: Lifetime Learning Publications.

WHEELWRIGHT, S. C., and S. MAKRIDAKIS (1980). *Forecasting Methods for Management,* 3rd Ed. New York, NY: John Wiley and Sons.

CHAPTER **2**

Model Building
and Forecasting

This chapter treats the remaining two phases of the forecasting process depicted on the Flowchart.

- The *specification phase* consists of the modeling and forecasting activities associated with developing a theory of demand.

- The *evaluation phase* consists of the preparation, presentation, and important tracking functions that must accompany a forecasting effort.

THE SPECIFICATION PHASE

The specification phase deals with activities in the forecasting process that are necessary for

- Developing a theory of demand.
- Dealing with data.
- Selecting the appropriate forecasting techniques.

Developing a Theory of Demand

A *demand theory* provides the basis for determining the specific factors that have influenced demand in the past and are likely to affect demand in the future. In addition to the demographic and economic factors previously mentioned, factors such as income, price, market potential, and habit are also known to influence many purchasing decisions and, hence, the total quantity of sales.

Once a demand theory is formulated, it is necessary to determine if any additional data can be gathered that can be used to measure the influence of the relevant demand

11

factors. A preliminary analysis can then be made to determine if models can be implemented in a cost-effective way.

Typically, time series models involve relatively little investment of a forecaster's time and money. Econometric models are generally more costly and may not be more accurate (Armstrong, 1978); yet they have explanatory power, which is missing from most time series models. Jenkins (1979) presents a number of multivariate transfer function and intervention models that are based on time series (Box-Jenkins) methods and that have some explanatory capabilities.

Econometric models are generally more useful for planning and policy analysis, because they allow for consideration of alternative scenarios. This additional information is costly, since it may take considerable time and effort to develop these models.

Dealing with the Data

The selection of an appropriate forecasting technique depends crucially on the data for which the forecast is being developed. Of primary importance here is the identification and understanding of the historical patterns of the series. An understanding of the time series provides information necessary to establish parameters on two of the decision criteria shown as row headings in Table 2.1:

> Row 1. The pattern of data that
> can be recognized and
> handled.

> Row 2. Minimum data
> requirements.

If the trend, cyclical, seasonal, or irregular (nonhorizontal) patterns can be recognized, then techniques that are capable of handling those patterns can be readily selected. Are the data smooth or irregular? Are there any apparent errors in data collection or input that would distort an analysis? These might be evident as data gaps, discontinuities, or "odd-ball" data values.

Analysis of this sort will determine the starting point for the forecast; i.e., selection of the specific technique or modeling approach to use. Not all techniques are appropriate for all forecasting situations: if the data are very smooth, a time series approach might be quite adequate; for more volatile data, this approach will probably not be as useful. In many business applications, the selection of the best approach might be a quite subtle procedure.

Inherent in the selection of a given technique is an assumption that the forecaster will be able to discern which patterns in the data are representative of those that will most likely occur throughout the forecast period, and which patterns are not. Thus the amount of relevant historical data and the patterns of the relevant data serve to reduce the number of projection techniques that can be considered useful.

Table 2.1 Comparison of analysis and forecasting techniques.

	Qualitative					Quantitative — Statistical								Quantitative — Deterministic		
	Delphi method	Market research	Panel consensus	Visionary forecast	Historical analogue	Summary statistics	Moving average	Exponential smoothing	ARIMA (Box-Jenkins)	TCSI decomposition (Shiskin X-11)	Trend projections	Regression model	Econometric model	Anticipation survey, Intention-to-buy	Input-output model	Leading indicator
Pattern of data that can be recognized and handled — Horizontal	Not applicable					×	×	×	×	×	×	×	×	×	×	×
Trend	Not applicable					×	×	×	×	×	×	×	×	×	×	×
Seasonal	Not applicable					×	×	×	×	×		×	×			
Cyclical	Not applicable					×			×	×		×	×			
Minimum data requirements	Not applicable					5 Points	5 – 10 Points	3 Points	3 yrs. by mo.	5 yrs. by mo.	5 Points	4 yrs. by mo.	4 yrs. by mo.	2 yrs. by mo.	> 1000	5 yrs. by mo.
Time horizon for which method is most appropriate — Short term (0 – 3 mos.)		×	×			×	×	×	×	×	×	×	×	×	×	×
Medium term (3 mos. – 2 yrs.)	×	×	×	×	×	×	×	×	×	×	×	×	×	×	×	×
Long term (2 yrs. or more)	×		×	×	×	×					×	×	×		×	
Accuracy (scale of 0 to 10: 0 smallest, 10 highest) — Predicting patterns	5	5	5	5	5	2	2	3	2	7	4	8	2	2	2	2
Predicting turning points	4	6	3	2	3	NA	2	2	6	8	1	5	7	8	0	5
Applicability (scale of 0 to 10: 0 smallest, 10 highest) — Time required to obtain forecast	4	8	4	3	5	1	1	1	7	5	4	6	9	5	10	3
Ease of understanding and interpreting the results	8	9	8	8	9	10	9	7	5	7	8	8	4	10	3	10
Computer costs (scale of 0 to 10: 0 smallest, 10 highest) — Development	Not applicable					0	1	1	8	6	3	5	8	NA	10	4
Storage requirements	Not applicable					4	1	1	7	8	6	7	9	NA	10	2
Running	Not applicable					1	1	1	9	7	3	6	8	NA	0	NA

Note: This table is a subjective adaptation of material presented in Chambers, Mullick, and Smith (1974, pp. 63–70) and Wheelwright and Makridakis (1980, pp. 292–93).

Selecting Appropriate Forecasting Techniques

There are four additional selection-decision parameters shown in Table 2.1 that relate to the inherent characteristics of the techniques in terms of their

    Row 3. Capability of handling a
           given time horizon.

    Row 4. Accuracy.

    Row 5. Applicability.

    Row 6. Computer-related costs.

These four parameters differ from the first two, since they are influenced by the requirements, resources, and objectives of a project rather than by the nature of data.

The *time horizon* of a forecast has a direct bearing on the selection of forecasting methods, as can be seen in the table. The scaling from 0 to 10 in Rows 4–6 represents a subjective evaluation of forecasting methods. A score of 0 represents the low end of the range, and a score of 10 represents the high end of the range. As can be seen by the pattern of X's in the "Time horizon" rows, the longer the time horizon, the greater the reliance on qualitative methods. For the short and medium terms, there are a variety of quantitative methods that can be applied. The simpler moving average, exponential smoothing, and ARIMA models are relatively poor predictors of turning points and are primarily used in the short-to-medium terms. While econometric models are better at predicting turning points, it is important to remember that such prediction requires accurate forecasts of explanatory variables.

Judgments concerning *accuracy* are ultimately assessments of the relative precision attainable through the application of the various projection techniques. If it becomes apparent that the original objectives of precision cannot be attained, the "go–no go" decision must be reevaluated. If the objectives are attainable, then assessment of the relative precision of the techniques will lead to selection of appropriate data sources and specific projection techniques for the problem.

For example, if a forecaster's goal is to provide the most accurate one-year-ahead forecast, experience has shown that the ARIMA (Box-Jenkins) class of time series models fits a wide variety of data series, and can provide accurate short-term forecasts. In the short term, the inertia or momentum of an economic system often precludes dramatic change. If the goal is to provide a six-year-plan forecast (a long-term forecast), the forecaster will also want to have regression or econometric models available that relate the item to be forecast to the variables that cause it to grow or decline; over a six-year period, customers can find new suppliers, and their needs may change as their markets change; therefore, in the long term it is essential to relate the forecast item to what are known as its "drivers."

The improved accuracy that regression models bring to volatile periods consists of the accuracy with which the explanatory variables can be predicted.

If a certain amount of forecast error is allowable (if accuracy is not an overriding need), and when numerous forecasts are required, exponential smoothing models are frequently used in the short-to-medium terms. A forecaster is freed of the need to store great amounts of information, and the models can be run inexpensively.

The need for greater forecast accuracy in the short-to-medium terms, however, often leads the forecaster to consider the Box-Jenkins methodology for models of univariate (ARIMA) time series and multiple time series. And as the time period of the forecast extends to the medium and long terms (more than 12 months), trend projections, regression models, and econometric models become increasingly useful.

While regression and econometric models can also be used in the short term, their increased cost and complexity are seldom offset by increased accuracy over simpler ARIMA models, especially if economic or market conditions are stable. In more volatile times, the regression and econometric models become more valuable, since these models relate the forecast items to other pertinent explanatory variables. The forecaster who uses these models for the short term may have to explain why the forecast was different from the actual outcome but can take solace that the models make it quite easy to do this.

When sufficient data exist and the need for accuracy is great, as in predicting company revenues, the use of an ARIMA model as well as a regression or econometric model is recommended. This combination balances the generally superior short-term forecasting abilities of ARIMA models with the ability of regression and econometric models to relate the item that is being forecast to other forecasts of economic conditions, price changes, competitive activities, and other appropriate explanatory variables.

When methods are combined in this way, if both methods yield similar forecasts, an analyst is reasonably certain that the forecast is consistent with the assumptions made about the future and has a strong probability of being highly accurate. When the forecasts from the two or more methods are significantly different, the analyst should take this as a warning to exercise care. The role of judgment will be all the more critical, since the forecaster must decide to accept one or the other or perhaps some combination of the forecasts produced by the different methods. It is also important to advise the managers who have requested the forecast that the risks associated with the forecast are greater than normal.

The remaining two selection parameters in Table 2.1 (applicability and computer-related costs) are generally established by the forecaster and are based on knowledge of and experience with the time series, the techniques, and the forecasting process.

With the criteria now established for the selection parameters, the forecaster can proceed to reduce the list of potentially useful projection techniques even further. The knowledge of the data and of the operating conditions under which the forecaster must develop the forecast serve as primary inputs to the technique selection process. This knowledge must, however, be supplemented by a thorough knowledge of the techniques themselves.

THE EVALUATION PHASE

After selecting two or more techniques for serious consideration, the forecaster turns to the computer to fit models and to provide information upon which to judge the adequacy of a model. Model evaluation is referred to as *diagnostic checking*. There are a number of analytical tools available (e.g., residual analysis and forecast test results) for determining if it is possible to improve on the initial model. This may involve transformations of variables, addition of new variables, or selection of a different technique.

Generating Predictions and Confidence Limits with the Model

Having further reduced the alternatives through evaluations of historical performance and statistical significance tests, the forecaster now reaches the stage where actual forecasts are produced, tested, and approved. This effort is begun by generating predictions from the models that have survived the selection process.

One way a forecaster can test the validity of models is to use the models to generate predictions from data for prior time periods, for which the actuals are known. This is termed *ex post* prediction. In this way it is possible to monitor the performance of the models and determine the likely forecast accuracy. Table 2.2 illustrates a way of summarizing the results of the forecast test. If one error in a given year is unusually large, the median error may be preferred over the average absolute error.

The evaluation stage is normally iterative, inasmuch as one considers new variables and transformations of variables. Some techniques are rejected at this stage because of their

- Inability to provide statistically significant results.
- Inability to achieve the desired accuracy objectives.

In addition, the forecaster provides estimates of the reliability of the forecast in terms of forecast intervals or confidence limits at specified levels of confidence; alternatively, the reliability can be expressed as the likely percent of (amount of) deviation between forecast and actual results. For example, new car purchases for a year may be forecast to be ten million units plus or minus 700,000 units at the 90-percent confidence level; alternatively, experience has shown that in models of new car purchases the average annual percent of deviation (absolute value) between forecast and actual is 7 percent.

The Role of Judgment

The next step of the process deals with the relationship between the predictions from the various methods and the final forecast. The forecaster recognizes that the models

Table 2.2 Summary of one-year-ahead forecast errors from a hypothetical model with data from 1964 through 1979.

Historical fit	Percent error
1964–1974	− 8.7
1965–1975	− 1.2
1966–1976	+ 5.7
1967–1977	+ 3.8
1968–1978	− 2.3
Average absolute percent error =	4.3
Median absolute percent error =	3.8

Note: Percent error = (Actual − Forecast) / Actual.

are abstractions from the real world. Judgment is, by far, the most crucial element when trying to predict the future. Informed judgment is what ties together the forecasting process and the projection techniques into a cohesive effort that is capable of producing realistic predictions of future events or conditions.

Informed judgment, therefore, plays a significant role in minimizing the uncertainty associated with forecasting. Automated processes for model building are sometimes used in computing future demand from a set of key factors. However, such an approach is unlikely to replace the reliance upon sound judgment substantially. Judgment must be based on a comprehensive analysis of market activities and a thorough evaluation of basic assumptions and influencing factors.

Statistical approaches can provide a framework of information around which analytical skills and judgment can be applied in order to arrive at and support a sound forecast. To quote from Butler et al. (1974, p. 7):

> In actual application of the scientific approaches, judgment plays, and will undoubtedly always play, an important role. . . . The users of econometric models have come to realize that their models can only be relied upon to provide a first approximation—a set of consistent forecasts which then must be "massaged" with intuition and good judgment to take into account those influences on economic activity for which history is a poor guide.

The limitations of a purely statistical approach should be kept clearly in mind. Statistics, like all tools, may be invaluable for one job but of little use for another. An analysis of patterns is basic to forecasting, and a number of different statistical procedures may be employed to make this analysis more meaningful. However, the human element is required to understand the differences between what was expected in the past and what actually occurred, and to predict the likely course of future events.

Forecast Presentation, Monitoring, and Documentation

After deciding on the forecast values, it is common to prepare a *forecast package* for approval by the forecast user or by the forecaster's superiors. This package will highlight the forecast values, changes from prior forecasts, forecast methods, assumptions, and reasonableness checks.

Monitoring procedures are established after the approval of the forecast. Often referred to as *tracking,* the objective of monitoring is to be certain that the forecast remains relevant, to signal the need for forecast revision, and to provide information useful in determining the magnitude and direction of forecast revisions. To this end, assumptions and results are monitored continuously.

The forecast package described above is part of an overall documentation effort. Documentation makes meaningful results-analysis possible. Lacking it, the evaluation that the forecaster makes may be limited to a "numbers" exercise. Documentation will also help the forecaster to learn from past experience by identifying problem areas requiring further analysis. Also, managers and users will appreciate the additional insight that cannot be conveyed by tables of numbers alone.

Forecast Risk

Since future events and conditions cannot be predicted consistently with complete accuracy, the end product of the forecasting process can best be described as *giving advice*. In most cases, that advice is provided in the form of a single *best-bet* figure that represents either the value of a time series at a specific time or the cumulative value of the series at the end of a specific period of time. The best-bet figure can probably be best described as the median of a hypothetical frequency distribution. It is the point at which there is an even chance that the future outcome will fall above or below it.

The best-bet approach is a widely accepted method of providing advice about the future, because it facilitates precise planning and decision making. Its primary weakness, however, is that planning and decision-making processes assume that the forecast describes the future precisely, when in fact it cannot perform such a feat.

If the forecasting process is performed in a logical, systematic way, and is complemented by sound judgment and appropriately selected techniques, then the advice can probably be best provided to decision makers in the form of *range forecasts;* these are different ranges (different upper and lower bounds) about the estimate of some future point, and specification of the risk associated with each of those alternatives.

Decision making involves the assessment and acceptance of risk. Therefore, forecasters can best assist decision makers by providing forecast levels and associated probability statements that indicate the chance of each of those levels being exceeded. This does not mean that the forecaster takes a "shotgun approach" to predicting the future by incorporating the extreme alternatives at either end of the range. It simply means that the forecaster should provide the best-bet figure and state the associated risk levels on each side of the best bet. If the view of the future is presented in this format, the decision maker has much more information on which to assess risks associated with decisions.

MODEL BUILDING AND FORECASTING CHECKLIST

_____ Have you stated the *problem,* and determined the *objectives* of the model?

_____ What forecasts are needed, and how are they to be used, in terms of

- Variables to be forecast?
- Time horizon?
- Accuracy requirements?
- Level of aggregation?
- Frequency or number of forecasts?

_____ What explanatory factors are assumed to influence the variable(s) to be forecast?

_____ What types of specifications (quantitative models) are being considered?

_____ Have you dealt with available data that relate to

- Accuracy needs?
- Availability constraints?
- Storing and computing costs?

_____ Are the computational and model-building skills available in your organization?

_____ Have appropriate analyses and forecasting techniques been evaluated in the context of the problem at hand?

_____ Can your forecasting models generate confidence limits along with the forecasts?

_____ Have model predictions been compared at various lead times with actual data (forecast performance)?

_____ What kinds of adjustments or judgmental factors ("add factors") have been incorporated in the model(s) to account for unanticipated changes in the forecast?

_____ What steps have been implemented in order to obtain your management's acceptance of your forecasts?

_____ Are any reviews under way to evaluate the effectiveness of the overall forecasting process?

SUMMARY

Before a model can be specified, each technique must be rated by the forecaster in terms of its

- General reliability and applicability to the problem at hand.
- Relative effectiveness when compared to other appropriate techniques.
- Relative performance level (attainment of accuracy).

To summarize the specification phase, the forecaster must

- Perform a general analysis on the time series.
- Perform a screening procedure that reduces the list of all available modeling techniques to those that are capable of handling the data that must be considered.
- Perform a detailed examination of the techniques that are still considered appropriate.
- Make the final selection of two or more techniques that are considered to be the most appropriate for the given situation.

In the evaluation phase, the forecaster

- Determines the adequacy of the model.
- Determines the need to transform variables, add new variables, or select different techniques.
- Generates model predictions and confidence limits.
- Selects the forecast values.
- Provides reliability estimates.
- Presents a forecast package.

It must be emphasized that a superior knowledge of your time series, of the parameters of the forecasting process, and of the available techniques is still not enough to isolate and identify the one and only appropriate modeling approach. It should be noted that the identification of a *single* technique is neither the desirable nor the attainable objective of the technique selection process.

USEFUL READING

ARMSTRONG, J. S. (1978). Forecasting with Econometric Methods: Folklore versus Fact. *Journal of Business* 51, 549–64.

BUTLER, W. F., R. A. KAVESH, and R. B. PLATT, eds. (1974). *Methods and Techniques of Business Forecasting*. Englewood Cliffs, NJ: Prentice-Hall.

JENKINS, G. M. (1979). *Practical Experiences with Modeling and Forecasting Time Series*. Jersey, Channel Islands: GJ&P (Overseas) Ltd.

Part 2

Forecasting with Multiple Regression Models

CHAPTER **3**

Developing Functional Relationships

This chapter considers why a forecaster might want to consider building multiple linear regression models in addition to the simple linear regression models described in *The Beginning Forecaster*. Multiple regression models can:

- Provide superior forecasts (once again, the accuracy of the forecasts of the independent variables is crucial).

- Provide forecast users with a model that is encompassing and appealing, since more than one variable is taken into account.

- Be constructed with an approach whereby one independent variable is chosen to explain the trend in the dependent variable and one or more other variables are chosen to explain the deviations about the trend.

In this chapter, we will revisit one of the forecasting case studies presented in *The Beginning Forecaster* and show how the model can be improved with the addition of one or more new independent variables.

A FORECASTING PROBLEM

In *The Beginning Forecaster,* a forecasting problem was presented in which the goal was to provide accurate forecasts of demand volumes in general, and main telephone gain in particular. Telecommunications companies find these forecasts are very important factors in determining construction (investment) programs, departmental budgets, and the size of installation and repair forces.

25

The Variable To Be Forecast

A *main telephone* is defined as a telephone that is connected directly with a central office (location of a telephone switching machine), whether the telephone is on an individual or party line. Only one telephone for each customer on each line is considered a main telephone.

Main telephone gain (or, more simply, main gain) is a highly seasonal time series in which about 81 percent of the total variation in the monthly series is attributable to seasonality (Figure 3.1). For modeling purposes, seasonality can be removed by an appropriate seasonal adjustment program, or, by taking differences. To minimize calendar and other irregular variation, *quarterly* data were used in this study; differences of order 4 were taken to make the series stationary and free of seasonal variation. The differenced data showed a pronounced cyclical pattern in which the peaks and valleys generally corresponded to the business cycle (Figure 3.2).

Main gain represents the growth of main telephones in service and can be viewed as a difference of order 1 of main telephones in service. A difference of order 4 of main gain is then equivalent to a difference of order 1 followed by a difference of order 4 of main telephones in service. (Adjustment of in-service counts to compensate for errors in processing and transcription are ignored here.)

A previous case study *(The Beginning Forecaster,* Chapter 20) showed that quarterly main gain could be forecast on the basis of a housing-starts time series (information that construction of a housing unit has begun). Main gain lagged housing starts by about two quarters, and a model fitted by a robust regression procedure provided the most accurate forecasts. For the 1973–1978 period, the model produced six one-year-ahead forecasts with an average absolute error of 7.8 percent and five two-year-ahead forecasts with an average absolute error of 11.7 percent.

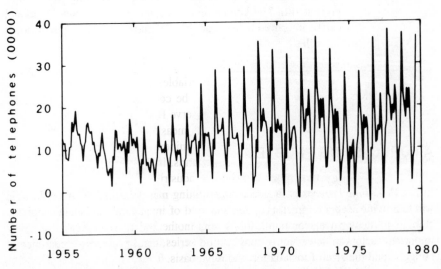

Figure 3.1 Time plot of a monthly gain in a main telephone series.

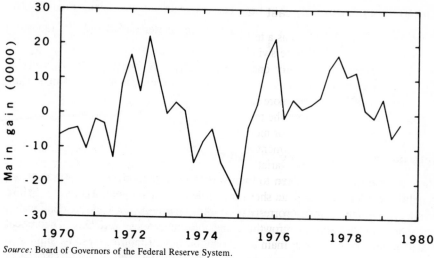

Source: Board of Governors of the Federal Reserve System.

Figure 3.2   Time plot of the differences of order 4 of the quarterly main gain, showing variation due to the business cycle. This time series is denoted as GAIN.

Identifying Variables for the Model

The case study that has just been described will be extended to an additional variable in a multiple linear regression model. A *multiple linear regression* model takes the form

$$Y = \beta_0 + \beta_1 X_1 + \cdots + \beta_k X_k + \varepsilon,$$

where Y denotes the *dependent* variable; $\beta_0, \beta_1, \ldots, \beta_k$ are the *regression coefficients;* and $X_1, \ldots, X_k$ are the *independent* variables; and ε is an error term.

The additional economic variable that will be considered for predicting main telephone gain is the growth in the Federal Reserve Board Index of Industrial Production. The FRB index, a cyclical indicator, measures changes in the physical volume or quantity of output of manufacturers, minerals, and electric and gas utilities. The index does not cover production on farms, in the construction industry, in transportation, or in various trade and service industries.

Experience has shown that superior forecasting models often result from forecasting similar "types" of series (e.g., one kind of in-service data versus another kind of in-service data, one gain series versus another gain series). Since the FRB index can be viewed as an "in-service type of series," differences of orders 1 and 4 were taken to have all series on a comparable basis. Thus, the time series that will be used in the multiple linear regression models are denoted as

GAIN: Differences of order 4 of quarterly main telephone gain (Figure 3.2).

DFRB: Differences of orders 1 *and* 4 of the quarterly FRB Index of Industrial Production (Figure 3.3).

HOUS: Differences of order 4 of the quarterly U.S. housing-starts series (Figure 3.4). Like main gain, housing starts represent the *growth* in housing stock.

Determining Lead–Lag Relationships

There is little reason to believe that the variations in the independent variables are exactly coincident with the variations in the dependent variable. Therefore, before continuing with the regression, it is important to determine if a lead–lag relationship exists between the dependent variable and the independent variables. Theoretical considerations may suggest appropriate relationships. Empirical results, similar to the testing of two- and three-quarter lags for the model with housing starts, may also be used to develop lead–lag relationships.

   Cross correlation is a way of determining lead–lag relationships in stationary *residual* series. The procedure can be viewed as shifting one series with respect to the other and calculating a correlation coefficient. The average relationship for each

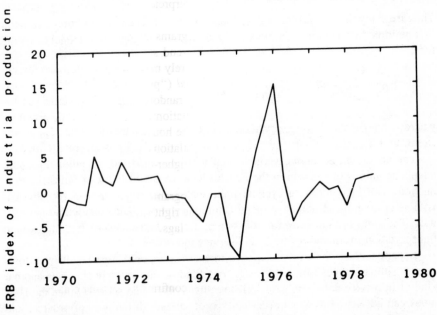

Figure 3.3 Time plot of the first- and fourth-order differences of the quarterly FRB Index of Industrial Production. This series is denoted as DFRB.

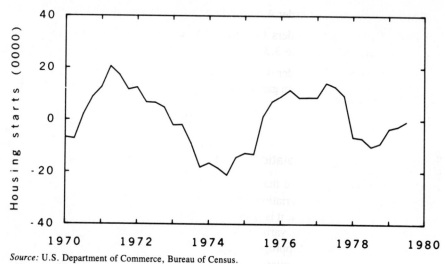

Source: U.S. Department of Commerce, Bureau of Census.

Figure 3.4   Time plot of the differences of order 4 for the quarterly U.S. housing starts series. This series is denoted as HOUS.

positive or negative shift is plotted as a spike in a diagram called the *cross correlogram*. The most significant spike represents the greatest correlation between the two series, and the corresponding lag can be interpreted as the lead or lag between the two series.

It is important to note here that cross correlograms are not valid tools when both series still contain significant seasonal and/or trend structures. In such cases, the pattern in the cross correlogram tends to be severely masked by the dominant trend or seasonal patterns; hence the need for *residual* ("pre-whitened") series in constructing cross correlograms. While not exactly random, differenced series can be cross correlated to find approximate lead–lag relationships, which can be tested in alternative regression models, as was done for the housing starts model.

Because fewer terms are involved in the calculation of a cross correlogram when lags are great, the spikes must be higher at higher-order lags to be considered significant.

Figures 3.5 and 3.6 are plots of cross correlograms—one for each of the independent variables plotted against main gain. The right side of the origin represents "leads" and the left side of the origin represents "lags." From the plots the following conclusions can be reached:

- DFRB is coincident with GAIN.

- HOUS leads GAIN by two quarters (this confirms the results obtained from the prior case study).

These conclusions are based on the cross correlogram plots and on prior research and knowledge of the nature of the data. From past experience, it is known that main

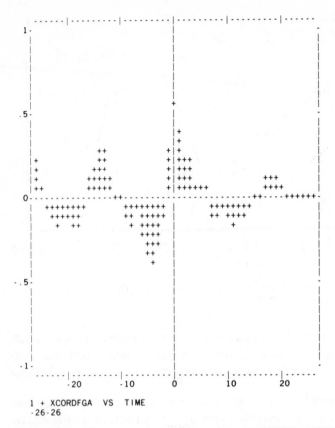

Figure 3.5 Plot of the cross correlogram between DFRB and GAIN, showing coincident correlation.

gain is either a coincident or leading indicator of economic activity. It is not likely that main gain would show a lead–lag relationship in excess of four quarters, however desirable. This tends to restrict attention to the center portion of the cross-correlogram.

The FRB index is a measure of economic production activity. Since there is generally a lag between the determination to change production levels and the implementation of such changes, the FRB index is likely to lag main gain. Since the differences of orders 1 *and* 4 of the FRB index represent growth rates (similar to acceleration, in this case) of the series, the differenced FRB series is likely to be coincident with the differenced main-gain series.

Since additional variables are added to the model to improve its forecasting capability, it makes sense to cross correlate the residuals of the current model with a new residual series (detrended, deseasonalized) to determine if the prior lead–lag relationships still apply. Any correlation among the independent variables may cause shifts in the lead–lag relationships in the model.

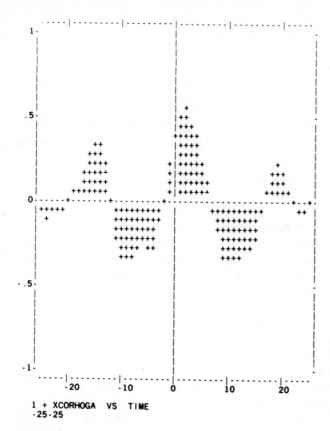

Figure 3.6 Plot of the cross correlogram between HOUS and GAIN, showing HOUS leading by two quarters.

Using Dummy Variables to Eliminate the Effects of Outliers

The main gain series was influenced by nationwide strikes in 1968 and 1971, and these caused the gain in the corresponding quarters to be unusually low. Since outliers are known to cause modeling problems, *robust regression* was used in the original case study to minimize the impact of these unusual values. In multiple linear regression models, *dummy variables* may also be used to eliminate the effects of outliers. (Dummy variables are treated in detail in Chapter 12.)

In this instance, two dummy variables are required. Each has zero values for all quarters except the one in which the strike occurred: the first dummy (D_1) has a value of 1 for the second quarter of 1968 and the second dummy (D_2) has a value of 1 for the third quarter of 1971. Since the dummy variables are used in order to explain the data in these two quarters perfectly, the outliers will not affect the estimates of the model parameters adversely.

Model Building and Forecast Evaluation

Four models will be evaluated:

- A simple linear regression model of GAIN against housing starts (HOUS, lagged by two quarters) estimated by ordinary least squares: Model A.
- Same as Model A, but estimated by a robust regression procedure: Model B.
- Model A, including the FRB index (DFRB) and dummy variables (D_1, D_2) to account for strikes in 1968 and 1971: Model C.
- Model C fitted by a robust regression procedure without dummy variables: Model D.

Since the objective is to produce the best forecasts, the focus of the evaluation will be the forecast test rather than model statistics. The mean absolute one-year-ahead forecast errors range from 6.3 percent for Model D to 9.6 percent for Model A. Model D is superior on the basis of one-year-ahead accuracy, and Model C is superior on the basis of two-year-ahead accuracy. Table 3.1 compares these models.

While the *minimum mean absolute error* in a forecast is a reasonable criterion for selecting the best forecasting model, it is also worth comparing models in terms of the largest annual absolute error contained in them. On this basis, Model D is superior since its largest one-year-ahead error (12 percent) and largest two-year-ahead error (15.2 percent) are less than any of the competing models (Table 3.1).

Apparently, few additional unusual values, beyond the 1968 and 1971 strikes, are impacting the housing starts, FRB index, or main gain series. As a result, the

Table 3.1 Comparison of the annual percent errors in forecasts (1973–1978) in four models for main telephone gain.

Model*	One year ahead		Two years ahead	
	Mean absolute	Largest annual	Mean absolute	Largest annual
A	9.6	19.9	14.1	21.3
B	7.8	24.1	11.7	20.8
C	6.5	12.9	8.0	16.5
D	6.3	12.0	8.6	15.2

*Model A (OLS): $\text{GAIN}_t = f(\text{HOUS}_{t-2})$
Model B (robust): $\text{GAIN}_t = f(\text{HOUS}_{t-2})$
Model C (OLS): $\text{GAIN}_t = f(\text{HOUS}_{t-2}, \text{DFRB}_t, D_1, D_2)$
Model D (robust): $\text{GAIN}_t = f(\text{HOUS}_{t-2}, \text{DFRB}_t)$

ordinary least squares *(OLS)* Model C (with dummy variables) produces almost the same results as the *robust* Model D (without dummy variables). The fitted models are:

MODEL C:

$$\text{GAIN}_t = 14{,}746 + 568.5\,\text{HOUS}_{t-2} + 13{,}185\,\text{DFRB}_t - 246{,}220\,D_1 - 225{,}270\,D_2.$$

MODEL D:

$$\text{GAIN}_t = 13{,}453 + 561.6\,\text{HOUS}_{t-2} + 13{,}376\,\text{DFRB}_t.$$

The coefficient of the housing starts variable in Model C is 1 percent greater than in Model D. The coefficient of the DFRB variable is 1 percent less in Model C. The coefficients of the dummy variables show the impacts of the two strikes. The 1968 strike (D_1) caused a reduction in demand of 246,220 main telephones and the 1971 strike caused a reduction of 225,270 main telephones.

The addition of the DFRB series has resulted in a reduction in the mean absolute one-year-ahead-forecast error from 7.8 percent to 6.3 percent. The largest annual error has been significantly lowered from 24.1 percent to 12.9 percent. The mean absolute two-year-ahead error has been reduced from 11.7 percent to 8.6 percent (8 percent with Model C). The largest two-year-ahead error has been reduced from 20.8 percent to 15.2 percent. In this example, the addition of one more independent variable in a multiple linear regression model has substantially improved forecast accuracy.

These forecast errors result from using the model in a forecasting mode with *actual* values for independent variables (*ex post* predictions). Thus the forecast errors reflect performance that is exclusive of inaccuracies in the independent variables. A more realistic evaluation would include *forecasts* of the independent variables (*ex ante* predictions). There might have been even larger forecast errors if predictions of the independent variables were inaccurate. Of course, it is also possible that inaccurate predictions of the independent variables would cancel out other influences and result in better forecasts than if the variables were correctly forecast. However, this might not give you much confidence in your model.

PREPARING DATA FOR MODELING

Some lessons learned from the previous case study include:

- Differencing operations may be necessary to put time series on a "comparable" basis.

- Modeling assumptions need to be verified (through residual analysis, for example) to validate statistical hypotheses.
- Other influences (calendar effects), not captured or not capable of being captured by the model, may be present.
- A multimethod approach helps assess model adequacy and enhances credibility of modeling results.

The remaining part of this chapter will deal with some premodeling issues you need to consider prior to embarking on a forecast modeling effort.

Removing Trend by Differencing

It has been noted that differencing is required to

- Put time series on a comparable basis for cross-correlation studies and regression modeling.
- Render time series stationary prior to time series modeling.

Differencing is obtained by subtracting a lagged version of itself from the original time series, Y_t. For example, a first difference or *difference of order 1* of the time series Y_t results in a new time series Z_t defined by

$$Z_t = Y_t - Y_{t-1}.$$

Similarly, a *difference of order 4* (or fourth-*order* difference) is given by

$$Z_t = Y_t - Y_{t-4}.$$

Here, Z_t is the difference between a value of the time series and the value of the series four periods earlier. For quarterly data, this represents a year-over-year change.

The most commonly used differencing operation for monthly data is the *difference of order 12* (or twelfth-*order* difference):

$$Z_t = Y_t - Y_{t-12}.$$

Again, this is a year-over-year change for monthly data.

Since year-over-year changes should be essentially free of seasonal patterns, the differences of orders 4 and 12 are used to change seasonal time series to new time series that are free of seasonality. Notice that this operation is not the same as a seasonal *adjustment* or the *removal* of seasonality from a series. A differenced series does not have the same scale as the original series.

Another point worth noting is that a fourth-*order* (or twelfth-*order*) difference is *not* the same as a "fourth" (or "twelfth") difference. A "fourth difference" is a first difference repeated four times; that is, let

$$Z_t = Y_t - Y_{t-1},$$

which is a *first difference*. The *second difference* is the first difference of Z_t; that is,

$$
\begin{aligned}
Z_t^{(2)} &= Z_t - Z_{t-1} \\
&= (Y_t - Y_{t-1}) - (Y_{t-1} - Y_{t-2}) \\
&= Y_t - 2Y_{t-1} + Y_{t-2}.
\end{aligned}
$$

To carry this out two more times, let $Z_t^{(3)} = Z_t^{(2)} - Z_{t-1}^{(2)}$, and $Z_t^{(4)} = Z_t^{(3)} - Z_{t-1}^{(3)}$; then the "fourth difference" turns out to be

$$Z_t^{(4)} = Y_t - 4Y_{t-1} + 6Y_{t-2} - 4Y_{t-3} - Y_{t-4}.$$

This is clearly not the same as the fourth-*order* difference,

$$Z_t = Y_t - Y_{t-4}.$$

Transforming Data in Order To Fit Assumptions

Regression and time series modeling are based on a number of statistical assumptions that will be made more explicit in later chapters. Transformations of data are often required to realize these assumptions. Assumptions that form a basis for selecting an appropriate transformation include:

- The approximate *normality* of the error distribution in a model.
- The constancy or *uniformity of the variability* in the errors.
- The applicability of *linear* forms for regression or time series (ARIMA) models.

A useful family of transformations on a (positive) variable Y are the power transformations devised by Box and Cox (1964):

$$
\begin{aligned}
W &= (Y^\lambda - 1)/\lambda && \text{for } \lambda \neq 0, \\
&= \ln Y && \text{for } \lambda = 0.
\end{aligned}
$$

These transformations were discussed in *The Beginning Forecaster*, Chapter 10. The parameter λ is estimated from the data, usually by the method of maximum likelihood. This procedure involves several steps:

- A value of λ is selected from within a reasonable range bracketing zero, say $(-2,2)$. A convenient set of values for γ is $\{\pm 2, \pm 1\frac{1}{2}, \pm 1, \pm\frac{2}{3}, \pm\frac{1}{2}, \pm\frac{1}{4}, 0\}$.
- For the chosen λ, evaluate the likelihood

$$L_{max}(\lambda) = \frac{-n}{2} \ln \hat{\sigma}^2(\lambda) + (\lambda - 1) \sum_{i=1}^{n} \ln Y_i$$

$$= \frac{-n}{2} \ln (\text{Residual SS}/n) + (\lambda - 1) \sum_{i=1}^{n} \ln Y_i,$$

where n is the total number of observations, and SS denotes "sum of squares."

- Plot $L_{max}(\lambda)$ against λ over the selected range and draw a smooth curve through the points. Figure 3.7 shows a plot of $L_{max}(\lambda)$ for the revenue series in a telecommunications forecasting problem used throughout this book for illustrative purposes.

- The value of λ that maximized $L_{max}(\lambda)$ is the *maximum likelihood estimate* $\hat{\lambda}$ of λ. Table 3.2 shows the maximum likelihood estimates of the four series in the telecommunications example.

- For applications, round the maximum likelihood estimate $\hat{\lambda}$ to the nearest value that makes practical sense. This should have a minor impact on the results, especially if the likelihood function is relatively flat (as it is in many cases).

- Determine an approximate $100(1-\alpha)$-percent confidence interval for λ from the inequality

$$L_{max}(\hat{\lambda}) - L_{max}(\lambda) \leq \tfrac{1}{2}\chi_1^2(1-\alpha),$$

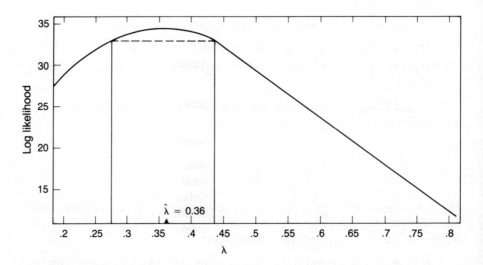

Figure 3.7 Plot of the logarithms of the likelihood function $L_{max}(\lambda)$ versus λ. Associated 95-percent confidence limits for the maximum are shown.

Table 3.2 Estimate of λ in the Box-Cox transformation
for the telecommunications example.

Series (Code)	$\hat{\lambda}$	Transformation close to
Toll revenues (REV)	0.36	Cube root ($\lambda = 0.33$)
Toll messages (MSG)	0.43	Square root ($\lambda = 0.50$)
Business telephones (BMT)	0.57	Square root ($\lambda = 0.50$)
Nonfarm employment (NFRM)	0.93	No transformation ($\lambda = 1.0$)

where $\chi_1^2 (1 - \alpha)$ is the percentage point of the χ^2 distribution with one degree of freedom, which leaves an area of α in the upper tail of the distribution (Appendix A, Table 3).

- The confidence interval can be drawn on the plot of L_{max} (λ) against λ by drawing a horizontal line at the level

$$L_{max} (\hat{\lambda}) - \tfrac{1}{2}\chi_1^2 (1 - \alpha)$$

of the vertical scale. The two values of λ at which this cuts the curve are the end points of the approximate confidence interval. Figure 3.7 also shows the 95-percent confidence interval for λ for the revenue data (REV).

Dealing with Seasonality

The principal reason for the removal of seasonal variation from a time series is the need for

- Recognizing important nonseasonal movements, such as turning points and other cyclical events in economic series.
- Developing regression relationships in econometric modeling.

The U.S. Census Bureau's X-11 procedure (Shiskin, et al., 1967) is probably the best known and most widely used seasonal adjustment procedure available today. A very recent method is the SABL procedure (Cleveland, et al., 1979), which was discussed extensively in *The Beginning Forecaster,* Chapter 19. Both the X-11 and SABL procedures consider an additive decomposition of the form

$$Y = T + S + I,$$

where T represents the long-term trend, S represents the seasonal component, and I represents the irregular or noise component. By removing (or subtracting off) one or more of these *components,* the forecaster obtains an *adjusted* series, which can be used to assess the important *components of variation* in a time series.

Both the X-11 and SABL procedures carry out the decomposition with a combination of filtering and regression techniques. SABL was designed to keep what seemed good in X-11 and to make changes where improvements were needed. The areas of improvement included:

- Graphics.

- Expanding the class of models by introducing power transformations.

- Smoothing (filtering procedures).

- Handling outliers (robust estimation).

- Calendar coefficient estimation.

In recent years, there have been other advances in the methodology for analyzing seasonal time series. Pierce (1980) summarizes some recent research on seasonal adjustment problems and procedures that has appeared in the literature. It is beyond the scope of this book to cover all these methods; later chapters on time series modeling will, however, touch again on various aspects of seasonal variation.

Adjusting for Calendar Effects

Besides the well-documented calendar adjustment (trading-day) procedure of the X-11 program (Shiskin, et al., 1967), there are few treatments of this important component in any forecasting books (see, for example, Makridakis and Wheelwright, 1978, section 4–5). Some recent advances have been made in the SABL program (Cleveland and Devlin, 1980), where the calendar component is given in terms of seven day-of-week coefficients $\alpha_1, \ldots, \alpha_7$, such that $\Sigma_{i=1}^{7} \alpha_i = 0$. The α's are estimated from the aggregated monthly series. For example, for the monthly main-telephone gain series and the airline data (Box and Jenkins, 1976, series G), the α's from the SABL estimation procedure were found to be the following.

	Main gain (1969-1979)	Airline
(Monday)	−24044	−0.0262
(Tuesday)	3002	−0.0284
(Wednesday)	10981	−0.0568
(Thursday)	8609	0.0075
(Friday)	191	0.0248
(Saturday)	880	0.0393
(Sunday)	381	0.0398

Figure 3.8 shows the main gain data and the four components (trend, calendar, seasonal, and irregular) determined by the SABL procedure. Calendar effects adjustment has been a neglected area; some recent experience with the SABL program has shown that calendar effects can be significant and should be adjusted in many monthly time series.

TOTAL MAIN GAIN

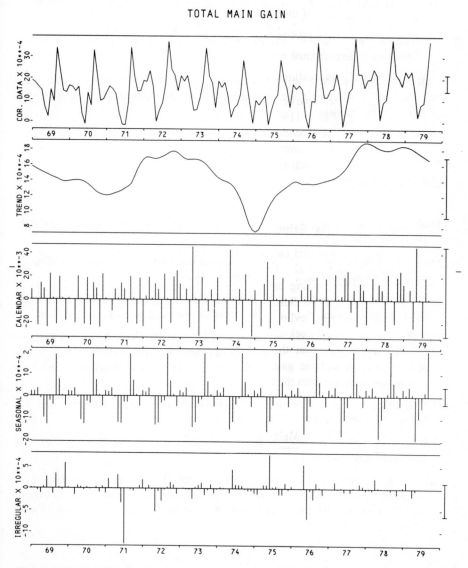

Figure 3.8 Monthly data and four components of the SABL decomposition of these data.

SUMMARY

The telecommunications case study examined in this chapter has shown that

- The forecasting accuracy of simple linear regression models may be enhanced by the addition of one or more variables. Since housing starts are only one of the factors causing growth in main telephones, the addition of the FRB Index of Industrial Production results in a more complete explanatory model and a more accurate forecasting model.

- Accurate forecasts of the independent variables are a prerequisite for improved forecasting performance.

In the next chapter, these same models will be evaluated in terms of summary statistics, and in Chapter 5 residual analyses will be performed to complete the evaluation of the models.

USEFUL READING

BOX, G. E. P., and D. R. COX (1964). An Analysis of Transformations. *Journal of the Royal Statistical Society* B 26, 211–52.

BOX, G. E. P., and G. M. JENKINS (1976). *Time Series Analysis, Forecasting and Control,* Revised Edition. San Francisco, CA: Holden-Day.

CLEVELAND, W. S., and S. J. DEVLIN (1980). Calendar Effects in Monthly Time Series; Detection by Spectrum. *Journal of the American Statistical Association* 75, 487–96.

CLEVELAND, W. S., D. M. DUNN, and I. J. TERPENNING (1979). SABL—A Resistant Seasonal Adjustment Procedure with Graphical Methods for Interpretation and Diagnosis. In *Seasonal Analysis of Economic Time Series,* A. Zellner, ed. Washington, DC: U.S. Government Printing Office.

MAKRIDAKIS, S., and S. C. WHEELWRIGHT (1978). *Forecasting, Methods and Applications.* New York, NY: John Wiley and Sons.

PIERCE, D. A. (1980). A Survey of Recent Developments in Seasonal Adjustment. *The American Statistician* 34, 125–34.

SHISKIN, J., A. H. YOUNG, and J. C. MUSGRAVE (1967). *The X-11 Variant of the Census Method-II Seasonal Adjustment Program.* Technical Paper No. 15, U.S. Department of Commerce, Bureau of the Census, Washington, DC: U.S. Government Printing Office.

CHAPTER 4

Building Regression Models

Linear regression models can be used to

- Predict a dependent variable from one or more (related) independent variables.

- Describe a functional relationship among regressor variables in which the estimated coefficients provide an econometric interpretation.

- Develop plans for policy evaluation where the model serves to express various policy alternatives.

This chapter provides the statistical background necessary for building, interpreting, and evaluating linear regression models. The material is basic to most of the forecasting techniques used in practice.

MULTIPLE LINEAR REGRESSION

In the multiple linear regression model, the *regression function* $\mu_{Y(X)}$ takes the form

$$\mu_{Y(X)} = \beta_0 + \beta_1 X_1 + \cdots + \beta_k X_k \, ,$$

where $X_1, \ldots, X_k$ are k *independent variables* and $\beta_0, \ldots, \beta_k$ are called *regression coefficients*. This model arises when the variation in the dependent variable Y is assumed to be affected by changes in more than one independent variable. Thus the *average value* of Y can be made to depend on $X_1, \ldots, X_k$. The dependence on X is henceforth suppressed in the notation; let $\mu_{Y(X)} = \mu_Y$. In this case, one speaks of a *multiple regression* of Y on $X_1, \ldots, X_k$.

Standard Assumptions

Conventional *normal regression* theory is based on the following assumptions:

- The mean μ_Y of Y is *linear in the β's;* that is,

$$\mu_Y = \beta_0 + \beta_1 X_1 + \cdots + \beta_k X_k,$$

- where $\beta_0, \beta_1, \ldots, \beta_k$ are called the regression coefficients.
- The variance of Y has the same value, σ^2, for all values of $X_1, \ldots, X_k$.
- Y is normally distributed.

Consider a sample (or time series values y_t, $t = 1, \ldots, n$) in matrix form:

$$
\begin{array}{ccccc}
y_1 & x_{11} & x_{12} & \cdots & x_{1k} \\
y_2 & x_{21} & x_{22} & \cdots & x_{2k} \\
y_3 & x_{31} & x_{32} & \cdots & x_{3k} \\
\cdot & \cdot & \cdot & & \cdot \\
\cdot & \cdot & \cdot & & \cdot \\
\cdot & \cdot & \cdot & & \cdot \\
y_n & x_{n1} & x_{n2} & \cdots & x_{nk}
\end{array}
$$

The n independent values $(y_1, \ldots, y_n)$ of Y, observed together with the values of the corresponding independent variables, are used to estimate the regression coefficients $\beta_0, \beta_1, \ldots, \beta_k$, and the error variance σ^2.

Multiple linear regression may be used to fit a polynomial function:

$$\mu_Y = \beta_0 + \beta_1 X + \beta_2 X^2 + , \cdots + \beta_k X^k.$$

By letting $X_1 = X$, $X_2 = X^2, \ldots, X_k = X^k$, you obtain the original formulation. Often X is used to represent a time scale (weeks, months, or years). It is worth emphasizing that the regressors may be any functional form; there need only be linearity in the parameters.

Other functional forms for the regression function include

$$\beta_1 \sin(\alpha t) + \beta_2 \cos(\alpha t),$$

or

$$\beta_1 \exp(\alpha_1 t) + \beta_2 \exp(\alpha_2 t).$$

The formal theory of normal multiple linear regression is very extensive and is dealt with in great detail in general statistics texts. Only the interpretation of those aspects of the theory relevant to forecasting problems is treated in this chapter.

General theoretical developments of this important topic are found in Draper and Smith (1981), Rao (1973), Searle (1971), and Seber (1977).

In *The Beginning Forecaster,* summary statistics for simple linear regression models were discussed. These statistics included the t statistic, F statistic, Durbin-Watson or DW statistic, and the R-squared statistic. In this chapter, the F and R-squared statistics are generalized to be applicable for models with multiple regressors. Additional statistics, the $\bar{R}$-squared and the "incremental" F statistics, are introduced for the first time, since they only have meaning in the context of multiple linear regression models.

Goodness of Fit

In general, a measure of the effectiveness of the regression fit can be obtained by calculating the *multiple correlation coefficient R,* or its square, which is given by

$$R^2 = \frac{\text{Regression sum of squares}}{\text{Total sum of squares}}$$
$$= (T - Q) / T,$$

where $T = \Sigma (y - \bar{y})^2$ is the total variation, and $Q = \Sigma(y - \hat{y})^2$ represents the unexplained variation.

Hence R^2 is the proportion of the variation of Y that has been explained by including particular independent variables. It is also commonly referred to as the *coefficient of determination.*

To compare models with a different number of independent variables, the *corrected R^2, adjusted for degrees of freedom,* is used. This is given by

$$\bar{R}^2 = 1 - \left\{ \frac{(n - 1)(1 - R^2)}{n - k - 1} \right\},$$

where k = number of independent variables.

Unlike R^2, $\bar{R}^2$ can decrease when a variable is added. In fact, for $k \geqslant 1$, $R^2 \geqslant \bar{R}^2$ and, moreover, $\bar{R}^2$ can be negative.

An overall test of significance of the regression can be carried out by calculating the F statistic:

$$F = \frac{(T - Q)/\text{Regression degrees of freedom}}{Q/\text{Residual degrees of freedom}},$$
$$= \frac{\text{Regression sum of squares/Regression degrees of freedom}}{\text{Residual sum of squares/Residual degrees of freedom}}$$
$$= \frac{\text{Mean square due to regression}}{\text{Mean square due to error}}.$$

An examination of the F table (Appendix A, Table 4) shows that, for most practical problems, an observed value of approximately 4 or more probably points to *statistical significance* for the appropriate degrees of freedom, provided the normality assumptions are valid.

In general, if the F statistic is significantly greater than unity, it indicates that the data do not support the null hypothesis of zero values for the regression coefficients. Then one is inclined to accept the alternative hypothesis that there is a regression relationship with at least one of the β_1, β_2, . . . , β_k different from zero.

Significance of Regression Coefficients

The t statistic measures the statistical significance of the regression coefficient for an independent variable. The t ratio follows a Student's t distribution that looks very similar to the bellshaped normal distribution. However, a t distribution is shorter and fatter, and its variance ($= v/(v - 2)$) is larger than that of the standard normal ($= 1$). For each positive integer v, called *degrees of freedom,* there corresponds a different t distribution.

With $n < 30$, the observed t value should be greater than approximately 2.0 in absolute value for significance at the 95-percent level (Appendix A, Table 2). When this is the case, one rejects the null hypothesis that the regression coefficient is zero. A statistically significant value not equal to zero is said to exist for the coefficient. This cannot be inferred as proof of a cause-and-effect relationship, however. Each variable can, for example, be related to a third (possible causative) factor and only be related coincidentally to each other.

You may test if a specific independent variable X_i is necessary, by using the t test on the estimated regression coefficient b_i. A nonsignificant t value implies that, given the effects of all other independent variables, X_i does not explain a significant amount of additional variability in Y. Cases can occur in which each regression coefficient b_i is not significant, and yet the regression as a whole is significant, as indicated by an F test. Hence, in multiple linear regression, an F test should always be performed.

In addition to summary statistics and significance tests, it is important also to consider

- The forecasts given by the model for future periods. Are they reasonable?
- The comparison of forecasts with actuals. Is the accuracy level acceptable?
- The residual pattern over the fitted and forecast periods. Does it appear to be random?
- Confidence limits for the residuals and their cumulative sum over the forecast period. These are useful for monitoring the forecasts as actual results become

available. Patterns of overforecasts, underforecasting, or of too many values falling outside the limits suggest that the model needs to be reevaluated. This subject was treated in detail in *The Beginning Forecaster,* Chapters 14 and 22.

An Incremental *F* Test

It is often important to test to see if the inclusion of an additional variable significantly improves the fit of a linear regression model. For example, you may want to examine to see if the inclusion of the FRB Index of Industrial Production in a model relating main gain to housing starts results in a significant reduction in the sum of squares due to errors.

In multiple linear regression problems, the significance of the regression coefficient cannot, in general, be tested with *t* tests on a one-by-one basis. This is so because individual regression coefficients are correlated among themselves and the *t* ratios are not independently distributed with a Student's *t* distribution. It is then necessary to perform an incremental *F* test.

The numerator of this *F* statistic, *F**, is the change in the sum of squares due to error in the old model minus the sum of squares due to error in the new model, divided by the difference in error degrees of freedom. The foregoing quantity is divided by the mean square error of the new model:

$$F^* = \frac{[\text{Residual SS}_{old} - \text{Residual SS}_{new}/(\text{df}_{old} - \text{df}_{new})]}{\text{Residual SS}_{new}/\text{df}_{new}}.$$

When only one variable is added to the model at a time, the incremental *F* test is equivalent to a *t* test. If the *t* statistic for the new variable is significant, so is the incremental *F* statistic. However, when several variables are added at a time, the group of variables may be significant even though one or more *t* statistics may appear insignificant. For example, in Chapter 12, we show how indicator variables are used to account for seasonal variation. In this case, while one or more of the variables appear insignificant, the seasonal variation is described by the presence of *all* the indicator variables. The incremental *F* test will indicate whether the added variables as a *group* are statistically significant.

Table 4.1 summarizes the statistics and test results from the four models presented in Chapter 3. Since extreme residuals are downweighted in a robust regression, the *R*-squared and *t* statistics cannot be compared to corresponding ordinary least squares (OLS) statistics. By downweighting extreme observations, the residual variation will be greater with robust estimation than with OLS (where every observation receives equal weight). By its very nature, the violation of OLS assumptions implicit in the use of robust regression means that normality cannot be assumed, and that the standard statistical significance tests cannot be applied. Since robust

Table 4.1   Summary of model results from the four models for GAIN presented in Chapter 3.

Model	HOUS	DFRB	Coefficient estimates		R^2	DW	Standard deviation of residuals $\times 10^3$
			D_1	D_2			
A	581.1 [5.5]	—	—	—	0.29	1.16	80.3
B	729.1 [NA]	—	—	—	NA	—	60.2
C	568.4 [7.0]	13,185 [6.3]	−246,220 [−4.2]	−225,270 [−3.8]	0.65	1.33	57.9
D	561.6 [NA]	13,376 [NA]	—	—	NA	—	54.5

Notes: t statistics corresponding to the coefficient estimates are shown in brackets below the estimates.

[NA] = not available.

R^2, $\overline{R}^2$, F, and t statistics are unavailable for the robust regression models for theoretical reasons.

All F statistics are significant at the 5-percent level.

regression is intended to fit the bulk of the data, the parameter estimates can be checked for reasonableness, but a forecast test may be the preferred criterion for selecting a forecasting model.

EVALUATING REGRESSION MODELS

The adequacy of model assumptions can be examined through a variety of methods, mostly graphical, involving the analysis of residuals. This topic is treated in Chapter 5. Graphical methods also play a role in assessing

- How individual data values affect the estimation of least squares estimates of regression coefficients.
- How the selection of independent variables impacts the fitting process.

Regression Pitfalls

There are a variety of regression pitfalls that you should be aware of and that must be avoided if possible. These pitfalls include trend collinearity, overfitting, extrapolation, outliers, nonnormal distributions, multicollinearity, and invalid assumptions regarding the model errors (e.g., independence, constant variance, and, usually, normality).

When the dependent and independent variables are time series, there are many special pitfalls to be avoided. Suppose that you are interested in forecasting the cyclical variation in one series on the basis of predictions of a related cyclical time series, such as revenues and employment.

In a simple or multiple linear regression model, a very high value of the R-squared statistic may result from what is known as *collinearity*. This often occurs when both series have very strong trends. It is quite possible that the trends are highly correlated but that the cyclical patterns are not. The dissimilarities in cycle may be masked by the strong trends.

Similarly, when a regression model is performed on raw time series, it is not clear just what information will result. If both series have rising trend and corresponding strong seasonality, the regression will very likely show a very high R-squared statistic. Alternatively, if there is a strong underlying relationship between variables but their seasonal patterns differ, the regression may appear insignificant.

As a rule, data should be adjusted for those possible sources of variation in which one is not interested in order to study the relationships with respect to those forces whose effects are of primary interest.

In the case of a telephone revenue–message relationship, seasonality is not of primary interest, so you can use seasonally adjusted data. A high correlation now means that there is a strong trend–cyclical correlation. In order to determine whether there is strong cyclical relationship, the appropriate procedure is to fit trend lines and correlate residuals from the fitted values.

Overfitting

Another source of danger in regression is overfitting. This occurs when too many independent variables are used to attempt to explain the variation in the dependent variable. Overfitting may also arise when there are not enough data points. If the number of independent variables is "close to" the number of observations, a "good fit" to the observations may be obtained, but the coefficient and variance estimates will be poor. This results often in very bad estimates for new observations.

Extrapolation

In forecasting applications, regression models are frequently used for purposes of extrapolation; that is, for extending the relationship to a future period of time for which data are not available. A relationship that is established over a historical time span and a given range of independent and dependent variables may not necessarily be valid in the future. Thus extreme caution must be exercised in using correlation analysis to predict future behavior among variables. There may be no choice in some cases, but the forecaster should recognize the risks involved.

Outliers

Outliers are another well-known source of difficulty in correlation analysis. A single outlier can have a significant impact on a correlation coefficient. Figure 4.1 shows a scatter diagram of sixty values from a simulated sample, from a bivariate normal distribution with theoretical correlation coefficient $\rho = 0.9$. One point was moved to become an outlier. The empirical correlation coefficient is now calculated to be 0.84. The figure shows that, except for this single point, the scatter is quite linear, and in fact, with this outlier removed, the estimated correlation coefficient is 0.9 (Devlin, et al., 1975). Robust methods, which are discussed in *The Beginning Forecaster,* offer some protection in these instances as well as others in which there is nonnormality in the error distribution.

Multicollinearity

Multicollinearity arises when the independent variables in a regression model are highly correlated. Models with more than about five independent variables often contain regressors that are highly mutually correlated. In such cases, it may be

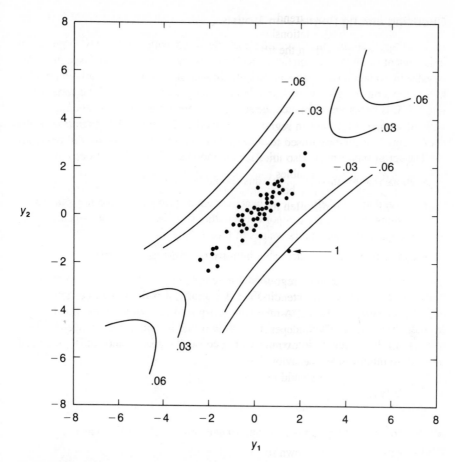

Figure 4.1 Scatter plot with influence function contours for a sample of bivariate normal data with the added outlier ($n = 60$, $\rho = 0.9$, $r = 0.836$).

profitable to seek linear combinations of these variables as regressors, thereby reducing the dimension of the problem. See, for example, Belsley, et al. (1980); Draper and Smith (1981); and Chatterjee and Price (1977, Chapter 7). See Chapter 14 and the time series forecasting example in Chapter 9 for additional approaches for identifying and treating this problem.

The standard assumptions discussed in the beginning of this chapter regarding errors (independence, constant variance, and normality) and the additivity of the model need to be evaluated. If, for a particular set of data, one or more of these assumptions fail, it may help to transform the data. Tests for invalid assumptions are discussed in Chapter 5.

Regression and correlation analysis are only tools and cannot replace thoughtful and thorough data analysis on the part of the forecaster.

Choosing among Regression Models

In some applications, the forecaster may be faced with a relatively large set of independent variables that, on theoretical or statistical grounds, should be considered of value in explaining a dependent variable. In such situations, it is clearly impossible to develop a meaningful regression relationship by incorporating all the variables of interest. It is necessary to reduce the set of independent variables for the regression without jeopardizing the usefulness of the model. Generally, interpretation of the coefficients is secondary to the need to explain variability with an overall model.

The set of independent variables for the models fall into three groups:

- Those that for theoretical reasons should be included.

- Those that are likely to be of benefit, such as proxy and "dummy" variables. Dummy variables are categorical in nature (yes–no, on–off, war–peace, . . .) and generally take on the values of 0 or 1.

- Others that may have desirable intuitive or statistical properties.

Given a class of models and model assumptions, the next step is to establish rules for accepting or rejecting variables. Generally, it is desirable to estimate lack of fit by an estimate of the expected sum of squared deviations of the fitted values from the true values. Then, a good procedure would be one that minimized the lack of fit. For a linear regression model with a constant error variance of σ^2, the ideal quantity to minimize is

$$\sum_{i=1}^{n} \text{Var}(\hat{y}_i),$$

which is σ^2 times (the number of independent variables). The variance σ^2 can be estimated as

$$s^2 = \frac{1}{n-k} \sum_{i=1}^{n} (y_i - \hat{y}_i)^2,$$

where k is the number of independent variables and n is the number of observations.

There are other criteria available for choosing k. These include Mallows's C_p, Anscombe's $s^2/(n-k)$ (as modified by Tukey), and Allen's PRESS (standing for "PREdiction Sum of Squares"). These criteria, which are discussed in greater detail in Draper and Smith (1981), Chatterjee and Price (1977, Chapter 9), and Mosteller and Tukey (1977, Chapter 15), should be applied with care; they can serve as a useful guide in making a sensible choice of k.

There are a variety of algorithms in use among regression practitioners for selecting subsets of independent variables (Draper and Smith, 1981; Mosteller and Tukey, 1977). Most of these approaches, while widely used, suffer from a number of theoretical and practical difficulties. Among these are methods using all possible regressions, backward elimination, forward selection, and stepwise regression.

When a method makes use of *all possible regressions,* the method utilizes all possible regressions for k variables. While everything is covered, the method can be expensive, time consuming, and usually unwarranted.

Backward elimination begins with all variables in the model. A partial F test is used to calculate the contribution of each variable. The variable with the lowest F value is removed, and the process is repeated until all variables have F values greater than a preselected value. With this technique, results for the full model are available, and the procedure is cheaper than the all-possible-regressions procedure.

The *forward selection* procedure starts by selecting a variable that best explains the variation in the dependent variable, by using the partial correlation coefficient. By using the residuals, a second variable is found that best explains the remaining variation in the dependent variable. The model is then reestimated with the new variable included. The calculations and selection procedure are repeated until no remaining variable has a partial correlation larger than a preselected value. While this procedure is economical and leads to a continuous improvement in the model, it ignores the possible reduction in importance of earlier variables.

The *stepwise technique* is similar to forward selection. At each step, however, the previously selected variables are examined with a partial F test to determine if any are not now contributing significantly. This method takes into account possible relationships between independent variables but requires more selection criteria. There are also weighted and resistant fitting variations to the stepwise method; however, these approaches should never be used without examining numerical and graphical outputs, such as the residual distributions and outliers, and reviewing regression coefficients at each step for possible changes in magnitudes and signs.

MULTIPLE REGRESSION CHECKLIST

_____ Is the relationship between the variables linear?

_____ Have linearizing transformations been tried?

_____ What is the correlation structure among the independent variables?

_____ Have seasonal and/or trend influences been identified and removed?

_____ Have outliers been identified and replaced when appropriate?

_____ Do the residuals from the model appear to be random?

_____ Are any changes in the variance apparent (is there heteroscedasticity)?

_____ Are there any other unusual patterns in the residuals, such as cycles, or cupshaped or trending patterns?

_____ Have F tests for overall significance been reviewed?

_____ Do the t statistics indicate any unusual relationships or problem variables?

_____ Can the coefficients be appropriately interpreted?

_____ Have forecast tests been made?

SUMMARY

This chapter has explained

- The multiple linear regression model and its assumptions.
- The interpretation of basic summary statistics.
- An overview of common pitfalls to be avoided in building multiple linear regression models. Detailed discussions of the pitfalls are presented in the appropriate chapters.
- An overview of regression models that automatically select subsets of independent variables. Exercise caution when using automated variable selection programs.
- Modeling assumptions implicit in formulating a forecasting model, including robustness considerations.
- The importance of the availability of relevant and appropriate independent variables for both the historical and future time periods.
- The importance of the interpretation of the results in the light of the assumptions.

The next chapter deals with some diagnostic tools, graphical as well as statistical, for dealing with the appropriateness of regression modeling assumptions.

USEFUL READING

BELSLEY, D. A., E. KUH, and R. E. WELSCH (1980). *Regression Diagnostics*. New York, NY: John Wiley and Sons.

CHATTERJEE, S., and B. PRICE (1977). *Regression Analysis by Example*. New York, NY: John Wiley and Sons.

DEVLIN, S., R. GNANADESIKAN, and J. R. KETTENRING (1975). Robust Estimation and Outlier Detection with Correlation Coefficients. *Biometrika* 62, 531–45.

DRAPER, N. R., and H. SMITH (1981). *Applied Regression Analysis*, 2nd ed. New York, NY: John Wiley and Sons.

RAO, C. R. (1973). *Linear Statistical Inference and Its Applications*. New York, NY: John Wiley and Sons.

SEARLE, S. R. (1977). *Linear Models*. New York, NY: John Wiley and Sons.

SEBER, G. F. (1977). *Linear Regression Analysis*. New York, NY: John Wiley and Sons.

CHAPTER 5

Analyzing Regression Residuals

Residual analysis is a process designed to reveal departures from the assumptions of a regression model, such as independence, constant variance, and normality. In this process, we try to determine what the "unexplained" variation might tell about the adequacy of the model. Quite often, transformations of variables or inclusion of new variables are suggested by residual plots. This chapter will:

- Briefly summarize the basics of residual analysis treated in *The Beginning Forecaster*.

- Provide new graphical aids (e.g., partial residual plots) appropriate for multiple linear regression models.

- Suggest a sequence of procedures to follow to improve the efficiency of the residual analysis process for multiple linear regression models.

TESTING MODELING ASSUMPTIONS

By analyzing regression residuals, you can determine if any of the modeling assumptions have been violated. A residual plot that has no visible pattern provides no evidence to reject the hypothesis that the residuals are independent, have zero mean, and have constant variance, for example.

Testing for Independence

Plots of residuals against time for the four models presented in Chapters 3 and 4 are shown in Figure 5.1. The large negative residuals in the second quarter of 1968 and the third quarter of 1971 in Model A are the results of strikes. In Model C, the introduction of dummy variables has accounted for the strike-attributable residuals of Model A. Note also that the range of variation in the residuals for Model C is considerably less than the range in Model A. Models B and D have been estimated by robust regression. The extreme residuals in these plots have been downweighted and the plots are of the adjusted residuals.

All four residual plots have an observable pattern of positively autocorrelated errors. Positive residuals tend to be followed by positive residuals, and negative residuals tend to be followed by negative residuals. The large positive residuals in the latter part of 1975, especially in the fourth quarter, are less pronounced in Model C than in Model A. The addition of the DFRB series as a variable in Model C has explained approximately one-half of the prior residual variation in the fourth quarter of 1975. The cyclical nature of the residual plots, and their approximate coincidence with U.S. economic recession and recovery periods, suggest that the addition of a cyclical economic variable may improve the models.

The run test for randomness was presented in *The Beginning Forecaster*, Chapter 15. The Durbin-Watson statistic provides a test for *first-order* autocorrelation in the residuals. A plot of autocorrelations of the residuals, called an *autocorrelogram*, when compared with the appropriate confidence limits, will suggest if there is any significant correlation among residuals *one or more lags apart*. If any of the tests or plots indicate significant serial correlation in the residuals, a revised model specification (new variables, transformations) may correct the problem. In other cases, the techniques presented in Chapters 13 and 23 should be considered.

Figure 5.2 shows plots of the autocorrelations of the residuals from the four models. All models show significant correlation at lag 1. Model A also shows significant correlation at lags 4 and 13. The correlation at lag 13 is a result of the two strikes, which occurred thirteen quarters apart. Following each strike, a recovery to a high occurs four quarters later. Also, it frequently seems to be the case that a peak occurs approximately four quarters after a trough has been reached (explaining the negative sign at lag 4). The robust regression in Models B and D seems to eliminate the autocorrelation at lags 4 and 13 of Model A. For the robust models, the effect of the adjusted residuals would show up in an autocorrelogram. The unadjusted residuals (Data – Robust fit) will be more extreme than those corresponding to an OLS (ordinary least squares) fit for the strike months. The addition of dummy variables to explain the data in the strike months (Model C) likewise tends to reduce autocorrelation.

About Constant Variance

Nonlinear residual patterns, such as a cupshaped residual pattern over time, or increasing or decreasing dispersion in the plot of the residuals against the predicted values, suggest the need to transform one or more variables. The logarithmic and square root transformations are the most frequently applied, though others, such as the *Box-Cox transformations*, may also prove useful.

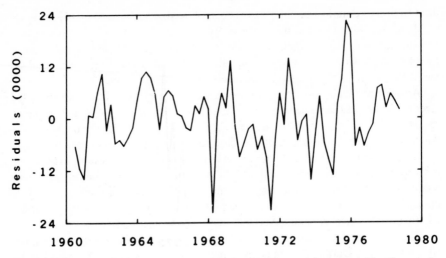

Figure 5.1(a)   Plots of residuals against time for the four models presented in Chapters 3 and 4.

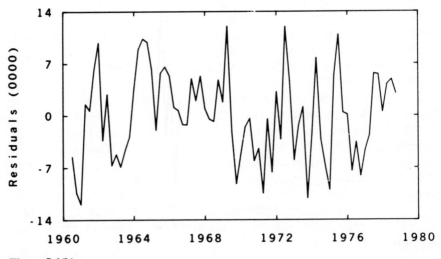

Figure 5.1(b)

Testing for Normally Distributed Residuals

When residuals are normally distributed, a very extensive (though not always realistic), often simple, and quite elegant set of statistical tests of significance can be applied. The *quantile-quantile* (Q-Q) *plot,* for example, where the quantiles of the residuals are plotted against the quantiles of the normal curve, provides a convenient

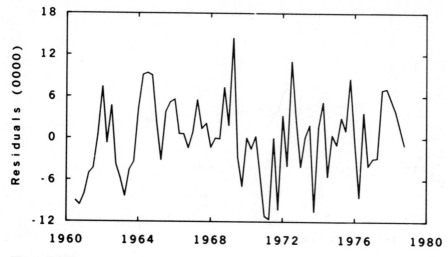

Figure 5.1(c)

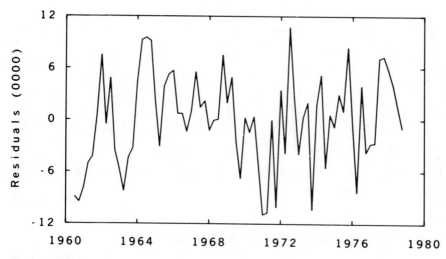

Figure 5.1(d)

way to test for normality. If the Q-Q plot is linear, the residuals are not unlike normally distributed errors. Points falling a significant distance from an otherwise linear relationship should be evaluated as potential outliers.

Transformations may be found that can render the distribution of the residuals approximately normal. In cases where normal residuals cannot be obtained, a robust regression method should also be considered since, in these cases, the usual tests of statistical significance have no meaning.

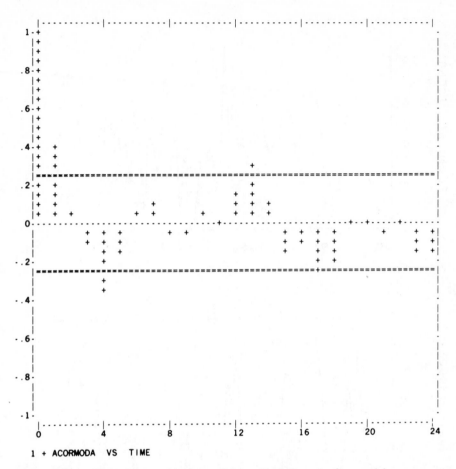

Figure 5.2(a) Plots of the correlograms of the residuals from the four models.

PLOTTING TECHNIQUES FOR RESIDUALS

There are many ways of graphically analyzing residuals, including

- Plots of residuals against time.
- Correlograms of residuals.
- Quantile-quantile plots.
- Plots of residuals against predicted values.
- Plots of residuals against independent variables.
- Plots of partial residuals against independent variables.

The first two items have already been discussed.

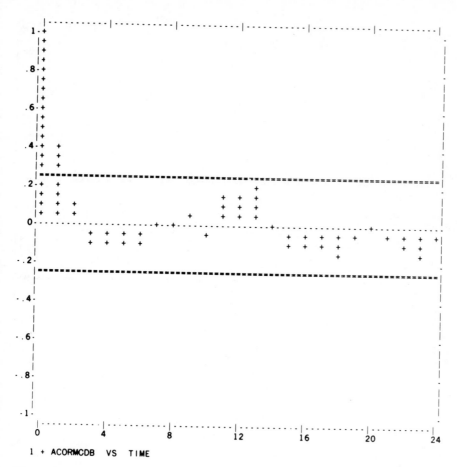

Figure 5.2(b)

Quantile-Quantile Plots

Figure 5.3 shows Q-Q plots of the residuals of the four models we have been considering, versus the normal curve. In Model A, the two strikes are readily apparent in the lower left quadrant of the plot. Since the residuals of this model are *not* normally distributed, the t test of the significance of the housing variable has little meaning. Such extreme outliers must be taken into account. In Model C, it is apparent that the dummy variables have solved the problem.

The objective of the Q-Q plot is to test for normality. Therefore, the unadjusted residuals are used for the robust models, because the adjusted residuals have been downweighted, and extreme residuals would not be apparent if these were used. If the Q-Q plot of the unadjusted residuals (Data − Robust fit) is linear, the forecaster, in general, can rely on OLS estimation.

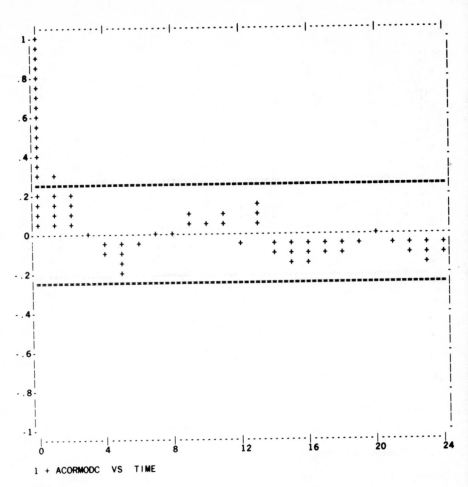

Figure 5.2(c)

Plotting Residuals against Fitted Values

A plot of the residuals against predicted (fitted) values is often helpful in deciding if a transformation of the dependent variable is appropriate. For example, if the plot shows increasing dispersion with increasing values of the dependent variable, a logarithmic or square root transformation of the dependent variable may help. The model is then reestimated, and the plot of the residuals against predicted values should then have constant variance.

Figure 5.4 plots the residuals against predictions for the four models. The plots do not indicate the need for transformations of the variables. There are no obvious patterns of increasing or decreasing dispersion over the range of the predicted values.

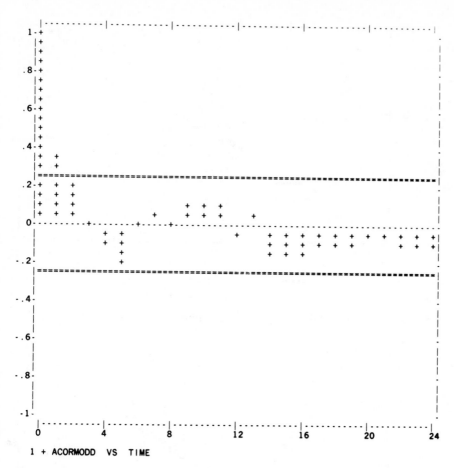

1 + ACORMODD VS TIME

Figure 5.2(d)

Plotting Residuals against Independent Variables

The residuals can also be plotted against each independent variable, in order to

- Detect outliers.
- Assess nonhomogeneity of variance.
- Determine if a transformation is required.
- Assess the importance of one variable in the presence of all the other independent variables in the model.

This graphical technique is particularly useful in the early stages of modeling.

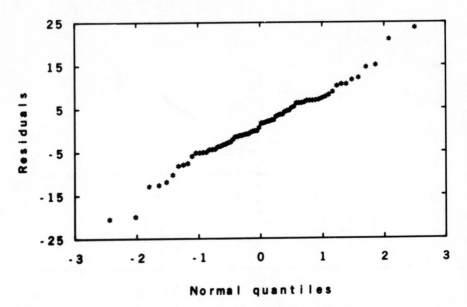

Figure 5.3(a) Normal Q-Q Plots of the residuals from the four models.

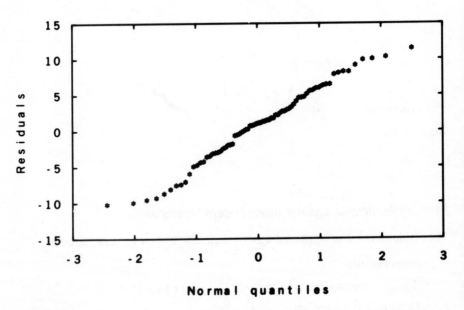

Figure 5.3(b)

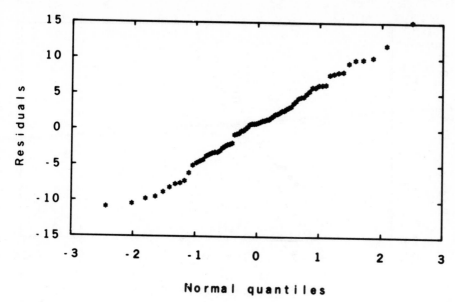

Figure 5.3(c)

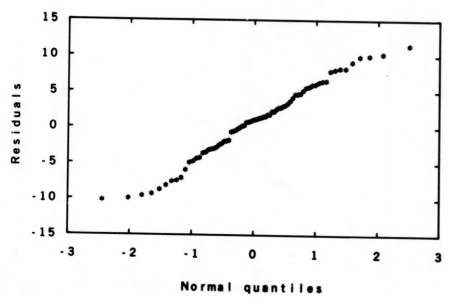

Figure 5.3(d)

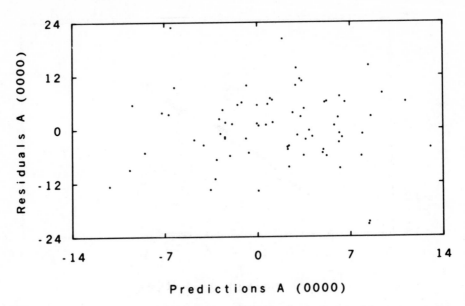

Figure 5.4(a)   Plots of the residuals against predictions for the four models.

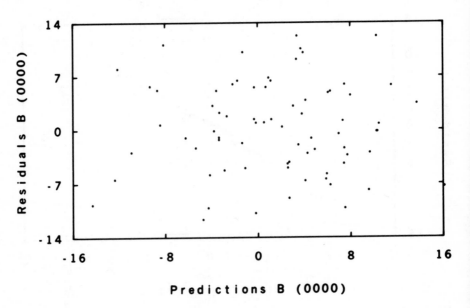

Figure 5.4(b)

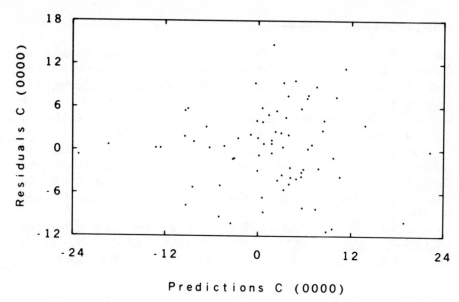

Figure 5.4(c)

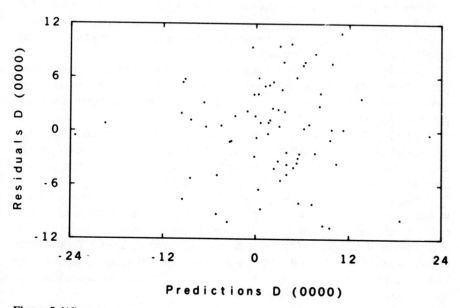

Figure 5.4(d)

Using Partial Residual Plots

Larsen and McCleary (1972) have shown that plots of *partial residuals* can be very informative in regression analysis. The partial residual plot allows you to determine each independent variable's ability to "explain" the dependent variable uniquely, given that all the other independent variables are already in the model. It also lets you assess the importance of the nonlinearity of a particular variable, and helps in precise selection of the appropriate transformation.

In the case of a very close fit, the partial residual plot might mask nonlinearities and outliers. Also, if the independent variables are highly correlated with each other (an undesirable condition in OLS estimation), the partial residual plot loses, to some extent, its ability to explain the impact of each independent variable on the dependent variable.

Partial residuals are obtained in the following manner. First, the complete model is fitted, including the independent variable X_i. The usual residuals (Data $-$ Fit) are then calculated. The value $\hat{\beta}_i X_i$ is added to the usual residuals. Consequently,

$$\text{Partial residual} = \text{Data} - \sum_{j \neq i} \hat{\beta}_j X_j .$$

Since the parameters have been estimated by considering all the independent variables, there is no bias in their estimates owing to a missing variable. The partial residual plot for an independent variable now includes the usual residuals plus any contribution provided by the variable X_i.

The plots of the partial residuals against each of the independent variables identify those variables that have the largest apparent dependence. These variables are retained in the model. Those with only slight dependence are candidates for elimination or replacement.

Figure 5.5 shows a plot of the usual residuals against the independent variable X_1. No violations of the model are apparent, and there seems to be little correlation with X_1. The partial residual plot depicted in Figure 5.6 shows a slight positive correlation with X_1, given that X_2, X_3, X_4, and X_5 are in the model.

A plot of the usual residuals against X_2 will show increasing variability for this variable, and little correlation. The partial residual plot in Figure 5.7 shows a definite negative correlation between X_2 and Y. It also shows increasing variance, but of a lesser amount. It is often the case that the original plots of Y against X_i show little or nothing about the relationship of X_i and Y, given that all the other variables are in the model $(Y|X_1, X_2, \ldots , X_{i-1}, X_{i+1}, \ldots , X_k)$. The partial residual plot tells much more.

Figure 5.8 plots the usual residuals against X_3. This suggests the need for a quadratic term in the model. The partial residual plot in Figure 5.9 suggests that a logarithmic term rather than a quadratic term will give better results, and this would require a transformation.

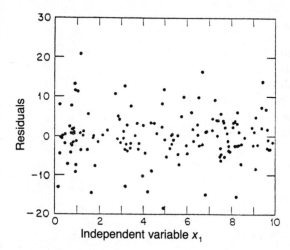

Figure 5.5   A plot of usual residuals against the independent variable X_1.

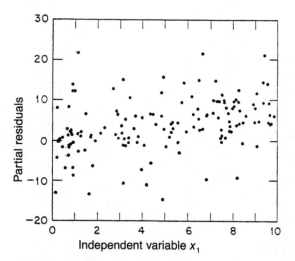

Figure 5.6   The partial residual plot shows slight positive correlation with X_1, given that X_2 through X_5 are accounted for in the model.

Figures 5.10 and 5.11 show the partial residual plots against the HOUS variable and the DFRB variable, respectively (since the patterns are similar, only those for Model C are shown). Both plots show positive and essentially linear relationships with considerable variability. There are more high-leverage points in the DFRB variable than in the HOUS variable. These plots show no need to transform the variables.

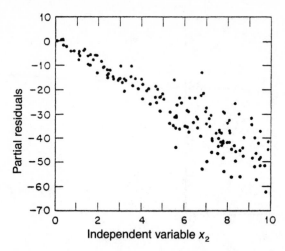

Figure 5.7   The partial residual plot shows negative correlation of X_2 with the dependent variable.

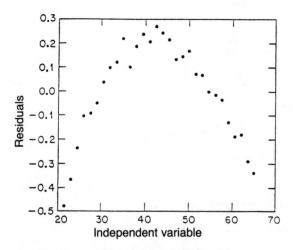

Figure 5.8   Plot of usual residuals against X_3.

In What Sequence Should Residuals Be Analyzed?

How should the forecaster sort out all these different ways of analyzing residuals? First, the summary statistics (F, t, incremental F, and R-squared statistics) should be reviewed. If the model passes these tests, residual analysis can begin. A recommended sequence of residual analysis begins with a plot of the residuals against the predicted values of the dependent variable. This plot is reviewed for constancy

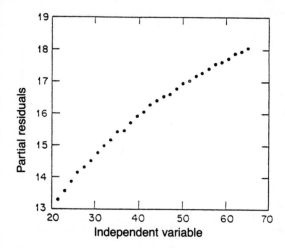

Figure 5.9   The partial residual plot suggests a transformation is required.

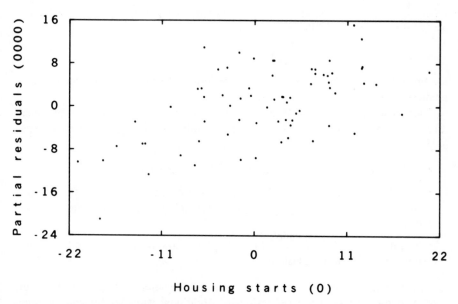

Figure 5.10   The partial residual plot against housing, the differences of order 4 of the quarterly housing starts series.

of variance among residuals. Patterns of increasing dispersion, with increasing magnitude, as mentioned earlier, suggest that logarithmic or square root transformations of the dependent variable should be made as a first attempt. The model is then

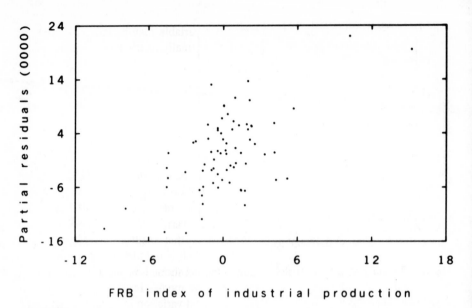

Figure 5.11   The partial residual plot against DFRB, the differences of order 1 and 4 of the FRB Index of Industrial Production.

reestimated to accomplish this, and the residuals are plotted once again against predicted values.

Once a satisfactory transformation of the dependent variable has been obtained, if one is indeed required, partial residual plots against each independent variable can be generated. At this time

- Delete variables with little correlation. Reestimate the model.

- Transform independent variables that exhibit nonlinear relationships (one at a time). After each transformation, reestimate the model and generate a new set of partial residual plots.

- Progress to additional analyses if all plots show linear and significant (not horizontal) relationships.

For time series models, plot the residuals and the autocorrelogram of the residuals. Test for serially correlated residuals. Additional variables for consideration may be suggested by these plots. The techniques discussed in Chapter 13 may be required if the serial correlation problem remains.

The next step is to generate quantile-quantile plots of the residuals versus the quantiles of a standard normal distribution. This will highlight potential outliers (review each end carefully) and indicate if there are departures from normally distributed residuals. Outliers should be investigated and replaced, if replacement is appropriate. Transformations of the dependent variable may be required. Alternatively, robust regression may be appropriate if normally distributed residuals cannot be obtained.

REGRESSION BY STAGES

One approach for identifying variables that should be included in a model is to build a regression model in stages. Using Tukey's notation (Mosteller and Tukey, 1977) for convenience, let $Y_{.1}$ denote the residuals after fitting a model with X_1 only. Instead of plotting the residuals $Y_{.1}$ against a new variable X_2, you can first regress X_2 on X_1 and denote with $X_{2.1}$ the residuals of X_2 that result after fitting X_1. Then $X_{2.1}$ represents the *additional* information in X_2 that is not already captured by X_1.

Now you can plot $Y_{.1}$ against $X_{2.1}$, and this will show if X_2 has a strong apparent dependence with Y, given that X_1 is already in the model. In other words, this plot will show the relationship between the unexplained variation in Y and the additional information in X_2. In the case where X_1 and X_2 are highly correlated with Y and with each other, the plot of $Y_{.1}$ versus $X_{2.1}$ will show little correlation, since X_2 adds little, given that X_1 is in the model.

Next, X_3 is regressed on X_1 and X_2; the residuals that result are denoted by $X_{3.12}$. At the next stage, the residuals of the model containing both X_1 and X_2 (thus, $Y_{.12}$) are plotted against $X_{3.12}$. Those variables that show high dependence are included in the model and those with little dependence are reserved for future use.

In general, let $X_{i.REST}$ denote the residuals of X_i on the rest of the independent variables. By regressing Y on $X_{i.REST}$, the same regression coefficient will result for X_i that results from the regression of Y on all the X's. The remaining regression coefficients are, of course, different. By displaying $Y_{.REST}$ (the fit of Y on *all* X_i) against each $X_{i.REST}$, it is possible to see the impact of each variable and, in some cases, the impact of individual points on the estimate of the regression coefficient.

In order to examine how to combine or compare coefficients in different regressions, Tukey uses the "minvar modification" of the fitted coefficient b_i, denoted by $b_{i.REST}$. For each i, $b_{i.REST}$ is that linear combination of fitted regression coefficients, including unity times the ith coefficient b_i, which has minimum estimated variance. These coefficients can be compared with the estimated variance of b_i itself. As a set of ratios, they may point to a useful combination of independent variables and to important dependencies among estimated regression coefficients (Mosteller and Tukey, 1977, section 13H).

The regression-by-stages approach is illustrated in the following example. A more traditional approach to this example will be developed in Chapter 9 as a demand-analysis case study. For the purposes of the present discussion, a regression relationship will be sought between the percent extension development in residences (extension telephones in households divided by main telephones in households) by geographic area, by income adjusted for the cost of living, by the percent of white-collar employees, and by the percent of households owning more than one automobile.

In a purely exploratory approach, you might consider all possible combinations of the independent variables and the dependent variable. In this example, interest lies in the relationship of the dependent variable to income, employment, and automobile ownership. Therefore, the first step is to display the frequency distributions or box plots (*The Beginning Forecaster,* Chapter 7) of the variables, and the second step is to generate scatter diagrams between the percent extension development and the independent variables. The frequency distribution of "percent of development" is plotted in Figure 5.12 and shows a long tail at the high end. At the low end are apparent observations showing 0–4-percent extension development. In actuality, these are transcription errors, which need to be removed prior to modeling. Since these small percentages are not totally unrealistic, they could remain undetected in an actual modeling procedure, and they have been left in as "undetected" outliers for this example; the impact of these outliers on the OLS parameter estimates will thus be partially assessable.

The scatter diagram between residence extension development and income in Figure 5.13 shows an essentially linear relationship, with most of the observations clustered below \$20,000 income. There is some indication of increasing variability with increasing income, and so a square root transformation of the dependent variable was taken in an attempt to obtain constant variance, and to "bring in" the high-leverage points in the upper righthand corner of the plot.

A scatter diagram of the square roots of extension development versus income showed some improvement in variance, but the points corresponding to very high income appeared to have bent downward, suggesting some nonlinearity.

Plotting the logarithms of the income variable (X_1) against the square roots of percent development appeared to exhibit the most promising relationship in terms of linearity, constancy of variance, and reduction in high-leverage points. A regression was performed and a plot of residuals $Y_{.1}$ against predictions was made. No obvious problems were apparent, and this was confirmed by the near straight-line configuration in a Q-Q plot of the residuals versus the normal quantiles.

The next variable that was entered in the regression was the percent of white-collar employment (X_2). This variable was regressed on X_1; then the residuals $Y_{.1}$ were plotted against the residuals $X_{2.1}$ (Figure 5.14). This plot shows the incremental explanatory power of X_2, given that X_1 is in the model. There is a slight linear positive relationship, with considerable variability. This variable was added into the model, the regression was performed, and the residuals showed no further problems.

The percent of households owning more than one automobile (X_3) was then regressed on X_1 and X_2. Plotting the residuals $Y_{.12}$ against the residuals $X_{3.12}$ showed a slight positive relationship, with even greater variability. Given that X_1 and X_2 are already in the model, X_3 provides relatively little additional information.

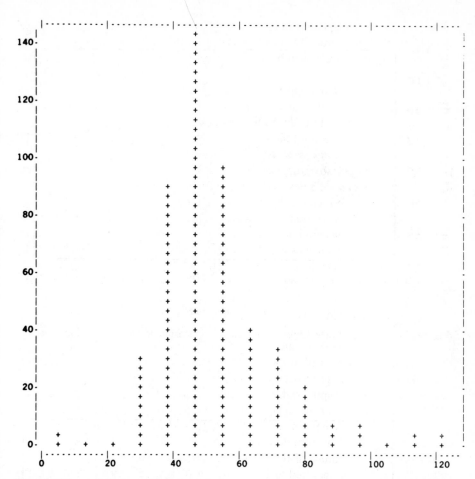

Figure 5.12   A frequency distribution of the percent of extension development, showing a long tail in the distribution at the high end.

The final model included all three variables; a Q-Q plot (Figure 5.15) of the residuals depicts a linear pattern. (In Chapter 10, OLS and robust regression will be used to estimate a model for this example, and the results with the three outliers will be compared for the two estimation procedures).

Table 5.1 summarizes the model results. R^2 increases and MSE (mean squared error) decreases as X_2 and X_3 are added to the model. All parameters are significantly different from zero (t test), and the F statistic is significant in all cases. The model for white-collar employment (X_2) shows significant correlation ($R^2 = 0.62$ percent) with income (X_1). The percent of households with more than one automobile (X_3) correlates less well ($R^2 = 0.54$) with income and employment. In this example, the

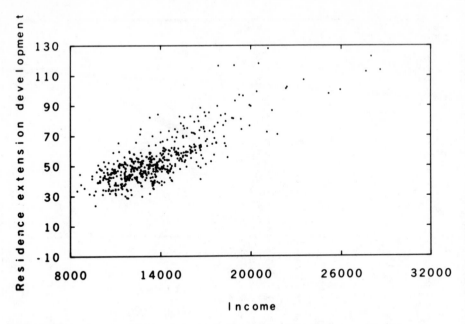

Figure 5.13   A scatter diagram between residence extension development and income.

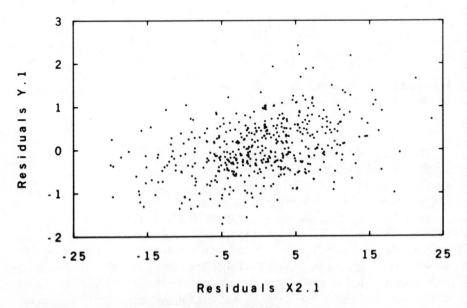

Figure 5.14   A plot of residuals $Y_{.1}$ against residuals $X_{2.1}$.

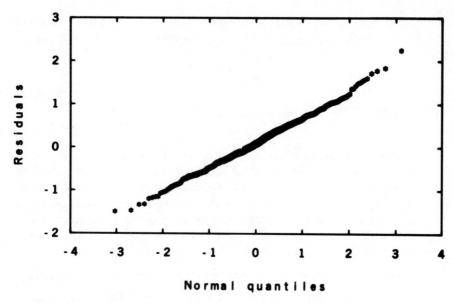

Figure 5.15 A normal Q-Q plot of the residuals in the full model.

Table 5.1 A summary of results from the final model.

Residuals included in model*	R^2	MSE	t-statistic X_1	t-statistic X_2	t-statistic X_3
$Y_{.1}$	0.61	0.42	27.3	—	—
$Y_{.12}$	0.67	0.36	11.0	9.2	—
$Y_{.123}$	0.69	0.34	7.5	8.6	4.9
$X_{2.1}$	0.62	54.1	27.6	—	—
$X_{3.12}$	0.54	59.2	11.9	3.0	—

Final model

$$Y = -13.33 + 1.93(\log X_1) + 0.032(X_2) + 17(X_3)$$
$$[5.9] \qquad [7.5] \qquad\quad [8.6] \qquad\quad [4.9]$$

*Y = Percent residence extension development
X_1 = Income
X_2 = Percent white collar employment
X_3 = Percent households with more than one auto

Note: All F statistics are significant at the 5-percent level.

first variable X_1 explains a large percent of the variation in Y. The addition of X_2 and X_3 raises R^2 from 0.61 to 0.69 and results in a 19-percent reduction in mean squared error.

SUMMARY

Residual analysis is perhaps the single most valuable diagnostic tool for evaluating regression models. Fortunately, much of the residual analysis can be carried out effectively by using graphical techniques. Residual analysis is concerned with revelation of departures from the assumptions of a model and suggestion of corrective steps. In building multiple linear regression models, the following sequence of steps is recommended. (It is understood that the model is reestimated after each transformation, new variable, etc.)

- Estimate the model, review the summary statistics, and calculate residuals.
- Plot the residuals against the predicted values to determine the need to transform the dependent variable.
- Plot partial residuals against independent variables to determine the need to transform or eliminate certain independent variables. Alternatively, use regression-by-stages to decide what variables and transformations are appropriate.
- In time series models, test for serially correlated residuals by plotting residuals against time and by plotting autocorrelograms of residuals. Correct or minimize the extent of the problem.
- Generate Q-Q plots to check for nonnormality in residuals.
- Consider robust regression if the OLS model assumptions are not appropriate.

If the model has survived all these tests,

- Generate forecast tests (see *The Beginning Forecaster*) to determine how well the model would have predicted recent (actual) performance.
- Consider using actuals as independent variables, to test the model.
- Consider using the independent variables to make appropriate forecasts at the time the regression is run; this will test how accurately the model forecasts these variables.

USEFUL READING

LARSEN, W. A., and S. J. McCLEARY (1972). The Use of Partial Residual Plots in Regression Analysis. *Technometrics* 14, 781–90.

LEVENBACH, H., and J. P. CLEARY (1981). *The Beginning Forecaster*. Belmont, CA: Lifetime Learning Publications.

MOSTELLER, F., and J. W. TUKEY (1977). *Data Analysis and Regression*. Reading, MA: Addison-Wesley.

CHAPTER **6**

Trend Models for Long-Term Forecasting

Many annual time series in socioeconomic systems are steadily increasing functions of time. This chapter treats some long-range forecasting methods based on trend extrapolation.

- For most trending data it is common practice to fit a simple curve, such as an exponential or straight line, through the data.

- For data tending to some saturation level, the logistic curve has found widespread application.

This chapter discusses these models by considering percent changes or relative growth rates as the dependent variable in a regression model involving magnitudes and reciprocals of magnitudes, thereby placing these models in the framework of multiple linear regression.

TREND-CURVE FITTING

The literature dealing with trend curves and long-range forecasting is quite extensive (Armstrong, 1978; Draper and Smith, 1981; Granger, 1980; Gregg et al., 1964). Most approaches use mathematical formulæ to fit successive values of the variable, from which projections are then made by extrapolation. Other techniques utilize models based on transformations of the data, which are then used for curve fitting.

Once the data are plotted, it is often evident that a transformation of the data suggests a simple curve. Taking logarithms, square roots, or reciprocals sometimes produces relatively linear patterns, so that a straight line may be adequate.

There are several dangers inherent in using trend curves as forecasting tools. While a given curve may yield a very good fit in terms of summary statistics, there is little guarantee that such a pattern will continue into the future. The forecaster should be aware that extrapolations need to be subjected to rational assumptions about the future course of events, such as saturation of a market versus continued growth. It may also happen that both of two curves give good results as fitting models, but may yield vastly different forecasts.

Exponential Growth

Exponential growth patterns appear frequently in practice. In numerous planning situations, the long-term growth rate is of interest and it is not desirable to incorporate the short-term business cycle fluctuations in the forecast. For example, the extremely long lead-time involved in planning for electrical generating capacity, particularly electricity generated in nuclear plants, would make the long-term forecast more important in such planning than the short-term forecast.

Exponential growth corresponds to a steady, constant growth rate. A semilogarithmic plot of the data would give a straight line. An *exponential curve* is given by the equation

$$Y_t = \exp(a + bt),$$

where b represents the *growth rate*.

By taking natural logarithms, it becomes evident that the curve represents a straight line:

$$\ln Y_t = a + bt.$$

From a statistical modeling viewpoint, these two curve specifications must be carefully distinguished, because of the different error structures that each may be assumed to present. In statistical terms, the exponential *curve* can be specified as an exponential *model;* that is, as

$$\mu_Y = \exp(\alpha + \beta t);$$

in terms of

$$\text{Data} = \text{Model} + \text{Error},$$
$$Y_t = \exp(\alpha + \beta t) + \varepsilon_t,$$

where ε_t represents a random error term.

Alternatively, a linear model,

$$\ln Y_t = \alpha + \beta t + \varepsilon_t^*,$$

represents exponential growth in terms of the *logarithms* of the data and an error structure ε_t^*. These two specifications arise from different statistical structures, and the differences can have a significant impact on the interpretation of results.

Both models can be fitted with conventional methods; however, the inferences and statistical properties of the estimates are different. As a general rule, it is simpler to "linearize" a problem as much as possible before using estimation methodologies to fit parameters. In the above example, the second model is linear in its parameters, thereby simplifying the problem of statistical fitting and inference.

FITTING CURVES BY USING THE TECHNIQUE OF LEAST SQUARES

The most commonly used curve-fitting technique is the technique of least squares. Curve-fitting with least squares is essentially a mathematical technique in which a vector of parameters, $\alpha = (\alpha_1, \alpha_2, \ldots, \alpha_k)'$, in a model $f(t;\alpha)$, is estimated by minimizing the sum of the squares of the deviations of the function from the data points. Thus, for a discrete time series, Y_t is assumed to be of the form

$$Y_t = f(t;\alpha) + \varepsilon_t, \qquad t = 1, 2, \ldots, n$$

where t represents time, $\alpha = (\alpha_1, \alpha_2, \ldots, \alpha_k)'$ is a vector of model parameters, and all $\{\varepsilon_t\}$ are independent, identically distributed, normal, random errors with zero means and a common variance, σ^2. The fitted model is given by

$$\hat{Y}_t = f(t;\hat{\alpha}),$$

where $\hat{\alpha} = (\hat{\alpha}_1, \hat{\alpha}_2, \ldots, \hat{\alpha}_k)'$ comprises the least squares estimates of $\alpha_1, \alpha_2, \ldots, \alpha_k$, respectively. In practice, the determination of the appropriate *functional relationship* $f(t;\alpha)$ is not always straightforward (nor is the estimation of $\hat{\alpha}$).

Saturating Growth

Among the trend-growth models that have saturation levels, the logistic model and modified exponential model are probably the most effective. The *logistic* growth curve is an S-shaped curve that has found widespread use in scientific and business applications. There are a variety of estimation methods in the literature (Oliver, 1964).

The logistic growth curve has the formula

$$Y_t = M/(1 - ke^{-at}).$$

It can be seen that if the saturation level, M, is known, or can be assumed to be known, the formula can be rewritten as a linear function of time, so that (with $\ln = \log_e$),

$$\ln\left(1 - \frac{M}{Y_t}\right) = \ln k - at.$$

In this linear form, the least squares approach can be assumed. In general, however, M is unknown and must be estimated from the data. With an unknown saturation level, the model is no longer linear, so that a nonlinear estimation technique must be employed.

There is a situation in which saturation can be linearly estimated. It can be shown that the percent changes or relative growth rate ("logarithmic derivative," or $d \ln Y_t = Y_t'/Y_t$) is a linear function of the magnitude Y_t. Since $Y_t'/Y_t \approx (Y_t - Y_{t-1})/Y_t$, the percent changes can be represented as

$$(Y_t - Y_{t-1})/Y_t = a - \frac{a}{M_1} Y_t.$$

By letting $b = a/M_1$, the model can be approximated as a simple linear regression model,

$$(Y_t - Y_{t-1})/Y_t = \alpha + \beta Y_t + \varepsilon_t.$$

This was the basis for studying saturation in population growth by Hotelling (1927) and serves as the motivation for the approach developed in the next section.

Another elementary but useful growth curve is the *modified exponential curve*, given by the formula

$$Y_t = ke^{-at} - M_2.$$

Basically, this curve can take on both the saturating and nonsaturating forms. The modified exponential model has a saturation level designated by M_2. The curve can be linearized to give

$$\ln(Y_t + M_2) = \ln k - at,$$

which can then also be formulated as a simple linear regression model, if M_2 can be assumed to be known.

RELATIVE GROWTH-RATE MODELS

Modeling time series data that are steadily increasing through time is an important part of many business and government planning organizations. This section presents an empirical approach for analyzing and projecting such trending time series. The basic idea is to consider changes in growth or, more specifically, the pattern of *percent changes* in the data over time.

Modeling Changes of Percent

Percent changes are sensible quantities to study as dependent variables, since they are independent of the units of the data. In addition, changes of percent for the linear, exponential, modified-exponential, and logistic growth-curves are elementary functions of the magnitude of the time series. Such considerations have led to the formulation of a class of relative growth-rate models for steadily trending data (Levenbach and Reuter, 1976). These models are capable of describing saturating as well as nonsaturating growth.

The most general model in this class describes the percent changes of the variable in terms of its magnitude and the reciprocal of its magnitude. Its mathematical form is known as the Ricatti curve. Thus the complete statistical model for empirical, relative growth rates, R_t, is given by

$$R_t = \alpha + \beta Y_t + \frac{\gamma}{Y_t} + \varepsilon_t, \qquad \varepsilon_t \sim N(0,\sigma^2).$$

The Ricatti Class

All models within the Ricatti class possess monotonically increasing solutions that depend on coefficients. The modified exponential, logistic, and Ricatti curves can be saturating or nonsaturating for certain ranges of values of α, β, and γ. The nonsaturating logistic curve and the Ricatti curve have limited application, since they have the shape of the tangent function. These functions, when extrapolated, approach a vertical asymptote, and thus rise too rapidly to be realistic.

The solutions of the growth rate models are known in closed form. It is well known that the exponential curve comes from constant growth rates

$$R_t = a.$$

The modified exponential is given by

$$R_t = a + \frac{c}{Y_t}.$$

The logistic curve is described by

$$R_t = a + bY_t$$

For linear growth, the percent changes are proportional to the reciprocal of the magnitude:

$$R_t = \frac{c}{Y_t}$$

A general Ricatti curve gives relative growth rates by combining exponential, modified exponential, linear, and logistic growth rates:

$$R_t = a + bY_t + \frac{c}{Y_t}.$$

The corresponding statistical model has the form:

$$R_t = \alpha + \beta Y_t + \frac{\gamma}{Y_t} + \varepsilon_t, \qquad \varepsilon_t \sim N(0,\sigma^2).$$

The coefficients α, β, and γ can be estimated by linear regression techniques.

There is an additional quantity in the growth curve (called the "integration constant") that can be fitted by a one-parameter nonlinear optimization technique, assuming that the remaining coefficients in the model are estimated from the linear regression model. Empirical growth rates (dependent variable) can be calculated by using either "first differences divided by the magnitude of the variable" or "first differences of the logarithms." There appears to be little practical difference between the two estimators. Approximate confidence limits on the forecasts for Ricatti models are available in Levenbach and Reuter (1976).

In order to see how these models can be used, consider the data depicted in Figure 6.1. These data appear to be growing like a purely exponential trend-curve. Now consider the percent changes associated with the data, which are plotted in Figure 6.2. They show an increasing trend, which implies the growth is even faster than exponential growth, percent changes of which would be constant over time. Thus, a model such as the modified exponential model shown in Figure 6.2 would provide a more realistic description of the trend. Figure 6.2 also shows the corresponding fitted, modified-exponential trend-curve.

An Extended Model

Note that in Figure 6.2 the differences between the empirically determined percent changes and the fitted percent changes appear to be increasing with time. In practice, this increasing variability may be accounted for in part by an extended model.

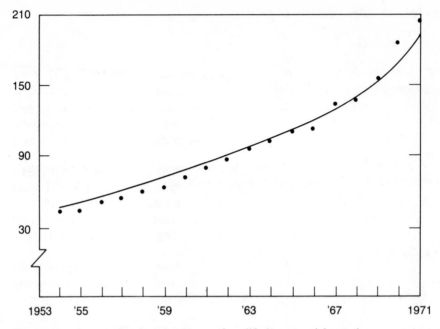

Figure 6.1   Data conforming to a pattern of modified exponential growth.

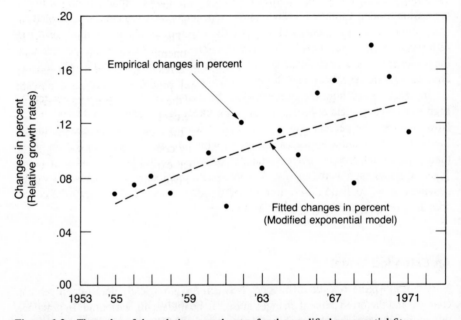

Figure 6.2   Time plot of the relative growth rates for the modified exponential fit.

The *extended model* has the form

$$R_t = \left(\alpha + \beta Y_t + \frac{\gamma}{Y_t} \right) f(X_t) + \varepsilon_t,$$

where R_t is determined empirically and $f(X_t)$ is some appropriate function of an exogenous or independent variable. Some simple choices in the discrete case include $f(X_t) = X_t$, $X_t - X_{t-1}$, or $(X_t - X_{t-1})/X_t$. Solutions for these models are similar in form to those for the basic trend model, except that time, t, is replaced by a function of X_t.

A pair of highly correlated and trending time series have been selected to illustrate the use of the extended model. These series, shown in Table 6.1, depict world energy consumption and world energy production for the years 1958 to 1969.

Here the *percent change of consumption* is expressed as a function of the *percent change of production*, in contrast to simply correlating the *magnitudes* of the variables directly.

Figure 6.3 represents a plot of the relative growth rates (percent change) of world energy consumption and world energy production that shows patterns that are nearly identical. Such a close relationship is to be expected, in this example, and it suggests an extended exponential model of the form

$$R_{Y_t} = \delta R_{X_t} + \varepsilon_t,$$

in which the percent changes between Y_t and X_t are directly proportional.

The exponential trend model fitted to the data for world energy consumption generated the predictions $\hat{Y}_t^{(1)}$ shown in Table 6.1. The residual pattern from this fit was fairly random. In addition, the modified exponential model was tried, which produced a slight improvement in fit, $\hat{Y}_t^{(2)}$, with a residual pattern similar to the exponential fit. The extended exponential model produced a rather significant improvement in fit. The extended model decreased the square root of the average of the square of the residuals (SASR) by about 50 percent.

In practice, it is necessary to have forecasts of the exogenous variable available in any extended model. However, further modeling could perhaps produce a close relationship of the exogenous variable to another explanatory variable, thereby developing a chain of closely interacting models involving only a few variables in each link.

SUMMARY

The trend models described in this chapter are useful for applications requiring forecasts for three to ten or more years. The relative growth-rate models

Table 6.1 World energy production and world energy consumption—million metric tons of coal equivalent, 1958–1969.

Year	Production, X_t	Consumption, Y_t	Exponential trend, $\hat{Y}_t^{(1)}$	Modified exponential trend, $\hat{Y}_t^{(2)}$	Extended exponential, $\hat{Y}_t^{(3)}$
1958	3826.2	3719.3	3692.5	3755.9	3708.5
1959	4056.8	3940.5	3879.6	3915.1	3940.9
1960	4297.5	4233.5	4076.2	4087.2	4183.9
1961	4273.8	4188.8	4282.7	4273.2	4159.9
1962	4512.1	4417.8	4499.7	4474.3	4401.0
1963	4792.4	4714.4	4727.7	4691.5	4685.2
1964	5087.8	4984.0	4967.3	4926.3	4985.3
1965	5325.1	5220.7	5219.0	5180.1	5226.9
1966	5626.2	5510.5	5483.5	5454.4	5534.1
1967	5761.1	5610.5	5761.3	5750.8	5671.9
1968	6141.0	6013.1	6053.2	6071.1	6060.7
1969	6521.9	6406.2	6360.0	6417.3	6451.4
		SASR*	77.7	73.9	33.2

*SASR = Square root of the average of the square of the residuals.
Source: United Nations Statistical Yearbook, 1969.

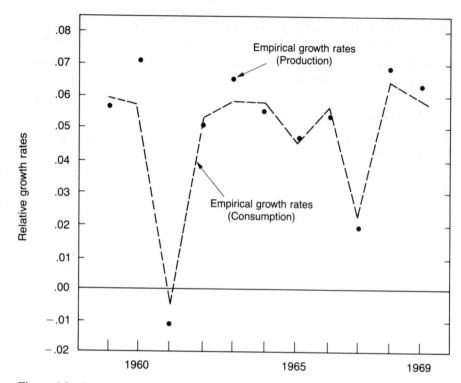

Figure 6.3 Empirical growth rates of annual world energy production and world energy consumption.

- Can be used to describe either saturating or nonsaturating forecasts that agree with assumptions about long-term growth prospects.

- Express the dependent variable in terms of changes in percent rather than magnitudes, thus providing not only a natural expression for the data but also a unifying modeling-construct for trending data.

USEFUL READING

ARMSTRONG, J. S. (1978). *Long-Range Forecasting: From Crystal Ball to Computer.* New York, NY: John Wiley and Sons.

DRAPER, N. R., and H. SMITH (1981). *Applied Regression Analysis,* 2nd ed. New York, NY: John Wiley and Sons.

GRANGER, C. W. J. (1980). *Forecasting for Business and Economics.* New York, NY: Academic Press.

GREGG, J. V., C. H. HASSEL, and J. T. RICHARDSON (1964). *Mathematical Trend Curves; An Aid to Forecasting*. ICI Monograph. Edinburgh, Scotland: Oliver and Boyd.

HOTELLING, H. (1927). Differential Equations Subject to Error and Population Estimates. *Journal of the American Statistical Association* 22, 283–314.

LEVENBACH, H., and B. E. REUTER (1976). Forecasting Trending Time Series with Relative Growth Rate Models. *Technometrics* 18, 261–72.

OLIVER, F. R. (1964). Methods for Estimating the Logistic Growth Function. *Applied Statistics* 13, 57–66.

Part 3

Demand Analysis and Econometrics

CHAPTER **7**

Elements of
Demand Analysis

The specification stage of model building was discussed in Chapter 2, and the need to develop a theory of demand for the item to be forecast was emphasized there. The demand theory is important since it suggests the variables that should be considered in the model. It is difficult to gain acceptance of a model when the variables do not "make sense." A number of basic elements of demand analysis are treated in this chapter. These include:

- Key determinants of demand.
- The price–quantity relationship.
- The rationale for relative price as a demand variable.

DETERMINANTS OF DEMAND

Economists have long attempted to determine what causes people to behave as they do in the marketplace. Over the years, one aspect of this research has evolved into a theory of demand (Samuelson, 1978). *Demand* expresses the inverse relationship between price and quantity; it shows the maximum amount of money consumers are willing and able to pay for each additional unit of some commodity, or the maximum amount of the commodity they are willing and able to purchase at a given price. There may not be enough of the commodity available to satisfy the demand. Economists concern themselves not with a *single* item purchased by members of a group (a market) but rather with a *continuous flow* of purchases by that group. Therefore, demand is expressed in terms of the *amount desired* per day, per month, or per year.

There are a number of *determinants of demand*. Demand varies with tastes, total market size, average income, the distribution of income, the price of the good or service, and the prices of competing and complementary goods.

Tastes

Some changes in taste are passing fads, such as the demand for hula hoops; others are more permanent, such as the demand for private automobiles rather than public transportation. Advertising is quite often intended to bring about a change in tastes. But whatever the causes, whenever tastes change, the demand for some goods increases, and the demand for others decreases.

Market Size

If a specific forecasting problem suggests that the size of the market should be modeled, suitable factors are population and the number of households.

All other things being equal, one would expect the demand for a good to increase in proportion to the growth in the total population or certain age groups within the population. Of course, these people must have the ability to pay for the good.

Income

The demand for a good generally increases as real income increases. When people are poor, food, clothing, and shelter account for most of their expenditures. As households become wealthier, more income is left over to be spent for additional items, such as durables (appliances and furnishings), housing improvements, and services. Also, it is important to consider real or constant dollar income when considering income as a determinant of demand. If inflation eats up all the increases in current dollar income, the household is no better off than it was before; there is no additional money available to purchase nonbasic goods or services.

Distribution of Income

Average household income may or may not be a very good measure of wealth, depending on the distribution of income. For example, in some population areas there are few rich people and many poor people; consequently one must consider the distribution of income as well as its average value when determining demand for an item.

Price

The Law of Demand states that demand for a good declines as the price of the good increases. Goods and services are desired to satisfy wants and needs, however, and since there are often alternative or competing goods available, a rise in the price of one good will cause some people to substitute other goods or services.

Goods or services can be considered as complementary or competitive commodities. *Complementary commodities* are used together to achieve a result. An increase in the price of one good will result in a decrease in the demand for its complement. For example, gasoline and automobiles are complementary. When the price of gasoline increased significantly, the demand for automobiles tended to decline—especially the demand for less fuel-efficient models.

There are also many goods or services that are *competing* to satisfy the same needs or desires. Consequently, a decrease in the price of one of these goods will cause a decrease in demand for its competitor.

All of the above determinants are acting simultaneously to establish the demand for a product. Because of this simultaneity, it is not possible to develop a simple theory of demand if all the variables are allowed to change at once.

THE DEMAND CURVE

To circumvent simultaneity, one can use a technique employed in all scientific research: assume that all but one of the determinants of demand are held constant. It is then possible to measure the effect of one independent variable, such as price, with all others held constant. A different determinant, such as income, can then be made to vary, and changes in the quantity of the product for which there is demand can be measured. This is simple in theory, but in the real world it is not possible to hold other determinants constant. Empirical verification of demand must take this into account.

The demand for a product will increase as the price of the product decreases, all other determinants held constant. As price falls, a product becomes cheaper relative to its substitutes, and thus it becomes easier for the product to compete for the consumer's dollar. The relationship, known as the *demand curve*, is plotted in Figure 7.1.

Movement along the Curve

When the quantity of a demanded product changes because of a change in price, this is referred to as *movement along the demand curve*.

When the entire curve shifts to the right or left because of a change of something other than price, it is referred to as a *shift in the demand curve*. The solid line in Figure 7.2 shows the original demand curve of Figure 7.1. A shift to the right, as shown by the dotted line, could result from an increase in income to a higher level, I_1. The curve could also shift as a result of an increase in the price of a competitive product, a favorable change in tastes, an increase in population, and a change in the distribution of income, for example.

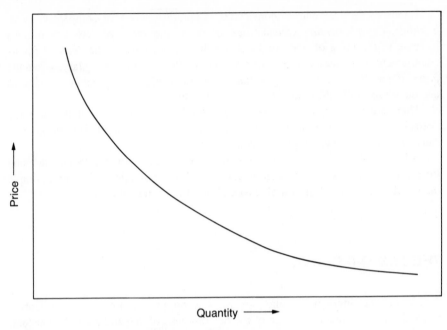

Figure 7.1   A theoretical demand curve for a product.

Some groups of consumers, having certain product preferences, will be able to buy more of a product if their income increases as a result of an increase in government transfer payments or favorable changes in tax policy.

Demand theory recognizes that there are also certain goods that are considered to be less preferable by a consumer allocating an increased income. Examples of these goods are potatoes or frozen foods, which a consumer willingly substitutes with more expensive food, given the opportunity. An increase in income will result in a decrease in the quantity of these goods that is demanded.

Relationship to Supply

The demand curve is only one-half of the story. One must also be concerned with supply. Let us discuss briefly the notion of supply and its relationship to demand and price. Many of the forecasts from econometric models in the past several years were incorrect because of a failure to correctly model the supply side of the equation and its impact on price.

Supply is the maximum amount of an item a producer or manufacturer is willing and able to sell (i.e., produce) at a given price, or it shows the minimum price a producer will accept to provide one additional unit of an item. It is stated in terms of quantity per unit of time (day, month, year).

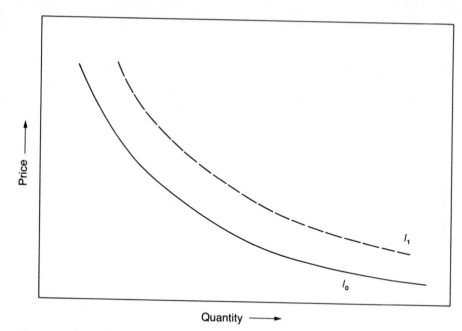

Figure 7.2   A shift in demand caused by a change other than price.

The factors that impact supply include the goals of business firms, the state of technology, the price of a product, the prices of alternative products, and the costs of the factors of production.

It is rather obvious that the goal of a business firm is a critical consideration of supply. The objective of most companies is to maximize profits within certain constraints. However, some firms choose to market only products that result in high rates of return on invested capital, rather than to market a product for which there is a larger demand but a lower rate of return. Some firms do not want to take risks and have smaller production runs than could otherwise be achieved.

At any given time, what is produced and how it is produced is a function of the technology that currently exists. As knowledge expands, it often becomes possible to manufacture a product at costs that are substantially lower than in an earlier period. The invention of the transistor is a case in point. The mass production of the transistor and, later, large-scale integrated circuits made it possible to manufacture many electronic devices at substantially reduced costs. Electronic devices such as radios, calculators, and televisions now have a mass market instead of a select market because of technological changes and consequent lowered prices.

All other things being equal, the profitability of a higher-priced product exceeds that of a lower-priced product. In deciding what markets to enter, firms will seriously consider the price that they believe can be charged for the product. Production will be shifted to the highest-profit items to the extent feasible and these are often the higher-priced goods.

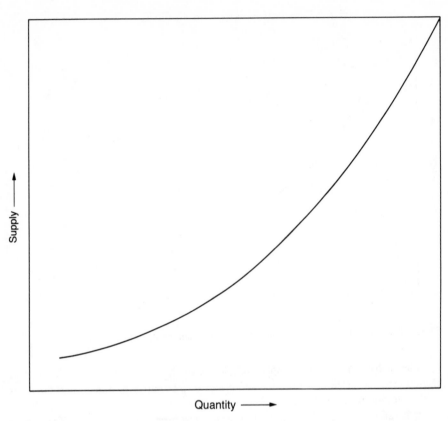

Figure 7.3   A theoretical supply curve.

Changes in the factors of production also affect profitability. These factors include land, labor, and capital. A rise in the cost of one factor—say labor—may result in the lower profit for a product. The firm may choose to produce less of that product and shift to a different product that requires less labor per unit of output. Changes in the cost of one factor relative to others will also change the methods of production.

The relationship between the quantity of a good that is supplied and its price is referred to as the *supply curve*. An example of a supply curve is shown in Figure 7.3. The relationship is upward-trending. It is assumed that all determinants of supply other than the price of the good are held constant. The assumption that the supply curve is upward-sloping has a great deal of intuitive appeal. Generally, if a firm can obtain a higher price for its goods or services, it will be willing to supply more of these.

Movement along the supply curve means that the quantity a firm is willing to supply is measured by the price of the good or service. A shift in the supply curve means that, at any given price, the firm is willing to provide either more or less of

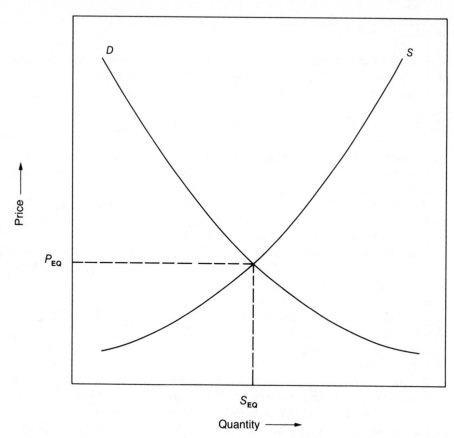

Figure 7.4   Intersection of demand and supply curves showing the equilibrium price for a product.

the product. The shift is a result of a change in one or more of the determinants of supply other than the price of the product. For example, the curve for the product might shift to the left if the state of technology for that product is not advancing as rapidly as for another product. That is, it may be becoming more expensive to continue providing this product than to switch to another for which new technology has substantially reduced the costs of production.

Equilibrium Price

The market price for a product is determined by the *intersection* of the supply and demand curves. This can be seen in Figure 7.4. An illustrative example shows why this is the long-run equilibrium price. Suppose the quantity that is demanded as a function of time, Q_t^D, has the following relationship with price, P_t:

$$Q_t^D = 3.0 - P_t.$$

The quantity that is demanded at any time is given by a constant amount (3.0), less the current price. As price increases, the quantity demanded decreases.

Now, suppose the quantity that is supplied, as a function of time, has the equation

$$Q_t^S = 0.5 + 0.5P_{t-1}.$$

The quantity that is supplied at any time is given by a constant amount (0.5), plus one-half times last period's price. If last period's price went up, a firm will be willing to increase the quantity it supplies.

Further, assume there is no capability of storage or inventory and that all of the product that is supplied will be consumed in each period. This might be the case for an agricultural commodity, for example.

Assume that last period's price for an agricultural commodity was $1.33. The supply equation would then indicate that the quantity supplied next period would be 1.17 units. (Partial units are of no concern because the data could be in thousands of units.) The demand equation would indicate that at that level of supply ($Q^D = Q^S$), the price would be $1.83. The change in price is +$0.50. For several more iterations (periods), the change in price will alternate in sign and successively diminish to some arbitrarily small amount. At equilibrium, the price is $1.66, and equilibrium demand is 1.34 units.

Figure 7.5 shows the "cobweb" pattern that leads to an equilibrium price and quantity. Starting with an initial quantity of 1.17, the price will rise from $1.33 to $1.83 by progressing vertically up to the demand line. At that price, the quantity that is supplied in the next period will increase to 1.41 by progressing horizontally to the supply curve. With the quantity of supply at 1.41 units, the price must be dropped to $1.59 to sell all the product. The progression to the equilibrium price looks like a cobweb.

In many cases, the marketplace actually works this way. For agricultural products and cattle, the supply is generally fixed and must be sold. The price is based on available supply. Farmers and cattlemen do base their adjustments of future supplies on current prices. It may take a period of time, perhaps a year or two, before a new equilibrium price is established. In regulated industries, prices must be approved by governmental agencies after the presentation of a rate case. This means that prices tend to remain fixed for a relatively long time, and the quantity that is demanded of a good or service must change in response to the new price, to establish an equilibrium level.

In practice, there may be a market-period equilibrium (all inputs are fixed and consumers have no chance for adjustment), a short-run equilibrium (in which some adjustments are possible), and a long-run equilibrium (in which firms have had a chance to adjust all production factors and consumers have had a chance to change their consumption habits.)

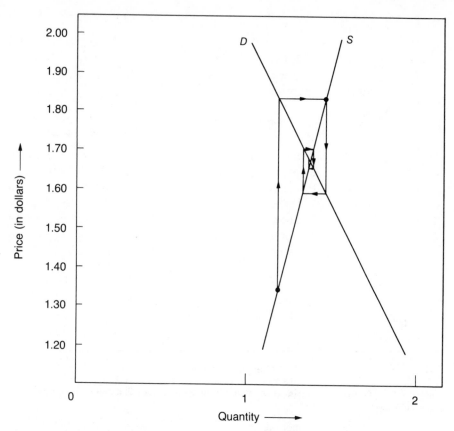

Figure 7.5   A cobweb pattern results from progression to an equilibrium price.

THE BUDGET LINE

To develop a formal theory of demand, it is useful to introduce the concept of the budget line. Let us restrict the analysis of this to the choice between two products that is made by a consumer with a fixed income. Assume that there are no savings and that all income is spent. The *budget line* is intended to show the quantity of each product that can be purchased with the fixed income. An example of a budget line is shown in Figure 7.6.

If monthly income is $1000 and the price of X is $100, then the consumer can buy 10 units of X per month. If product Y costs $50.00, the consumer can buy 20 units of it per month. Individual consumers can select any combination of quantities of X and Y that will total $1000 per month. The possible combinations that use all the income are represented by the budget line.

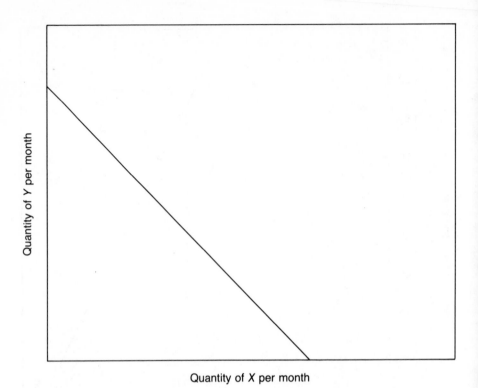

Quantity of X per month

Figure 7.6 The budget line shows the quantity of each product that can be purchased with fixed income.

If income were to decline and there were no changes in prices, the budget line would shift inward parallel to itself, as is shown in Figure 7.7. A household buys less of all goods. However, if income remains constant but the price of X increases, the budget line rotates downward, as is shown in Figure 7.8.

There are a number of rather basic conclusions that can be drawn from the budget line approach. If prices remain constant but income changes, the budget line moves parallel to itself. A household consumes either more or less of all products. Changes in prices of some goods relative to others changes the slope of the budget line. The household can buy either more or less of one product and the same quantity of another product. If prices of all goods go up by 10 percent and incomes go up 10 percent, the budget line remains the same. The household can buy no more units of either product than before. If all prices increase by the same percent and income remains constant, the budget line moves inward parallel to itself; that is, the household buys less of all goods with its income.

These conclusions are rather obvious, but the approach does cause one to think about the effects of *relative* prices and income levels on household demand.

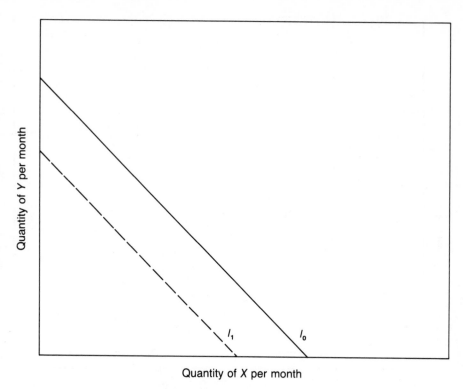

Figure 7.7   A shift in the budget line.

MARGINAL UTILITY

Another useful concept in explaining consumer behavior is the theory of marginal utility. This theory states that most choices are not "yes" or "no," but rather "more" or "less" (Phlips, 1974; Samuelson, 1978). Utility is the economists' term for satisfaction. Many times it is not possible to rank one's enjoyment or satisfaction quantitatively. For example, you do not hear someone say, "I enjoyed that cup of coffee 87.6 percent." Utility is an abstraction. It is a way of looking at the problem of relating the desirability of different products or services.

The theory states that in most cases it is the marginal or incremental utility of the last units consumed that determines the choice between more of good X or less of good Y.

For example, if you have already had five cups of coffee this morning, how much additional satisfaction will the sixth cup provide? The theory states that the utility

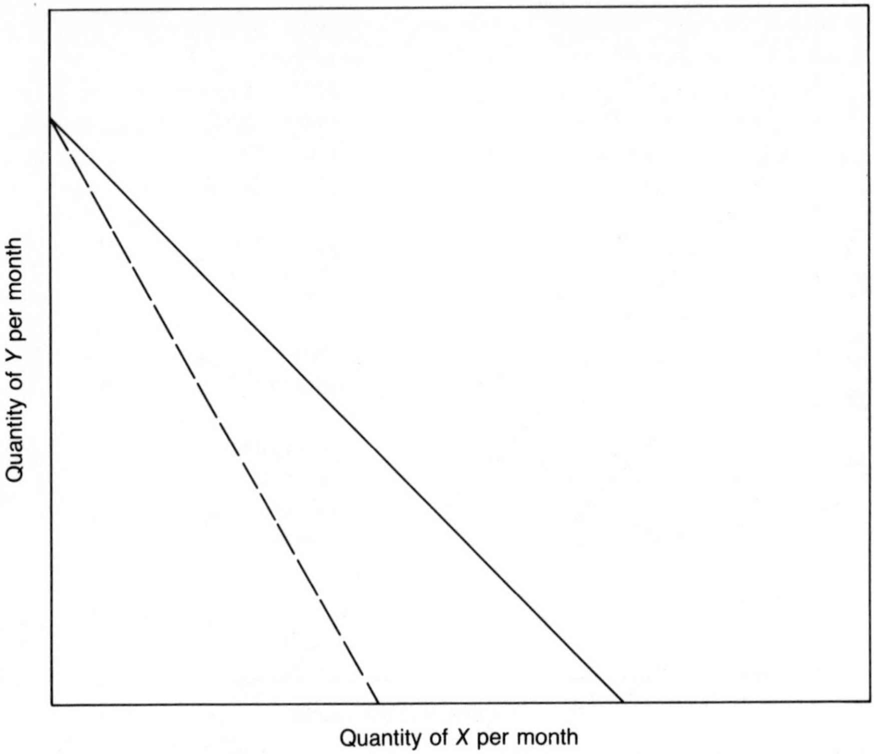

Figure 7.8   A rotation in the budget line.

or satisfaction of the last unit of a given product will decrease as more and more of that product is consumed. The consumption of all other goods or services is assumed to be held constant.

Maximizing Utility

If a household is trying to maximize its utility or satisfaction, then it will allocate its purchases so that the utility of the last dollar spent is the same for all products; that is, if the household can get more satisfaction from a dollar's worth of X than of Y, it will continue purchasing X until the marginal utility of X equals the marginal utility of Y.

The theory explains what appear to be anomalies in the marketplace. For example, water is essential to life and gold is not. Why is the price of water so low and the price of gold so high? The marginal utility theory explains this contradiction.

Since water is so readily available, the consumer is not willing to pay a great deal for the last gallon after consuming many gallons of water. However, if one had to choose between purchasing enough water to survive, or an ounce of gold, it is apparent that the price of water would exceed that of gold.

The question is not, "Do you want water or gold?," but rather, "How much are you willing to pay for a little more water or a little more gold after you have already purchased or consumed a certain amount of each commodity?"

To maximize utility or satisfaction, the following relationship must hold:

$$\frac{\text{Marginal utility of a unit of } X}{\text{Marginal utility of a unit of } Y} = \frac{\text{Price of a unit of } X}{\text{Price of a unit of } Y}.$$

Therefore, if the marginal utility of a unit of X is only 80 percent of the marginal utility of a unit of Y, a consumer should only be willing to pay 80 percent as much for a unit of X as for a unit of Y.

The equation indicates that, to maximize its total satisfaction, a household will allocate its expenditures in such a manner that the ratios of the marginal utilities of each pair of goods will be equal to the ratio of the prices. *Relative price* is an important consideration in determining household demand.

SUMMARY

This chapter serves as a prelude to demand modeling, in that it attempts to explain why certain variables of a demand model include:

- The price of a good.
- Prices of alternatives.
- Income.
- Habit.
- Advertising or promotion.
- Market size.

It is common practice to deflate prices and income by dividing the nominal values by appropriate indexes (e.g., Consumer Price Index, and GNP deflators) to yield *real* income and *real* price. The real or constant dollar figures remove the illusion of wealth resulting from inflation in incomes related to inflation in prices.

Supply considerations are mentioned to sensitize you to the importance of taking supply factors into account. Forecasts of prices used in demand models have frequently been underestimated because of dramatic price increases in material costs (e.g., of fuel, of commodities). The result of underestimating prices has been an overestimate of demand (e.g., of electricity, of fuel).

The demand curve shows the relationship between the quantity of something that is demanded and its price. The forecast of the quantity that is demanded at various price levels is important to businesses, since it permits them to maximize profitability by considering price and cost tradeoffs.

USEFUL READING

PHLIPS, L. (1974). *Applied Consumption Analysis*. New York, NY: American Elsevier Publishing Co.

SAMUELSON, P. (1978). *Economics,* 9th ed. New York, NY: McGraw-Hill.

CHAPTER **8**

Estimating Demand Elasticities

Two important determinants of a firm's profitability—indeed its survival—are cost and the demand for its products or services. Demand must exist or be created if the firm is to survive. It must also be high enough at least to cover fixed costs. Because of its key role, all corporate planning activities require a careful analysis of demand over time.

Forecasters are also concerned with the relationship between quantity demanded and price and can play an important role in helping their firms make pricing decisions by estimating price elasticities for products and services with their models.

This chapter addresses:

- The definition of elasticity.

- Arc versus point elasticity.

- What determines price elasticity.

- Estimating elasticity with regression models.

- Constant elasticity models.

- Short- versus long-term price elasticity.

PRICE ELASTICITY

As described in the previous chapter, the *demand function* describes the relationship between the quantity that is demanded of a good or service and all of the variables that determine demand. The *demand curve* is that part of the demand function that relates the price that is charged to the quantity that is demanded, *when all other variables are held constant*. While this curve is important, it fails to show how sensitive the quantity demanded is to price. The missing element is *elasticity,* and

105

it explains the responsiveness of changes in demand to changes in prices or any of the other variables.

While the notion of elasticity can be used to interpret relationships between the dependent variable and any or all independent variables, primary attention is often given to price elasticity. *Price elasticity, E,* is defined as the percent of change in the quantity demanded, Q, as a result of a given percent of change in price, P:

$$E = \frac{\text{Percent change in } Q}{\text{Percent change in } P} = \frac{\Delta Q/Q}{\Delta P/P}.$$

An important condition in the definition of elasticity is that all factors influencing demand other than *own-price* (price of the item under consideration) are held constant while own-price is varied.

Consider message toll service, a telephone service, as an example. The price elasticity will be a negative number, since fewer messages Q will be placed at higher prices P, with all else held constant. Also note that it is a dimension-free number.

In Figure 8.1, consider the equilibrium position at the intersection of supply curve S and the demand curve D at point E_1. If the supply curve is shifted up and to the left to S_2, the small price increase causes a substantial change in quantity demanded. In this case, demand is very sensitive to price. Figure 8.2 shows a different relationship between supply and demand. Again, equilibrium is reached at E_1, but when the supply curve moves from S_1 to S_2, the quantity that is demanded drops only a small amount from Q_1 to Q_2. (Of course, for these examples, the visual effect of a "large" or "small" amount would depend on the scaling that is chosen for the plots.) For actual applications, a quantitative measure of response is desired. A useful formula for actually measuring elasticity is the one that determines the *average elasticity* between points E_1 and E_2 on the demand curve:

$$E = \frac{\text{Change in } Q/\text{Average } Q}{\text{Change in } P/\text{Average } P} = \frac{\Delta Q/\frac{1}{2}(Q_1 + Q_2)}{\Delta P/\frac{1}{2}(P_1 + P_2)} = \frac{\Delta Q(P_1 + P_2)}{\Delta P(Q_1 + Q_2)}.$$

With this formula, the elasticity is the same whether the starting point is E_1 or E_2.

Arc Elasticity

Table 8.1 shows the data that go into the calculations of arc elasticity for three different hypothetical products. The elasticity for Product A equals the percent change in the quantity that is demanded (33 percent) divided by the percent change in price (-16.7 percent), or -2.0. For Products B and C, the elasticities are -1.0 and -0.5, respectively.

The demand for Product A is very responsive to price changes, since the elasticity exceeds unity in absolute value. This is referred to as an *elastic* product. Product B demonstrates *unitary elasticity,* since the percentage decrease in price

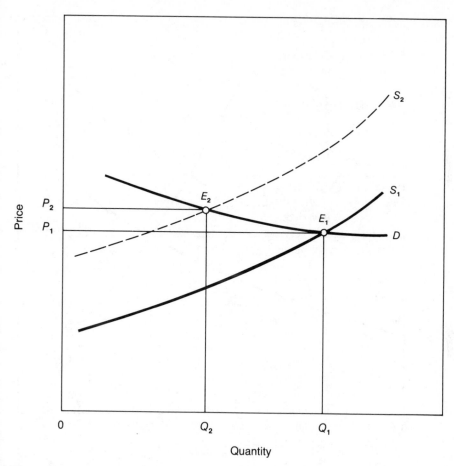

Figure 8.1 Intersecting supply and demand curves—the quantity demanded is *sensitive* to price changes.

equals the percentage increase in quantity demanded. Product C is referred to as *inelastic,* since the percent change in the quantity that is demanded is less than the percent change in price. These examples are all arc elasticities. *Arc elasticity* is a measure of the average elasticity over the range of the prices and quantities specified.

Point Elasticity

A second measure of elasticity is point elasticity. *Point elasticity* is calculated at a specific point on the demand curve, as is illustrated in Figure 8.3. At point 1, the elasticity is calculated from the tangent *(T)* to the curve at point 1. Figure 8.4 shows that there are several possible arc elasticities (1–4, 1–5, 1–2, 1–3) for the demand

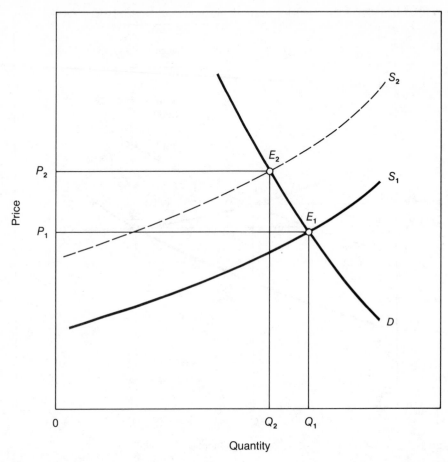

Figure 8.2 A different relationship between demand and supply—the quantity demanded is *insensitive* to price changes.

curve but only one point elasticity. If the arc in Figure 8.3 is successively shortened from b' to b'', the arc elasticity approaches the point elasticity.

A linear demand curve is shown in Figure 8.5. Except for a horizontal or vertical demand line, the point elasticity changes in value when calculated at different points on the demand curve. The percent change in quantity divided by the percent change in price is a constant, but the ratio P/Q will be different for each point on the line.

Determination of price elasticities is more than an academic exercise. Price elasticities tell how price changes affect total revenues. Depending upon the price elasticities, a price change will result in an increase in total revenues, no change, or a decrease in total revenues. If elasticity is unitary, total revenues are unchanged by price changes. If demand is elastic, total revenues decline if price is increased,

Table 8.1   Data for calculation of an arc elasticity.

Product description	Old amount	New amount	Change	Average	Change (percent)	E
PRODUCT A						
Price	$1.95	$1.65	−$.30	$1.80	−16.7	Elastic
Quantity	10,000	14,000	+4,000	12,000	+33	
PRODUCT B						
Price	$1.80	$1.40	−$.40	$1.60	−25	Unitary
Quantity	10,500	13,500	+3,000	12,000	+25	
PRODUCT C						
Price	$1.80	$1.20	−$.60	$1.50	−40	Inelastic
Quantity	10,800	13,200	+2,400	12,000	+20	

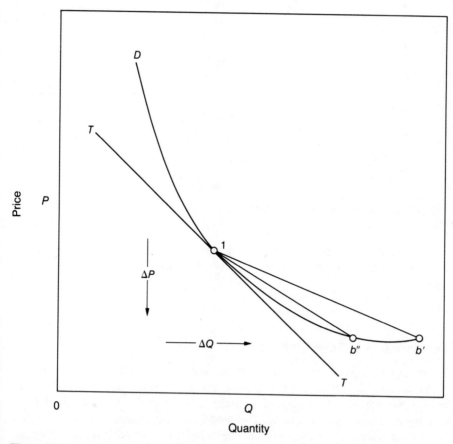

Figure 8.3   A point elasticity is calculated at a specific point on the demand curve.

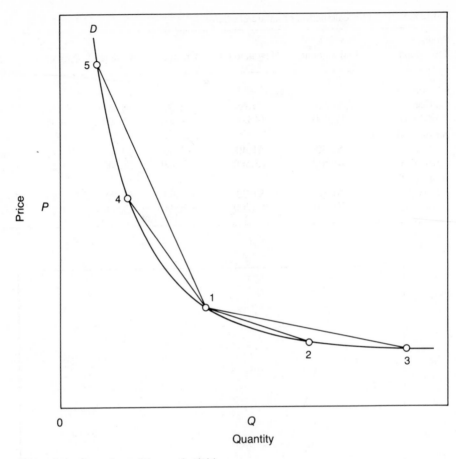

Figure 8.4 Several possible arc elasticities.

because the quantity demanded drops by a greater percent than the price increases. For inelastic demand, total revenues rise when price increases and drop when price decreases.

Determinants of Price Elasticity

To attempt to explain the determinants of elasticity at a given time requires an understanding of economics, psychology, consumer preference, and many other things. Nevertheless, it is possible to make some general statements about what determines demand and elasticity. The model-building process requires that you know what independent variables should be included in the model and have a feeling for their sizes and signs.

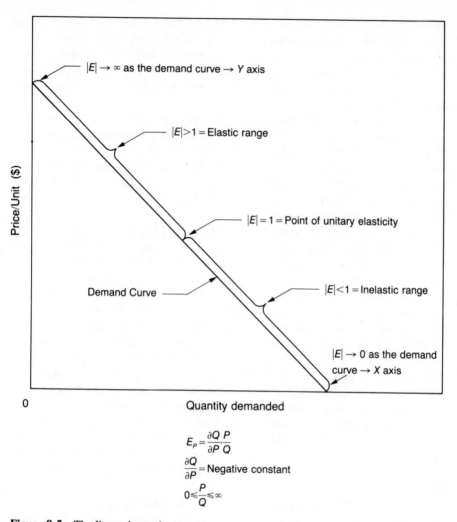

Figure 8.5   The linear demand curve.

In general, price elasticity is determined by at least four factors:

- Whether or not the good is a necessity.
- The number and price of close substitutes.
- The proportion of the budget devoted to the item.
- The length of time the price change remains in effect.

If the product or service is a necessity, its demand will be inelastic. Consumers will pay any reasonable price for a necessity. Lack of substitutes for a product will also cause demand to be inelastic. If a good is both a necessity and without a substitute, demand will tend to be very inelastic. If substitutes are available, consumers will switch their purchases to those substitutes that have not increased in price. If the proportion of income spent for a good is small, price changes may not have too great an impact on the demand for the good. If the proportion of the income is large, price increases will cause postponements in demand or reductions in the quantity demanded.

In the telephone industry, basic telephone service is considered to be a necessity for most businesses and households. There are certainly some substitutes for telephone service, such as mail service or telegrams, but these services are often considered inferior to verbal communication. Further, the cost of telephone service is not normally a major part of a family's income or a business's expense. For that reason, the average household or business is likely to absorb increased telephone costs so long as they are moderate.

"Vertical services," such as telephone extensions, premium sets, Touch-Tone, and Custom Calling (the last two being trademark names of the Bell System) may demonstrate a different demand pattern Taylor (1980). These items are not necessities, and price changes may cause significant reductions in the quantity demanded.

Finally, there is the amount of usage of a basic service that is generated by the customer—the number and length of calls. Some calls may be important enough that they will be made regardless of price changes. Other calls may be for convenience or casual purposes, and these calls can be postponed or eliminated. However, the determination of elasticity is even more complex in this area. Customers may shift their calling patterns and make telephone calls in the evening or on weekends when rates are less. They may change the number as well as the duration of calls, and they may dial the calls themselves instead of using an operator.

The longer a price change remains in effect, the more elastic the demand for a product. Consumers become aware of price changes and adjust their consumption habits to the new circumstances. "Elastic" as used here is a relative term. Demand becomes "more elastic" as time goes by; but it could still be inelastic—i.e., less than one.

Other factors influencing price elasticity are the frequency of purchases and the presence or absence of complements (e.g., automobiles and gasoline). Frequently purchased and relatively inexpensive products may be more inelastic than infrequently purchased expensive items.

Price Elasticity and Revenue

Price elasticities are used to show how price changes can affect total revenue. Since revenue equals unit price *times* quantity demanded, a price change can result in an increase, no change, or a decrease in total revenue (Figure 8.6).

Let us consider first the impact of own-price elasticity on revenue by means of an example, deferring the impact of cross-elasticity until later. Suppose a forecast for a service predicts $1000 in revenues for a particular future year. Assume the

Behavior of model	Unitary	Inelastic	Elastic
Price rise	Total revenue remains the same	Total revenue increases	Total revenue decreases
Price decline	Total revenue remains the same	Total revenue decreases	Total revenue increases
Gain versus loss	Gain = Loss	Gain>Loss	Gain<Loss

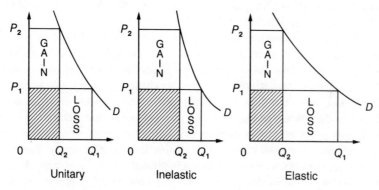

Figure 8.6   The relationship of price to revenue.

own-price elasticity is -0.2. Defining P = existing price, Q = forecast of demand at the existing price, and R = forecast of revenue at the existing price, then $R = P \cdot Q = \$1000$.

What will be the impact of a 10-percent price increase in the service under consideration, effective at the start of the future year? The new price is clearly just $1.10 \cdot P$, but quantity demanded would be somewhat less than before. With $E = -0.2$, a 10-percent price increase would result in approximately a 2-percent demand loss. Hence, the new demand will be approximately $0.98 \cdot Q$, and the revenue R' after the price change will be

$$R' = (1.1P)(0.98Q)$$
$$= 1.08\ R.$$

Thus the 10-percent price increase will increase revenues by only about 8 percent. In general, the own-price elasticity effect will cause an X-percent price increase to yield less than an X-percent increase in revenues. The "reprice" value of a rate or price increase is the incremental revenue that would result if there were no demand reaction. It would be $100 in the example. The amount by which revenues fall short of the reprice value, $20 here, is called *revenue repression*. A revenue repression factor may be defined as the ratio of revenue repression to reprice value. In this case, it is 0.2.

An elasticity of -0.2 implies a minimum revenue repression factor of 0.2; e.g., at least $20 for a price increase worth $100 on a reprice basis. The revenue repression factor often goes up with the magnitude of the price change.

To continue with the $1000 service example, suppose it has been determined that this service must yield $1100 to enable the company to earn its objective rate of return. A 10-percent rate increase will not be sufficient, owing to the revenue repression of $20 (revenue repression factor of 0.2). Evidently, to net $100 in incremental revenues, almost a 12.5-percent increase in price will actually be necessary. The reprice value ($1000 · 1.125) of such an increase is $1125. A repression of $25 (0.2 · $125) leaves a net of $1100.

Demand for a given good or service is frequently affected by prices of other goods and services as well as its own price. A cross-elasticity coefficient measures this interaction effect. The *cross-elasticity* measures the percent of change in the demand for good A as a result of a given percent change in the price of good B. If a good has a close substitute, a price increase for one will probably create increased demand for the other. When products are complementary (used together), a decrease in the price of one will lead to an increase in demand for both products. The cross-elasticity is positive for substitutes, and negative for complementary goods. As before, this is a dimensionless number, since it is defined in terms of percent changes.

As an example, let us use the price of WATS (Wide Area Telephone Service—a discounted long-distance offering based on a flat rate for specified hours of use; the elimination of the cost of itemizing individual calls permits a lower rate to be charged). One expects the price of WATS to affect MTS demand (Message Toll Service—the normal long-distance rates apply). An increase in WATS prices is expected to result in an increase in demand for MTS, resulting in a positive cross-elasticity. Thus MTS and WATS are examples of *substitute* services. Note, however, the asymmetry in the definition: the relationship between *MTS demand* and *WATS prices* is not necessarily the same as the corresponding relationship between *WATS demand* and *MTS prices*.

Consider now the relationship between demand for MTS and the price of basic local-telephone service. This would suggest that as the price of local service goes up, customers might be expected to cut back on toll calling, which is more discretionary. In this case local and MTS services are examples of *complementary services,* where the cross-elasticity is negative.

In summary, gaining knowledge of price elasticities, both own-price and cross-, for at least all of the major services and products of the business, is essential for intelligent and effective business forecasting and planning. While this is by no means an easy goal to achieve, it represents an area where much progress has been made in recent years and must continue to be made in the years to come.

OTHER DEMAND ELASTICITIES

Attention so far has been concentrated on price elasticity. However, there is an elasticity associated with each independent variable in the demand function. Income is another determinant of demand that also receives attention. For most goods and services, one would expect a positive relationship between demand and income. *Income elasticity* is the percent change in demand divided by the percent change in income. However, the elasticity is now positive instead of negative. If the demand for a good is income-inelastic, the increase in demand will not be proportional to the percent increase in income. As national income rises, for instance, a business firm will not experience a proportional growth in revenues, and its share of national income will decline.

Income elasticity is a two-edged sword. If the economy contracts and income declines, the revenues of an income-inelastic firm will shrink less than the revenues for an income-elastic firm. The firms whose goods are income-elastic are more concerned with anticipating the business cycle expansions and contractions.

EMPIRICAL ESTIMATION

The previous example has served to illustrate a very important principle: ignoring the impact of own-price elasticities in a rate- or price-planning process will lead to revenue shortfalls. Of course, compensating for this effect implies that the price elasticity coefficient must be estimated by one method or another. This quantification is frequently not an easy task.

If there has been at least one change in price, a manual calculation of elasticity can be performed. The change in quantity demanded that is owing to the price change will have to be subjectively estimated. After that, you should determine what the demand would have been had there been no price increase. This can be done by using actual data for the other determinants of demand, or by making a time series model where predictions are generated by using actuals up to the time of the price change. By whatever means seems most reasonable, you can subjectively determine what the quantity demanded would have been had there been no price increase. Since the actuals for past changes in price are available, the average elasticity can be calculated. If a price increase is expected in the future, the forecasts can be adjusted downward by using this manual elasticity calculation. Obviously, this is a rough estimate of elasticity and depends on the judgment of the analyst. However, the results can still be helpful to management and forecasters in predicting the impact of proposed price changes.

Using Regression Models

Since all of the independent variables in a demand function can change simultaneously, regression analysis is the primary tool for estimating elasticities. A model may take the form

$$Q_Y = Q(P_Y, I, P_X, A, B, \varepsilon, \ldots),$$

where the quantity demanded, Q_Y, is a function of the (deflated) price P_Y of the product, income I, the price P_X of a competing or complementary product, and other factors, which could include market potential, A, advertising, B, any of a number of other variables, and a random error term, ε.

In an application of demand models, either an additive or a multiplicative model may be appropriate in a given situation. Elasticities can be calculated from either model, but the calculation is different for the additive model than for the multiplicative model.

The *multiplicative* model has the simpler calculation and will be discussed first: this demand model has the form

$$Q_Y = \alpha P_Y^{\beta_1} I^{\beta_2} P_X^{\beta_3} A^{\beta_4} B^{\beta_5} e^{\varepsilon},$$

where e^{ε} is a *multiplicative* error term.

To estimate the coefficients by using ordinary least squares, it is convenient to take a logarithmic transformation ($\ln = \log_e$):

$$\ln Q_Y = \beta_0 + \beta_1 \ln P_Y + \beta_2 \ln I + \beta_3 \ln P_X + \beta_4 \ln A + \beta_5 \ln B + \varepsilon,$$

where $\beta_0 = \ln \alpha$.

The point elasticity for price (the partial derivative of $\ln Q_Y$ with respect to $\ln P_Y$) is represented by the coefficient β_1 of the demand function.

The conclusion drawn from a *multiplicative* demand function is that the elasticities are given directly by the regression coefficients. This is a *constant elasticity* model: regardless of the location on the demand curve, the elasticity is the same. This may or may not be a realistic assumption, in practice.

With any regression model, it is useful to see how stable the coefficients are over different regression periods. If the purpose of the model is to identify elasticities, the coefficients must remain relatively stable in order to satisfy the requirement that they be accurately determined. Multicollinearity (Chapter 14) among independent variables often causes the coefficients to change significantly over different time periods.

The elasticities should also make sense. If it is believed that income is positively related to demand, the coefficients of income should be positive. The price coefficient should be negative. If this is not the case, the models may be improperly specified, or the wrong independent variables may be included in the model. More likely, the independent variables are multicollinear.

The calculation of elasticity is more detailed for an additive demand model than for the multiplicative model. In an *additive* demand model,

$$Q_Y = \beta_0 + \beta_1 P_Y + \beta_2 I + \beta_3 P_X + \beta_4 A + \beta_5 B + \varepsilon .$$

The point elasticity for price (the partial derivative of $\ln Q_Y$ with respect to $\ln P_Y$) is given by $\beta_1 P_Y/Q_Y$. This can be demonstrated by using a simplified model,

$$Q_t = a - bP_t.$$

Let

$$Q_{t-1} = a - bP_{t-1},$$

and subtract Q_{t-1} from Q_t; then

$$Q_t - Q_{t-1} = (a - bP_t) - (a - bP_{t-1}) = -b(P_t - P_{t-1}),$$

or

$$\Delta Q = -b\Delta P.$$

Thus,

$$b = -\frac{\Delta Q}{\Delta P} .$$

To obtain the elasticity, you multiply by P/Q and obtain

$$\frac{\Delta Q}{Q_Y} / \frac{\Delta P}{P_Y} = -bP_Y/Q_Y .$$

Therefore, the price elasticity varies at every point (Q_Y, P_Y) along the demand curve, since the ratio of P_Y to Q_Y is different for every point on the demand curve.

A Revenue Model

In a revenue model, the same variable (price) may appear on both sides of the equal sign. Statistical bias may be introduced; therefore, extra care is required. Suppose that revenues are related to price P and income I in a multiplicative model,

$$Q_Y = \alpha P^{-\beta_1} I^{\beta_2} e^\varepsilon.$$

Then the revenue R is

$$R = QP$$
$$= \alpha P^{(1-\beta_1)} I^{\beta_2} e^{\varepsilon},$$

and

$$\ln R = \ln \alpha + (1 - \beta_1)\ln P + \beta_2 \ln I + \varepsilon.$$

The point elasticity is the coefficient of $\ln P$, namely $(1 - \beta_1)$.

The elasticity range for a model of quantity demanded versus a model of revenues is shown in Table 8.2.

The price elasticity for a revenue model is one greater than the elasticity of the corresponding model for quantity demanded. Therefore, it is possible for the price elasticity of a revenue model to be positive.

Figure 8.7 shows three different elasticities. In the market period all inputs are fixed and demand is perfectly inelastic. In the short run, there is some elasticity and the long-run elasticity may be perfectly elastic.

Long-Term Elasticity

The usual price elasticity referred to in demand models is a constant one-time elasticity. However, demand models that use lagged quantity as an independent variable allow for an increasing elasticity over time. Referred to as *dynamic* models, they are appropriate when a relatively long period of time passes before the restrictive effects have run their course. Consider a large computer system as an example. A large increase in rental rates may cause a customer to decide to replace the current system with a lower-priced competitive offering. However, it might take several years to replace the computer system, owing to the long planning, manufacturing, and installation intervals for such a large-scale project.

Consider a simplified demand for such a situation

$$\log Q_t = a \log Q_{t-1} - b \log P_t .$$

Table 8.2 Range of elasticity for a quantity model versus a revenue model.

Range	Quantity model	Revenue model
Elasticity	$-\infty < E < -1$	$-\infty < E < 0$
Unit elasticity	$E = -1$	$E = 0$
Inelasticity	$-1 < E < 0$	$0 < E < 1$

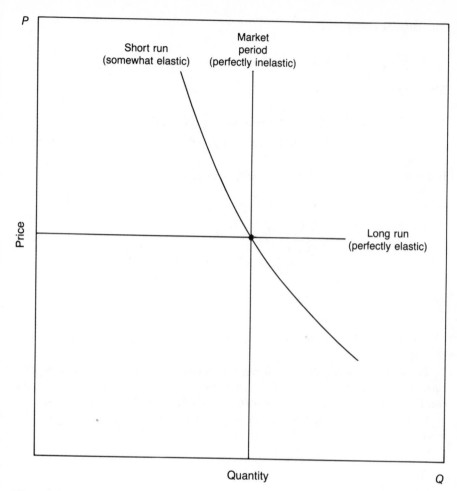

Figure 8.7 Three elasticities.

The current-quarter (short–term) price elasticity is equal to $-b$. Assume that price is increased at time $t = 0$ and held constant thereafter. The *long-term elastic effect* of the price change can be determined as follows. At $T = t + 1$,

$$\log Q_{t+1} = a \log Q_t - b \log P_{t+1} .$$

Since $\log P_{t+1} = \log P_t$,

$$\log Q_{t+1} = a(a \log Q_{t-1} - b \log P_t) - b \log P_t$$
$$= a^2 \log Q_{t-1} - b(1 + a) \log P_t.$$

Similarly, at $T = t + 2$,

$$\log Q_{t+2} = a \log Q_{t+1} - b \log P_{t+2} .$$

Since $\log P_{t+2} = \log P_t$,

$$\log Q_{t+2} = a(a^2 \log Q_{t-1} - b(1 + a) \log P_t) - b \log P_t$$
$$= a^3 \log Q_{t-1} - b(1 + a + a^2) \log P_t .$$

If the process is continued, the elasticity at each time is given by

$t = 0$	$E = -b$
$t = 1$	$E = -b(1 + a)$
$t = 2$	$E = -b(1 + a + a^2)$
$t = 3$	$E = -b(1 + a + a^2 + a^3)$
.	.
.	.
.	.
$t = n$	$E = -b(1 + a + a^2 + a^3 + \cdots + a^n).$

For demand models that use a lagged dependent variable, the *long-run price elasticity* can be shown to *equal the short-run elasticity divided by (1 minus the coefficient of the lagged dependent variable)* (Pindyck and Rubinfeld, 1976).

In this case, as t approaches infinity, the long-term elasticity is given by $-b/(1 - a)$, since

$$\sum_{i=0}^{\infty} a^i = \frac{1}{1 - a} .$$

Figure 8.8 shows two different elasticity profiles based on different values for a and b. The closer a is to 1.0, the longer it takes to reach the long-term elasticity. It is also apparent that the elasticity approaches its long-term value asymptotically. After an initial rise, further increases in elasticity are relatively minor. For this reason, it is more practical to consider the amount of time that elapses until 90 percent or 95 percent of the long-term effect is reached. For example, the time to reach 90 percent of the long-term effect is

$$\ln (0.10)/\ln a ,$$

where a is the coefficient of the lagged dependent variable.

Long-term elasticity can be extremely important, since it can be considerably larger than current-period elasticity. For example, if $a = 0.75$ and $b = -0.3$, the

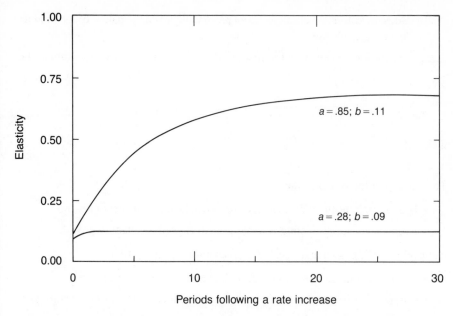

Figure 8.8   Two different elasticity profiles based on different values for *a* and *b*.

long-term elasticity $= -0.3/(1 - 0.75) = -1.2$. What appears to be highly inelastic in the short term can be elastic in the long term. If this distinction is not recognized, a company may continue to raise prices in the belief that demand is inelastic, and constantly find revenue shortfalls one or two years in the future.

DEMAND ANALYSIS CHECKLIST

_____ Have the determinants of demand been identified (price, market potential, income . . .)?

_____ Have the determinants of supply been identified?

_____ Are there sufficient data to build a regression model?

_____ Do the data have sufficient variability to estimate elasticities?

_____ Is there a multicollinearity problem?

_____ Is the model to be used for forecasting or explanatory purposes?

_____ Is an additive or multiplicative model more appropriate?

_____ Does the estimated elasticity seem reasonable given the nature of the product (necessity, substitute's price)?

_____ Lacking adequate data for modeling, can a manual elasticity estimate be made?

SUMMARY

The determination of elasticities

- Is an important function essential for understanding business growth.
- Is important for predicting revenue growth as well as quantity growth.

Demand modeling is an analytical as well as a forecasting activity.

- The ideal demand model will identify elasticities and forecast well.
- Caution is warranted when trying to use demand models for forecasting purposes.
- It is rarely the case that a good demand model will automatically be a good forecasting model.

USEFUL READING

PINDYCK, R. S., and D. L. RUBINFELD, (1976). _Econometric Models and Economic Forecasts_. New York: McGraw-Hill, Inc.

TAYLOR, L. D. (1980). Telecommunications Demand: A Survey and Critique. Cambridge, MA.: Ballinger Press.

CHAPTER 9

The Demand Analysis Process

This chapter presents a seven-step procedure for estimating elasticities in the demand analysis process. These steps are:

- Defining measures of demand.
- Identifying determinants of demand.
- Collecting data.
- Estimating model coefficients.
- Generating demand response functions.
- Producing forecasts.
- Tracking results.

DEMAND ANALYSIS AS A PROCESS

Historical demand data must be analyzed during periods when the causal factors have changed. For instance, to develop a demand model that assesses the impact a price change has had on demand for a product, the historical data must be analyzed during periods when the price has changed. Assuming that such data are available, the resulting model can be used to generate the price-demand curve that existed *during the historical or past period being analyzed*. If one assumes that the historical market structure will remain relatively stable into the future, it is reasonable to use this curve to estimate how demand will respond to *future* price changes.

If this assumption is unrealistic, the demand response predicted by the model should be modified by using testimonial data (i.e., surveys of consumers and/or market experts) and the judgment of the product/service manager about future demand or price. However, if historical price–demand data exist, a demand curve based on a model of these data is a useful starting point for estimating future price–demand relationships, regardless of the judgmental refinements that may later be required.

Demand analysis for a particular product or service can be divided into seven sequential steps. Each step includes a unique part of the analysis; however, execution of the steps is often an iterative process. A close correspondence between the demand analysis process and the general forecasting process should be noted (see the Flowchart).

Defining Measures of Demand

1. *Define one or more appropriate measures of demand.*

The most common measures for telephone *product* demand, for example, are in-service quantities and what is known as inward movement (i.e., sales). However, there is increasing need to model the total market size and the relative portion served by a particular product line. The most common measure of telephone *service* demand is usage. Examples would be total messages and the average length of a message. Quite often a useful surrogate for the demand for a group of products or services is its associated revenue divided by an appropriate price index. This ratio is an equivalent unit of demand that incorporates all of the internal cross-elasticities within a particular product line (e.g., PBX equipment) or a major service category (e.g., Message Toll Service).

Identifying Determinants of Demand

2. *Use economic theory and marketing knowledge to identify the most likely determinants of demand.*

The most important step in this stage is to develop a theoretically sound but not too restrictive framework, consistent with the data obtainable for analysis. In many cases, economic theory is inadequate or incomplete, thus complicating the specification problem.

Many aspects of this step are unique to each model; however, it is often useful to begin by partitioning the determinants of demand into controllable and noncontrollable groups. The determinants that are under management control are called decision variables or *marketing instruments* and include price, advertising, promotional campaigns, sales effort, distribution techniques, and so on.

The only marketing instrument whose influence on demand is strongly supported by economic theory is price. The *Law of Demand* states that an inverse relationship between demand and price will exist. However, the other marketing instruments, especially advertising and sales efforts, can logically be expected to influence demand in a positive way.

In most cases, the number of marketing instruments to be considered are severely limited by a lack of data. The prices of the product being modeled and potential cross-elastic product prices are the most common instruments for which historical data are available. However, with some work, reasonable historical indicators of advertising and sales efforts can occasionally be obtained.

If one or more of the marketing instruments did not change in value during the historical period, they can be immediately dropped from consideration. Although these instruments may be important determinants of demand, their influence cannot be measured through demand modeling if they did not change in value.

Once the marketing instruments have been identified, it is necessary to identify the noncontrollable determinants of demand. These determinants are called *environmental variables* and will usually determine the total market size and its fundamental growth rate. Quite often, one or two environmental variables can adequately explain most of the fundamental market movement. A few possible environmental variables are personal income, a business activity index, an unemployment rate, number of business establishments, and population. Some important but often unavailable environmental variables are competitive sales activity and prices. However, a surrogate for these influences can sometimes be developed.

As a start, you should include the most important four or five variables explicitly, listing other potentially useful variables for later consideration. Starting with a few variables avoids getting embroiled in serious statistical problems early in the analysis and allows for maximum flexibility of interpretation.

Collecting Data

3. *Collect historical data on demand and its likely determinants.*

Once the potential determinants of demand have been identified, you proceed to the data collection phase. Measurement always involves data, so you must be aware of data availability and limitations. The data can be collected in two basic formats. One is the time series format, in which demand quantities for a particular product/service in a particular geographical area are obtained for several successive periods of time (e.g., past months, quarters, or years). The other is the cross-sectional format, in which demand quantities for several similar products/services or a single product/ service in several geographical areas are obtained for one historical period.

These formats can be combined when the primary objective is to estimate *average* response functions (e.g., price–demand curves) across all cross sections. The combined format is called a "pooled" data base and can sometimes yield more accurate estimates of a response curve than either of the separate formats can. However, if the demand model is to be used to make forecasts over time in the various cross sections (e.g., cities, states), it is usually best to develop separate demand models each of which uses a separate time series data base for each cross section. If the model is needed for an *aggregate* forecast over all cross sections, a time series data base aggregated over all cross sections will often be appropriate.

Since it is not always clear in advance which type of data will produce the best model, the data gathered for a demand analysis should have both times series and cross-sectional dimensions, and it should be easy to aggregate the data across either dimension.

Estimating Model Coefficients

4. *Use statistical estimation procedures to identify and validate the most likely structure for the demand model.*

After data collection has been completed, you must specify plausible ways in which the potential determinants of demand can be analytically related to demand. For instance, one plausible relationship might be to express the *level* of demand as a linear combination of relative price, or a business activity index, or the number of business establishments demanding some product, and variables that account for seasonal and/or exceptional influences.

Another plausible relationship might be to transform all variables (logarithmically) so that the percent changes in demand are expressed as a linear combination of the percent changes in the determinants. In general, each plausible model should contain measures of relative price, relative ability to purchase, and potential market size.

In addition, cross-elastic prices, advertising expenditures, sales effort measurements, and indicators of seasonal variation or variation owing to special events should be considered. Also, analytic structures that allow for dynamic (lagged variable) relationships (i.e., short-run versus long-run responses) should be considered (see, for example, Parsons and Schultz, 1976; Phlips, 1974; Pindyck and Rubinfeld, 1976).

Demand theory appears to give little guidance as to model form, other than to indicate the imposed constraints that lead to easily interpreted coefficients. An *additive* demand model would imply that the elasticities are *changing*. On the other hand, *constant* elasticities are implicit in a *multiplicative* or *log-linear* demand model.

In theory, signs of parameter values can be assigned that are based on the nature of the product or service. There is, however, little theoretical information that will help to establish the magnitude of the coefficients. In practice, empirical studies may verify the signs and relative sizes of the coefficients. Such studies must of course be based on detailed knowledge of the product or the services as well as clear understanding of the underlying market conditions.

The end result of the estimation and validation step is the selection of a statistically and economically sound demand model that most reasonably explains the historical patterns of demand. This final model is generally a single equation that expresses demand as a function of its determinants and their respective elasticity coefficients.

In practice, it is sometimes possible to select a model simply on the basis that it has been used for some closely related empirical study. Other times the selection may be based on implicit assumptions about elasticities. An analyst always looks for evidence that certain model forms are more defensible than others, however.

Alternatively, you may decide to take an orthodox approach and reject the data rather than the theoretically specified model. A preferable alternative would be to modify the theoretically specified model in accordance with the evidence displayed by the data.

Each potential model structure will constitute a hypothesis of market behavior. Each hypothesis can then be accepted or rejected after standard statistical tests have been made. This process is not as open ended as it may appear, since only a small number of logical model structures will apply in any particular situation.

To test alternative models, the response parameters (e.g., elasticities) in each model must be computed from the historical data by using statistical estimation procedures. The most common procedure is multiple regression analysis. It can produce accurate estimates of the response parameters, and provides a variety of statistical measures useful in comparing alternative models.

Generating Demand Response Functions

5. *Use the demand model to generate price–demand and other demand response relationships.*

The demand curves and related elasticity coefficients of the marketing instruments can often be computed directly from the model structure. However, in some cases, the demand curves can only be determined by inserting simulated values of the marketing instruments into the model and observing the resulting demand response.

In many cases these procedures will be limited to the generation of price–demand and environmental variable response functions. However, as historical data for other marketing instruments become available, their response curves can also be incorporated into the demand model.

Producing Forecasts

6. *Use the demand model to generate conditional demand forecasts.*

To forecast with a demand model, you must obtain reliable forecasts of the determinants of demand and enter these forecasts into the demand model. The model will then generate a *conditional* demand forecast. That is, it will produce a demand forecast that depends on the accuracy with which the determinants of demand are forecast.

The demand model therefore translates the problem of forecasting demand to the problem of forecasting the determinants of demand. Although this may appear to be a questionable tradeoff, it often turns out to be very sensible. Some governmental and private organizations specialize in developing reliable forecasts of the environmental variables included in typical demand models, so accurate data for these variables can be obtained without difficulty. Also, product/service managers should be able to forecast the future values of their marketing instruments with reasonable accuracy, so they are also a potential source of reliable data. Indeed, a strength of demand models is that they provide a *systematic* way to combine the best available information on future product/service demand.

The inherent strength of the conditional demand forecast is its ability to quantify all the assumptions in a demand forecast rigorously. The response assumptions (i.e., elasticity coefficients) are an integral and readily identifiable part of the model structure. The input assumptions are simply the values selected for the forecasted determinants of demand. Thus if the demand forecast does not materialize, the reasons for this can often be isolated and explained with a minimum of supplementary analysis.

The forecast provided by the demand model will often *not* be the final forecast. Purely judgmental adjustments may need to be made: this is most common when *new* products/services are introduced, which are expected to be cross elastic with the forecast of demand. In this case, the original demand model forecasts can serve as a basis to track and measure the accuracy of the judgmental estimates of cross-elastic response.

Tracking Results

7. *Track model forecasts and actual product/service demand and use the differences between them to guide future refinements of the model and make preliminary elasticity estimates of new marketing instruments.*

This last step in the demand analysis process is one of the most critical. A demand model is not a "one-time shot." Although it is a valid representation of past and current market structures, it may have diminishing validity in the future. By regularly comparing the model forecast against actual demand quantities, you can use the pattern of differences to determine when the model structure needs modification.

Equally important is the use of the differences between a forecast and an actual outcome to make preliminary estimates of the elasticity coefficients of new marketing instruments. In many cases, the demand model will not contain any information about a particular marketing instrument. However, if the value of that marketing instrument changes in the future, the simultaneous change between what has been predicted and actual performance can often be used to estimate the elasticity coefficient of the responsible instrument. Preliminary estimates of self and cross elasticities of a price change can be made through this "tracking" procedure.

The execution of these seven steps is fairly straightforward for a trained analyst. However, the development of a valid model requires a large amount of creativity and a reasonable amount of time. Once the data are collected, an experienced analyst should be able to construct and document a valid model in a reasonably short time. However, take into account the iterative nature of certain steps, especially those pertaining to the collecting of data and testing of alternative model structures, since these may require additional time.

SPECIAL PROBLEMS TO WATCH OUT FOR

It has been noted that a problem with measuring the demand curve (quantity versus price) is that all the determinants of demand and supply will always be changing. If there are enough sufficiently varying data, regression methods can assist in measuring the demand curve.

Regression done with ordinary least squares (OLS) works best when there is no correlation among the independent variables. If the independent variables are correlated, the coefficient estimates produced by OLS analysis may not be valid. In practice, all independent variables have some correlation and one problem is to determine whether the correlations among independent variables are sufficiently strong to invalidate the regression results. Regression analysis therefore does have limitations, especially when *multicollinearity* exists among independent variables.

When multicollinearity exists, in effect, the OLS procedure cannot be used to determine which of the independent variables "explains" the variation in the dependent variable. As a result of multicollinearity it is possible to have a significant F statistic for the overall regression and yet to find that none of the coefficients of the individual independent variables are significant. Also, the signs of the coefficients may disagree with what theory indicates should be the way the market should function. For example, the model may suggest that quantity demanded increases as price increases, or decreases as income increases. One may not believe this to be true. Collinearity problems will be treated again in the case study in the next chapter and in Chapter 14.

Other problems to consider are *specification errors,* such as the omission of relevant variables, the inclusion of inappropriate functional forms, and the inclusion of irrelevant variables. All these point to the importance of starting first with simple (parsimonious) specifications, so that potential and existing modeling problems can be understood thoroughly and handled appropriately.

There are other statistical questions that need to be considered for demand models. The significance of an included variable can be assessed by testing whether an estimated parameter for such a variable is significantly different from zero. The model should have overall significance as well, and no excluded variables should be able to make a significant change in the model when they are tested. A careful residual analysis should always be made to verify the validity of underlying assumptions about data distribution.

MANAGEMENT CONSIDERATIONS

To derive full benefit from a demand analysis of a product or service, a forecast manager should be certain that the following activities have been done:

- Define the product and service categories for which market response measurements (elasticities, etc.) are desirable and attainable.
- Determine the data sources, computer facilities, and associated analytical software system required to generate the market response estimates.
- Determine the personnel requirements for data base management, analytical studies, documentation, and tracking of estimates.
- Devise an organizational structure that can accomplish the required analytical activities in the most efficient manner.
- Implement the demand analysis program in manageable phases, with initial emphasis on creation of a data base, management of the data, and development of demand models for tracking purposes.
- Integrate the demand analysis activity into the ongoing product/service management process.

Each of these actions requires a considerable creative effort. Without these efforts a variety of problems can arise.

Of foremost concern are the organizational implications of a mismanaged demand analysis effort. In a typical company it is conceivable that the full range of demand analysis responsibilities could be divided between several management groups or departments. As a result, some responsibilities may be unnecessarily duplicated (e.g., data base management could overlap aggregate demand modeling or forecasting) and other responsibilities may only be done for specialized purposes, on an *ad hoc* basis or not at all (e.g., price–quantity forecasting, tracking, demand modeling or forecasting of a product line, or economic and market research). This makes it very difficult to integrate a comprehensive demand analysis program into the product/service management process.

Another sort of mismanagement of demand analyses can come about when market response data are gathered only on a centralized basis, through various types of survey activities or centralized analyses; centralized gathering of data may inhibit measurement of market responses on a local or regional basis. There is ample evidence that market responses as well as environmental factors have significant variation from region to region and that making an ongoing analysis of local demand in many regions is the only practical way to monitor and measure these changing effects over time.

Another reason why demand analysis is often not used effectively is simply that data are inadequate. Historical data are needed to measure market responses to any decision a company's management makes. If these data have not been collected and maintained in a systematic manner, only a very limited form of demand analysis can be done. Because systematic collection of data is often not done, it is often difficult to obtain detailed historical data on individual product lines; then much effort has to be expended to obtain accurate data on a disaggregated basis.

SUMMARY

Demand analysis

- Is one of the most comprehensive methods for understanding and measuring the factors that influence demand.
- Is a widely used technique for developing estimates of historical price–demand relationships.
- Is sometimes called demand modeling since it produces a mathematical model that explains demand in terms of its causal factors.
- Is one important area where the analytical methods discussed so far can be applied. Linear and log-linear regression methods are the primary statistical tools used to quantify relationships postulated by the theory of demand treated in this chapter.

For estimating price elasticity, the forecaster can

- Use a demand analysis (theory of demand) framework to select the variables that should be included in the model.
- Select the appropriate model form (multiplicative or additive). The multiplicative model is the more frequently used of the two.
- Estimate the model by regression methods, provided that model assumptions have not been violated.
- Determine the elasticity directly in the case of the multiplicative model.
- Calculate a long-term elasticity if a lagged quantity is a variable in the model.

A primary reason for building a demand model is that it enables you to interpret key regression coefficients. The forecasts produced from these models are of secondary significance; although good forecasting properties are a valuable attribute of any model. However, the inability of a demand model to produce reasonably accurate forecasts may reduce its overall credibility.

USEFUL READING

PARSONS, L. J., and R. SCHULTZ (1976). *Marketing Models and Econometric Research*. New York, NY: North-Holland Publishing Co.

PHLIPS, L. (1974). *Applied Consumption Analysis*. New York, NY: American Elsevier Publishing Co.

PINDYCK, R. S., and D. L. RUBINFELD (1976). *Econometric Models and Economic Forecasts*. New York, NY: McGraw-Hill.

CHAPTER 10

Demand Analysis and Forecasting: Two Case Studies

This chapter presents two case studies that will illustrate many procedures described in earlier chapters.

- The first study measures the price elasticity of a telephone toll service; it forecasts a telephone toll-revenue series by taking price changes into account.

- The second study predicts the demand for extension telephones in residences in various geographic regions through use of available cross-sectional census and economic data.

Both examples are designed to illustrate:

- The application of multiple linear regression models for forecasting purposes.

- The evaluation of summary statistics.

- The value of doing a residual analysis.

A DEMAND MODEL FOR TELEPHONE TOLL SERVICE

Toll revenues generated through telephone use are a significant part of the revenue of telephone companies. The price elasticity of the service must be understood before an optimum pricing scheme can be established. The demand model should also be helpful in predicting future revenues.

Developing a Price Index

To measure price elasticity in the first of our case studies, it was necessary to develop a price index that, at any point in time, would measure the difference between price of a typical call and the price of that call in a given base period. The "basket," or distribution of calls used to develop the price index, will need to take into account the number of calls made between various locations, the time of day these are made, the extent of operator assistance, and the duration of the calls (conversation time). A *chain index* was used; such an index has a value of 1.0 in the base period; the basis for that period is the initial distribution of calls and the price per call. At the time of each price change, the percentage increase in prices, based on the *current distribution* of calls, is computed. The existing price index is then multiplied by the most recent percentage change in price. This form of the price index allows customers to change the distribution or pattern of calling over time in response to differential price changes (e.g., large increases in the price of operator-handled calls will encourage customers to place fewer operator-handled calls and more directly dialed calls).

Developing a Demand Model

The dependent variable that will be used in the regression is the revenue derived from the service divided by the price index. This yields a surrogate for "quantity" that has as its advantage that it measures other customer reactions to price changes than simple reduction in the number of calls. Since a customer can offset a price increase by reducing conversation time, by calling in discount periods, or by dialing calls directly instead of by using an operator, as well as by reducing the number of calls made, changes in revenues will be a more meaningful variable in the regression than changes in the number of calls.

Independent variables for consideration in the model are the number of telephones in service and personal income in constant dollars. Before it is included as a variable in the right-hand side of the formula for our model, the price index is deflated by dividing by the Consumer Price Index to yield a relative price index. Finally, as a measure of habit, the "quantity" in the prior period $(t-1)$ is included. An advertising variable could also be tried.

Six years of quarterly historical data are available for use.

Consider the tentative model that has been described:

$$\text{Quantity}_t = f(\text{Quantity}_{t-1}, \text{Price}_t, \text{Income}_t, \text{Telephones}_t).$$

A multiplicative model form (for constant elasticity) is considered; then logarithms are taken so that ordinary least-squares regression can be applied to estimate the parameters. (In the following formula, Q = quantity, P = price, I = income, and T = telephones; $\ln = \log_e$.) We have

$$\ln Q_t = \beta_0 + \beta_1 \ln Q_{t-1} + \beta_2 \ln P_t + \beta_3 \ln I_t + \beta_4 \log T_t + \varepsilon_t.$$

The coefficient of the price variable provides an estimate of the price elasticity of the service. Table 10.1 summarizes the results of the tentative model and Table 10.2 shows the correlation matrix of the variables. The model explains 90 percent of the variation about the mean of the "quantity" series. All of the independent variables except price are very highly correlated with the dependent variable; but the income, telephone, and lagged quantity variables are also highly correlated with each other. Plots of these series (Figures 10.1–10.3) show a strong linear trend and, as a consequence, a potential multicollinearity problem may exist.

Dealing with Multicollinearity

The OLS estimation procedure works best when there is zero correlation between pairs of independent variables. When two or more independent variables have similar linear trends, multicollinearity exists and it may be difficult to interpret individual parameter estimates. Chapter 14 discusses several approaches that may be tried to

Table 10.1 Summary of results for the demand model.

Variable (expressed in logarithms)	Coefficient	t Statistic
Constant	-2.93	-0.67
Lagged Quantity	-0.007	-0.03
Tels (T)	1.10	2.43
Income (I)	0.58	1.90
Price (P)	-0.29	-2.66
R^2:	0.90.	
Residual standard deviation:	0.032	

Table 10.2 The correlation matrix for the variables in the model.

1	Quantity (Q)	1.00				
2	Lagged Q	0.86	1.00			
3	Tels (T)	0.92	0.92†	1.00		
4	Income (I)	0.25	0.35	0.45	1.00	
5	Price (P)	0.90	0.87†	0.93†	0.47	1.00

†Indicates high correlation among independent variables. Variables are expressed in terms of logarithms.

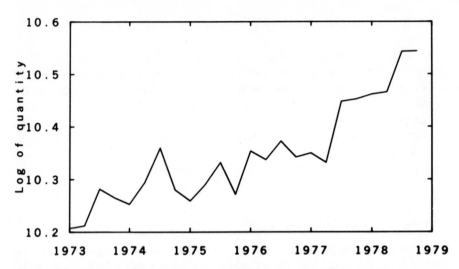

Figure 10.1 Time plot of the dependent variable for the toll revenues model—telephone toll revenues divided by the price index from first quarter 1973 through fourth quarter 1978.

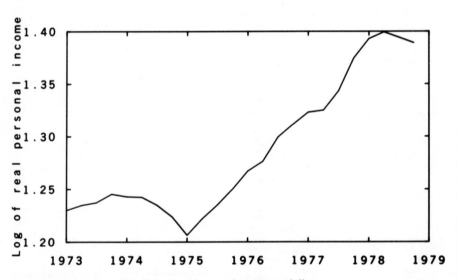

Figure 10.2 A time plot of personal income in constant dollars.

deal with multicollinearity. For the present case study, the problem is confirmed when the regression analysis is reviewed. The t statistic for the *lagged quantity* Q_{t-1} indicates that the parameter is insignificant, even though there is a correlation of 0.86 between "quantity" Q_t and *lagged quantity* Q_{t-1}.

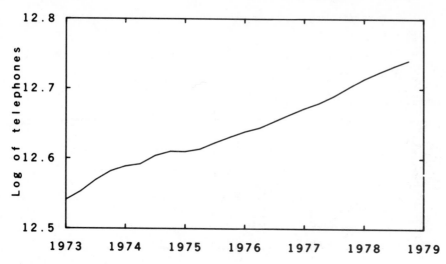

Figure 10.3   A time plot of the number of telephones in service.

Residual Analysis

The correlogram of the residuals (Figure 10.4) of the model indicates a significant correlation at lags 2 and 4. Even though the quantity series is dominated by trend, a seasonal pattern exists in the residuals. To correct this problem, a seasonal adjustment of the series can be made prior to modeling; if the seasonal pattern is not changing over time, it may be appropriate to make use of dummy variables (dealt with in Chapter 12) to account for the seasonal pattern.

Improving the Model

In our case study, dummy variables were introduced to account for seasonality and one of the independent variables was pruned from the model to reduce the multicollinearity problem. Two alternative models were then possible—the first model excluded the telephone variable T, and the second model excluded the habit variable (lagged quantity, Q_{t-1}). The regressions that result for both alternatives are summarized in Table 10.3. Both models have very high values for the R-squared statistic, significant values for the F statistic; and parameter estimates for lagged quantity, income, price, and telephones are all significant and of the expected sign. While only Dummy 3, the dummy variable introduced to represent the fourth quarter of each year, is individually significant, the incremental F test indicates that as a group the dummies are significant. For example, the mean squared error of Model 2 without the dummies is 10.43×10^{-4}, and with the dummies it is 5.01×10^{-4}. The incremental F test is computed as

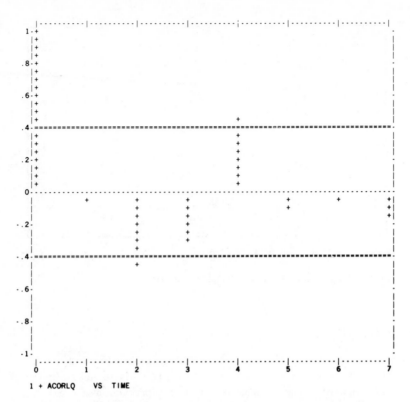

Figure 10.4   A correlogram of the residuals from the demand model.

$$F = \frac{(\text{Residual SS}_{old} - \text{Residual SS}_{new})/(df_{old} - df_{new})}{\text{Residual SS}_{new}/df_{new}}$$

$$= \frac{(10.43 - 5.01) \times 10^{-4}/(18 - 15)}{(5.01 \times 10^{-4})/15} = 5.4.$$

The incremental F statistic exceeds the critical value (3.29) for 3 and 15 degrees of freedom; this indicates that as a group the dummy variables are significant.

The Durbin-Watson statistic cannot be used to test for first-order residual autocorrelation in Model 1 because of the presence of the lagged dependent variable. The Durbin h statistic (see Chapter 13) can, however, be used as a test of this.

The Durbin h statistic is a large-sample statistic ($n > 30$); however, its value was so close to zero in the test that was made that the rejection of the hypothesis of first-order autocorrelation seems plausible.

The D-W statistic can be used for Model 2 and gives a value of 1.66 for 17 degrees of freedom, which is outside the range of positive autocorrelation (Appendix A, Table 5). Neither the h nor the D-W statistics will pick up the problem of serial correlation at lags other than one, and this indicates a need to plot the residual correlogram.

Table 10.3 Summary of regression results for alternative models.

Variable (expressed in logarithms)	Model 1		Model 2	
	Coefficient	t statistic	Coefficient	t statistic
Constant	4.63	3.2	−1.65	−0.7
Lagged Quantity	0.55	3.6	—	—
Tels (T)	—	—	0.98	4.6
Income (I)	0.76	3.7	0.64	3.3
Price (P)	−0.22	−2.6	−0.27	−3.7
Dummy 1	0.020	1.2	−0.006	−0.5
Dummy 2	0.016	1.0	−0.007	−0.6
Dummy 3	0.072	4.3	0.04	3.3
R^2:	0.95		0.96	
Residual standard deviation:	0.025		0.023	
Durbin-Watson statistic:	NA		1.66	

When correlograms of the residuals of each model were plotted, these confirmed a significant residual correlation at lag 2, with a value of approximately 0.40 for Model 1 and 0.48 for Model 2. An approximate test of significance is $2/\sqrt{n} = 2/\sqrt{24} \simeq 0.42$, where n is the number of observations—in this case 24 quarters. The residual autocorrelation at lag 2 is less than 0.42 in Model 1 but is greater than this in Model 2. A new independent variable might be found to explain this pattern.

For Model 2, residuals were plotted against the predicted values; this plot indicated a relatively large outlier at the fourth quarter of 1975. The second quarter of 1977 also showed a large negative residual. There were no obvious outliers in any of the variables that would suggest transcription errors within these time series.

A robust regression was performed to determine what impact these unexplained values had on the parameter estimates for Model 2. The price elasticity changed from −0.27 to −0.28 in the robust regression and the income elasticity dropped from 0.64 to 0.53: the relatively large residuals had not significantly changed the estimated price elasticity and had changed the income elasticity by less than one standard error. When a Q-Q plot was made, it did not indicate substantial deviations from normality.

Model Conclusion

While not perfect, the two models do provide an approximation of the price elasticity for telephone toll service. Model 1 estimates the elasticity to be −0.22 ± 2(.08) and Model 2 estimates it to be −0.27 ± 2(.07). The range of the price elasticity is −0.06 to −0.41 at the 95-percent confidence level. Once again, the problem of having only 24 observations for analysis results in rather imprecise estimates of the elasticity and argues against using 95-percent confidence limits; perhaps using 50-percent confidence limits would make more sense for this example. Using monthly rather than quarterly data would also improve the precision of the parameter estimates. Another alternative, pursued in Chapter 16, is the use of a *pooled* model: data for one geographic area are combined with data for adjacent areas to increase the degrees of freedom in a pooled regression model.

When Model 1 was compared with Model 2, both models indicated that a price elasticity of approximately -0.22 to -0.27 is the most likely value for a quarter in which price is changed. However, Model 1 is a dynamic model and indicates a long-term elasticity of $-0.22/(1 - 0.55) = -0.49$. This suggests that, all other variables held constant, the long-term response to a price change is about twice the short-term response. By conducting a controlled experiment, it would be possible to determine which model makes the best actual fit to empirically observed results.

The results of the first of our case studies should demonstrate that a high value for the R-squared statistic does not necessarily mean you have a good model. Problems of multicollinearity, serially correlated errors, and heteroscedasticity (not a problem here) could invalidate a model even if its R-squared statistic were high, and a high R-squared statistic does not necessarily mean that the model will forecast well; you must go beyond the R-squared statistic to develop worthwhile models.

Forecasting Results

Given the limited historical data, one-year-ahead forecasts were generated for 1977 and 1978; these indicated forecast misses of -3.4 percent and -6.2 percent for Model 1 and -1.2 percent and 1.6 percent for Model 2, respectively. In this example, Model 2 provided more accurate forecasts than the model incorporating a lagged dependent variable.

A CROSS-SECTIONAL MODEL FOR RESIDENTIAL EXTENSION TELEPHONES

In the second case study, an analysis was performed to help identify the potential for increased sales of extension telephones within residences in 470 geographic areas. A requirement was that the model should incorporate local economic and demographic data in a formulation understandable and acceptable to local sales personnel responsible for stimulating demand. Areas with below-average development, as predicted by the model, would be candidates for future sales campaigns.

Variable Selection

After experimentation with a variety of possible independent variables, on theoretical and statistical considerations several were ultimately selected. Median family income, adjusted for cost-of-living differences among the geographic areas, was an obvious candidate. Of the total work force in each area, the percent engaged in white-collar employment was selected, since these employees are generally intensive users of telephones on the job, and tend to take their telephone habits home with them. Of all households, the percent in which more than one automobile was owned was selected as an indicator of the propensity to consume rather than to save income. A number of other variables could also have served this purpose.

The Model-Building Process

As a first step in building a model, a plotting was made of the percent of residential extension telephones (residential extensions per hundred main residential telephones) by area. It showed a lack of symmetry in areal distribution, particularly so where percentages of extension telephones were high.

Scatter diagrams were next plotted to investigate relationships between the dependent and independent variables. In Chapter 5, we noted that, in a multiple linear regression model, inclusion of partial residuals plots may be helpful, since they show the relationship between the dependent variable and each independent variable, given that all the variables have been entered into the model.

A regression model containing the three independent variables and the partial residual plots was examined. The partial residual plots associated with income, employment, and automobiles are shown in Figures 10.5, 10.6, and 10.7, respectively. In Figure 10.5 most of the points are concentrated in the $10,000–$18,000 range, with a relatively small number of high-leverage points above $20,000. Figures 10.6 and 10.7 show a more uniform distribution. In all three plots the relationships appear to be positive and linear.

Diagnostic checks for constancy of variance and normality in the residuals were next made, and these completed the residual analysis. Figure 10.8 compares residuals against the values that were predicted in the model described above, and Figure 10.9 shows the normal Q-Q plot of the residuals of this model; Figure 10.8 reveals an apparent problem of increasing residual variance, and most of the residuals associated with development above 85 percent are positive; the normal Q-Q plot in Figure 10.9 shows that the high positive residuals form a longer tail than is present in the normal curve.

In most cases, you could stop at this point, since the departures from the assumptions of the model are not too severe. However, it was decided to go further to see if a more constant residual variance as well as a more linear Q-Q plot (one closer to normality) could be obtained, and if a better partial residual plot against income could be obtained by transforming the variables. For example, a transformation of the dependent variable might improve the Q-Q plot.

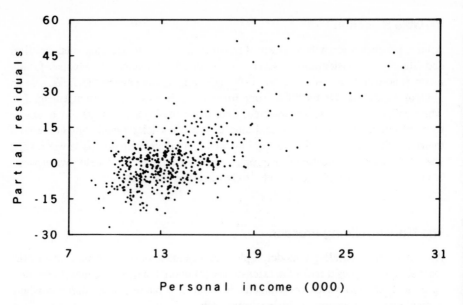

Figure 10.5   A plot of partial residuals against the income variable, for the extension telephone case-study.

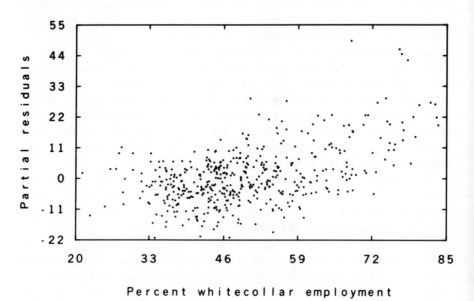

Figure 10.6   A plot of partial residuals against the employment variable.

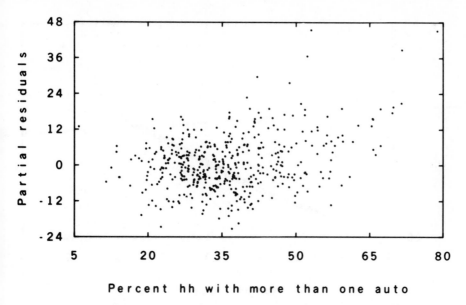

Figure 10.7   A plot of partial residuals against the automobile-ownership variable.

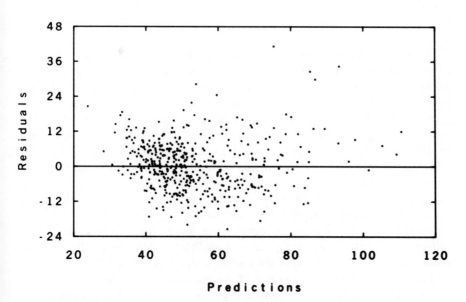

Figure 10.8   A plot of residuals against predicted values.

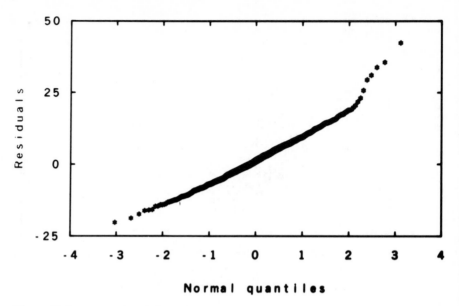

Figure 10.9 A normal Q-Q plot of the residuals.

Figure 10.10 shows a plot of the residuals against predictions for the model after a *logarithmic transformation* of the percentage development was taken. In this case, an obvious pattern of decreasing variance resulted, leading us to conclude that the logarithmic transformation is inappropriate.

Figure 10.11 shows a plot of the residuals against predictions for an alternative model in which a *square root transformation* has been applied to the dependent variable and a *logarithmic transformation* has been applied to the income variable. This is a two-step operation, the results of which are combined in the figure for convenience. (The logarithmic transformation was taken to obtain a linear relationship in the partial residual plot between the two time series.) Recall that the square root transformation is often helpful with data involving counts or percents. This plot is similar to Figure 10.8 but is an improvement, since Figure 10.11 shows a more uniform residual variance over the range of predictions. Also, the residuals are not as concentrated as the low end of the percent-of-development scale.

Figure 10.12 shows the Q-Q plot of the residuals of both models against the normal curve. A more linear pattern, and therefore a more normal pattern, exists. Figures 10.13, 10.14, and 10.15 show the partial residuals against (respectively) the logarithms of income, the percent of employment that is white-collar, and the percent of households with more than one automobile. Figure 10.13 is an improvement over Figure 10.5, since there are fewer high-leverage points far away from the bulk of the data. Figures 10.14 and 10.15 are slight improvements over 10.6 and 10.7, in

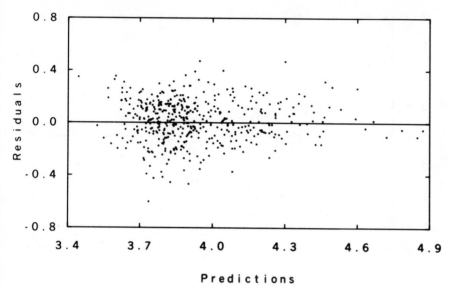

Figure 10.10   A plot of residuals against predictions after taking logarithms of the dependent variable.

that they appear to have a higher slope (and therefore a higher correlation) and fewer outliers. The refined model is an improvement over the original model.

In the early stages of development of the model for this case study, an analyst failed to notice that of the 473 geographical areas being examined, there were three areas for which no values (observations) were given in the dependent variable, yet there were values in the independent variables for these areas. The computer set the values for the missing observations equal to zero, by default. Later, a plot of residuals against the predictions offered by the final model (in which zeroes appeared as observed values for the three areas) demonstrated that the three residuals were in fact so large that they dominated all others and had the effect of lowering the R-squared statistic from 0.69 to 0.43. The analyst then made a Q-Q plot of the residuals against the normal curve, and it also made the appearance of the three outliers quite striking.

A Robust Alternative

Robust regression would offer some protection against outliers such as those just described, as is illustrated in Table 10.4. The left-hand part of the table shows that ordinary least squares (OLS) and robust regression analyses yield almost identical results, with no extreme values and approximately normal residuals. Comparison of the left-hand and right-hand sides show that the OLS results were distorted by the

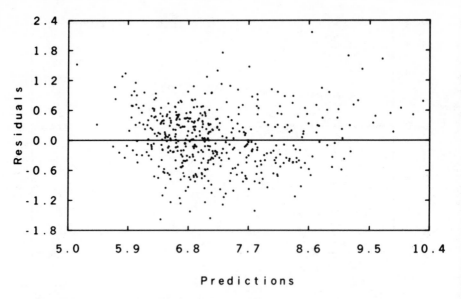

Figure 10.11   A plot of residuals against predictions after taking square roots of the dependent variable and logarithms of the income variable.

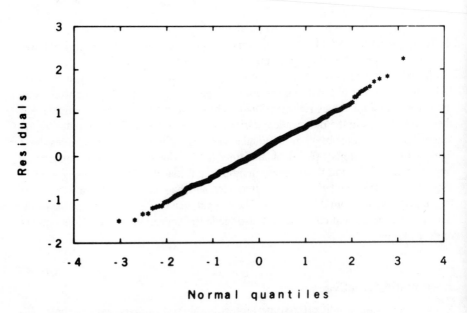

Figure 10.12   A Q-Q plot of the residuals for the refined model.

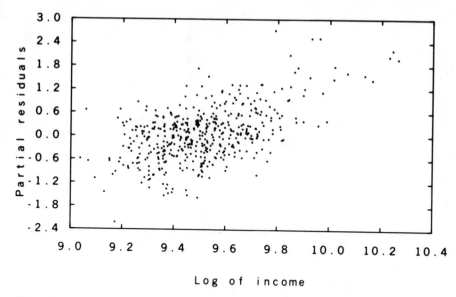

Figure 10.13 A plot of partial residuals against logarithms of the income variable for the refined model.

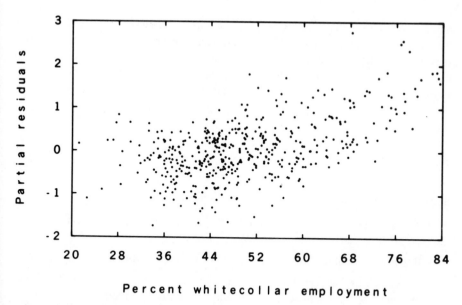

Figure 10.14 A plot of partial residuals against employment variable for the refined model.

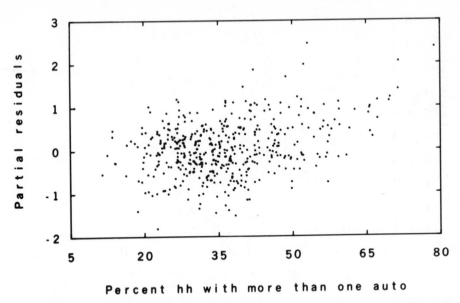

Figure 10.15   A plot of partial residuals against the automobile-ownership variable for the refined model.

outliers, while the robust results were not. Recall that only three of the 473 observations were extreme, yet these three altered the income coefficient and the constant term significantly. This example should suggest to you that a difference between OLS and robust regression coefficients of approximately one standard deviation is cause to review the OLS model and the original data in much greater detail.

SUMMARY

Some of the most important topics in demand modeling are:

- The data should be adjusted for outliers such as might result from unforeseen events (e.g., strikes) or missing observations.
- Replacement values for outliers should be estimated whenever possible.
- Relationships between the dependent and independent variables should be examined by means of scatter plots and partial residual plots.
- Transformations may be necessary to establish linear relationships between variables or to correct for increasing variability in the residuals of the model.
- If the primary purpose of the model is to generate a forecast and *not* to estimate elasticities, then autocorrelation or collinearity problems can be solved by using the values of percent changes, first differences, or deviations from trend.

Table 10.4 A comparison of OLS and robust regression results for the final model with and without the outliers.

Variable	No extreme outliers			Three extreme outliers		
	OLS		Robust	OLS		Robust
	Coefficient	Estimated standard deviation	Coefficient	Coefficient	Estimated standard deviation	Coefficient
Interest	−13.328	2.2627	−13.366	−9.6892	3.467	−13.366
Income	1.9275	0.2561	1.944	1.5500	0.392	1.944
White-collar employment	0.0319	0.0035	0.0308	0.0300	0.0057	0.0308
Households with more than one auto	0.0173	0.0037	0.0154	0.0171	0.0054	0.0155

- It is important to use residual plots as a means of checking model assumptions.
- The presence of autocorrelation in residuals suggests there may be some variation that is not explained by the model. This may be due to missing independent variables or to incorrect specification of the error structure.

USEFUL READING

BELSLEY, D. A., E. KUH, and R. E. WELSCH (1980). *Regression Diagnostics: Identifying Influential Data and Sources of Collinearity*. New York, NY: John Wiley and Sons.

DRAPER, N. R., and SMITH, H. (1981). *Applied Regression Analysis*. 2nd ed. New York, NY: John Wiley and Sons.

CHAPTER **11**

The Econometric
Approach to Forecasting

This book has emphasized the use of regression and time series methods to represent relationships among variables for a variety of forecasting applications.

- An economic relationship between a quantity of a commodity that is demanded in a market and the determinants of that demand (price, income, habit, etc.) was quantified by a demand model in the previous four chapters. For a demand model a single equation is used to express the relationship between economic variables.

- Economic models can be created, more generally, by groupings of relationships or equations. Such groupings describe an econometric system if the model that results can be used to estimate and test economic theories. This is the subject of the present and next five chapters.

SPECIFYING AN ECONOMETRIC MODEL

The role of econometrics is the empirical estimation and testing of economic models; *econometrics* is concerned with the quantification of relationships among economic variables. This requires a specification of a model in terms of mathematical equations. The variables in these equations need to be relevant to the economic problem and to be measurable as statistical data. With the application of appropriate statistical tools, the parameters in the econometric system can be estimated and tests of significance can be performed.

The *specification* of a regression model consists of a formulation of the regression equation(s), and of assertions about the right-hand side variables in those equations (regressors or explanatory variables), as well as error or disturbance terms. A *specification error* refers to an incorrect assertion about the form (linear, logarithmic) and the content (the variables that are selected) of a regression equation, or incorrect assumptions about the regressors and the disturbance term.

Specification of an econometric model requires proper consideration of economic theory, relevant variables, functional form, and lag structure.

Econometric Assumptions

The present state of the art of economic theory is such that there are few well known and well behaved utility functions that one can specify for use in estimating equations. As Phlips (1974, p. 93) suggests: "One therefore often proceeds otherwise, specifying directly the demand equations that seem appropriate for the problem at hand, and taking care to impose general restrictions which ensure their theoretical plausibility." These restrictions can include the sign and magnitude of coefficients and the lag structure.

As Phlips further notes:

> To identify a given relationship one has to take other possible relationships into account. To be able to identify a demand function with the help of statistical data, one has to realize that supply has simultaneously influenced the same data, for the simple reason that the observed prices are the results of the equalization of demand and supply.

The underlying assumptions behind an econometric system include these:

- Economic behavior can be described by a system of mathematical equations.

- The representation will capture the essential features of the economic relationships.

- Future values can be obtained for predictive purposes.

- As a policy tool, alternative economic scenarios can be developed by varying control variables.

Another basic assumption of an econometric model involves the specification of a disturbance or random error term. Relationships among economic variables will at best be only approximate, so the uncertainty inherent in the model is expressed through stochastic error terms (usually additive).

There are also two technical problems in constructing an econometric system of equations, which should be stressed:

- A mathematical problem of solving k equations in n unknowns ($k < n$).

- A statistical problem of estimating the parameters of an equation by using ordinary least squares (OLS).

In the practice of econometrics, several simplifying assumptions are generally made that must be kept in mind. The first is the *ceteris paribus* assumption, which states that all variables that are excluded from the model are held constant. Their mean effect is captured by the constant term of the regression equation and their variability is captured by determining the variance of the errors. This assumption is necessary but not generally realistic.

A second general assumption is the *aggregation assumption,* which assumes that, for example, the aggregate income elasticity equals the individual income elasticity. This assumption requires that all individual income elasticities be equal and that the income distribution remain fixed. Obviously, there will always be differences between a model based on aggregate data and one based on individual data. The unavailability of data about individual incomes forces an analyst to use aggregate data.

The aggregation assumption is also made when per capita specifications are used. One hopes that the division of aggregate income by population yields a per capita income that is representative of the individual incomes of the members of the population. When there is a wide income distribution, the aggregation assumption can become a problem. In most cases, there is little an analyst can do, except to be aware of the problem, since data for individual incomes are generally not available.

Explanatory Variables

Since the purpose of the independent variables in the model is to "explain" the variation in the dependent variable, you should consider variables that measure price, income, market size, advertising, and habit. The price variable often enters the equation in the form of a price index. It can be deflated by a measure of overall changes in prices to yield "real" price.

The income variable can be personal (before tax) or disposable income, nominal or real income, or aggregate or per capita income. There are also instances where consumers appear more concerned with a stream of income over time than with income at a particular point in time. A house may be purchased on the basis of current income and expected future income. In other cases, consumers may not change consumption patterns immediately after changes in income but change them more slowly over time. The lag terms in models are often used to estimate this gradual or *distributed* response to changes in the level of the independent variables.

Specification Errors

All errors arising from a misspecification of an econometric system are called specification errors. *Specification errors* can result from:

- The omission of a relevant explanatory variable.
- Disregarding a qualitative change in one of the explanatory variables (e.g., quality of a product).
- The inclusion of an irrelevant explanatory variable.
- Incorrect definition of a variable.

- Incorrect specification of the manner in which the error term enters the equation.

- Incorrect time lags.

Most commonly, specification errors involve omitting relevant variables and including irrelevant variables, and incorrectly specifying the functional form of the relevant variables. For example, a consumer price index may be a relevant but omitted variable in the demand equation for a consumer good. Likewise, temperature may be an included but irrelevant variable in this case. Specification errors of this kind could lead to estimated coefficients that have undesirable statistical properties, such as biasedness (see for example, Fisher, 1966; Johnston, 1973; Pindyck and Rubinfeld, 1976).

Other forms of misspecification are also important. In demand modeling, for example, it is important to correctly define quantity, price, income, and the market variable. Incorrect use of a price variable, instead of using "relative price," has been known to produce misleading results. The incorrect specification of the error term in a model can lead to vastly different inferences. For example, an additive error term in a log-linear model corresponds to a multiplicative error, not to an additive error in the original domain of the dependent variable.

Incorrect specification of the time lag in a model with lag-dependent variables can result from misspecifying the length or form of the time lag. Time lag specification occurs in the estimation of distributed lag models.

Problems of simultaneity will be discussed in Chapter 15 in connection with a "Connects-Disconnects" case study based on telephone data. Also to be considered are the specification errors that arise when error variances are not constant and normality assumptions are not satisfied.

ECONOMETRICS AND FORECASTING

Many econometric systems are often used for forecasting purposes. A variety of commercial timesharing vendors and academic institutions provide the business community with a plethora of economic forecasts from sophisticated, computerized econometric systems. These forecasts are an important asset to the decision maker in a rapidly changing economic environment.

While there is a widespread use of forecasts from econometric systems, there does not appear to be a universal acceptance that econometric systems produce consistently reliable and accurate forecasts. (See, for example, Armstrong, 1978; Granger and Newbold, 1977, Section 8.4.) There are often simpler approaches yielding more accurate projections. However, the role of econometrics in forecasting is to provide an economic rationale for the process along with a mechanism for projecting numerical "forecasts."

The following chapters will provide the basic information required for more forecasting applications. Chapter 12 provides examples of how dummy and lagged variables can be used to extend the capabilities of regression models to handle a variety of practical problem areas. Chapters 13 and 14 discuss approaches that can reduce problems of autocorrelation, nonconstant variance, and collinearity, which frequently plague econometric modeling. Chapter 15 introduces some simultaneous equation techniques, while Chapter 16 shows how to combine time series and cross-sectional data to improve the parameter estimates through pooled models.

SUMMARY

The description of economic behavior with systems of mathematical equations

- Has prompted forecasters to take considerable interest in the field of econometrics.
- Has led to the extensive use of econometric systems in forecasting and planning models.
- Has contributed to the study of how alternative economic assumptions affect policy making.

USEFUL READING

ARMSTRONG, J. S. (1978). Forecasting with Econometric Methods: Folklore versus Fact. *Journal of Business* 51, 549–64.

FISHER, F. M. (1966). *The Identification Problem*. New York, NY: McGraw-Hill.

GRANGER, C. W. J., and P. S. NEWBOLD (1977). *Forecasting Economic Time Series*. New York, NY: Academic Press.

JOHNSTON, J. (1963). *Econometric Methods*, 2nd ed. New York, NY: McGraw-Hill.

PHLIPS, L. (1974). *Applied Consumption Analysis*. New York, NY: American Elsevier Publishing Co.

PINDYCK, R. S., and D. L. RUBINFELD (1976). *Econometric Models and Forecasts*. New York, NY: McGraw-Hill.

CHAPTER 12

Using Dummies and Lagged Variables

In Chapters 7–10, a theory of demand was developed in which a single dependent variable was regressed on a set of independent variables that were assumed to explain the dependent variable. The forecaster must often apply special treatments, because these variables create significant statistical problems. Solutions to these problems include:

- The use of indicator variables for qualitative factors, and for seasonal and outlier adjustment.

- The use of lagged variables to describe effects that unfold over time.

USE OF INDICATOR VARIABLES

Indicator variables, better known as *dummy variables,* are useful for extending the application of independent variables to representation of various special effects or factors, such as

- One-time or fixed-duration factors or effects, such as wars, strikes, and weather conditions.

- Significant differences in intercepts or slopes for different consumer attributes, such as sex, race, and so on.

- Discontinuities related to changes in qualitative factors.

- Seasonal variation.

- The effects of outliers.

- The need to let the intercept or slope coefficients vary over different cross sections or time periods. This subject will be treated along with the pooling of time series and cross-sectional data in Chapter 16.

Indicator Variables for Qualitative Factors

In addition to quantifiable variables of the type discussed in earlier chapters, the dependent variable may be influenced by variables that are essentially qualitative in nature. Changes in government or public policy, wars, strikes, and weather patterns are examples of factors that are either nonquantifiable or very difficult to quantify. However, the presence of such factors can influence consumer demand for products and services. Dummy or indicator variables may then be used to indicate the existence or absence of an attribute or condition.

For example, suppose that for any given income level, the sales S of a product to women exceed the sales of the same product to men. Also, suppose that the rate of change of sales relative to changes in income is the same for men and women. A dummy variable can be included in the sales equation to account for sex. Let $D = 0$ for sales to men and $D = 1$ for sales to women. Then

$$S_t = \alpha_0 + \alpha_1 D_t + \beta_1 (\text{Income})_t + \varepsilon_t.$$

For this example the base or control condition will be "male" ($D_t = 0$). The prediction $\hat{S}$ of sales to men is therefore

$$\hat{S}_t = \hat{\alpha}_0 + \hat{\beta}_1(\text{Income})_t, \qquad \text{for } D_t = 0,$$

and the prediction of sales to women is

$$\hat{S}_t = \hat{\alpha}_0 + \hat{\alpha}_1 + \hat{\beta}_1(\text{Income})_t, \qquad \text{for } D_t = 1.$$

The coefficient $\hat{\alpha}$ is called the *differential intercept* coefficient. It indicates the amount by which sales to women exceed sales to men at a given level of income. The t test can be used to determine if $\hat{\alpha}_1$ is significantly different from zero. Figure 12.1 shows a plot of the two regression lines for the example just given.

Similarly, the mean sales of a product in one geographical area may show the *same* rate of change relative to an economic variable that the sales in another area show, yet total sales for each state may be *different*.

Models that combine both quantitative and qualitative variables, as both of the foregoing examples do, are called *analysis-of-covariance models*.

You must always be careful to introduce one less dummy variable than the number of categories represented by the qualitative variable. In the above case, the two categories (male, female) can be represented by one dummy variable.

Using Dummy Variables to Identify Different Slopes and Intercepts

In the above example, suppose you want to know whether the intercepts and slopes are different for women and men. This can be tested in a regression model of the form

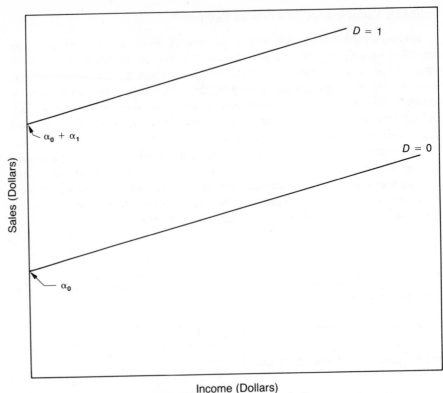

Figure 12.1 Plot of two regression lines.

$$S_t = \alpha_0 + \alpha_1 D_t + \beta_1(\text{Income})_t + \beta_2 D_t(\text{Income})_t + \varepsilon_t.$$

Then

$$\hat{S}_t(\text{Men}) = \hat{\alpha}_0 + \hat{\beta}_1(\text{Income})_t, \qquad \text{for } D_t = 0.$$

Likewise,

$$\hat{S}_t(\text{Women}) = (\hat{\alpha}_0 + \hat{\alpha}_1) + (\hat{\beta}_1 + \hat{\beta}_2)(\text{Income})_t, \qquad \text{for } D_t = 1.$$

The use of the dummy variable in the additive form allows you to identify differences in intercepts. The introduction of the dummy variable in the multiplicative form allows you to identify different slope coefficients. An example of the analysis-of-covariance approach will be discussed under pooling of cross-sectional and time series data in Chapter 16.

Using Dummy Variables to Measure Discontinuities

A change in a government policy or a change in a price may alter the trend of a revenues series. In the case of a price change, the preferable course of action is to develop a price index. Suppose that for any of a variety of reasons, such as lack of time or data, this is not possible. A dummy variable may be introduced into the model as follows. Let

$$Y_t = \beta_0 + \beta_1 X_t + \beta_2 D_t + \varepsilon_t, \qquad \text{where } D_t = 0 \text{ for } t < T^* \text{ and } D_t = 1 \text{ for}$$
$$t \geq T^*.$$

T^* is the time of the policy or price change, and D_t is a dummy variable with a value of zero for all time less than T^* and a value of one for all time greater than or equal to T^*; X_t is an explanatory variable.

In this example, the predicted values of revenues are:

$$\hat{Y}_t = \hat{\beta}_0 + \hat{\beta}_1 X_t, \qquad \text{if } t < T^*,$$

and

$$\hat{Y}_t = (\hat{\beta}_0 + \hat{\beta}_2) + \hat{\beta}_1 X_t, \qquad \text{if } t \geq T^*.$$

Another situation that often occurs is one involving a "yes–no" or "on–off" possibility. For example, the demand for business telephones is very strongly affected by presidential and congressional elections. These are held in even-numbered years: storefront campaign offices and candidates' headquarters are established; there is a large increase in the demand for business main telephones in September and October; then in November, the telephones that were related to the election campaign are disconnected. In odd-numbered years, local political elections are held; local politicians do not have the financial resources to establish as many campaign offices. Consequently, the impact on telephone demand is not as great in odd-numbered years. Aside from this, one can be sure that elections occur in odd- and even-numbered years alike. Therefore, it is possible to use a dummy variable that assumes that half the telephone gain attributable to an election occurs in September, and half in October, and that it all disappears in November. In this case, for a given year, the important consideration in assessing election-influenced telephone gain is whether there is or is not an election. This is what is meant by a yes–no, on–off, or categorical variable. The variable does not continue to take on different values for an extended time and in that sense is not quantitative in nature.

Using Dummy Variables for Seasonal Adjustment

There are numerous studies in which the analyst may decide to use dummy variables to account for seasonality in quarterly data. In these a dummy variable is used for the second, third, and fourth quarters. The seasonal effect of the first quarter is then captured by the constant term in the regression equation

$$Y_t = \alpha_0 + \alpha_1 D_{1t} + \alpha_2 D_{2t} + \alpha_3 D_{3t} + \beta_1 X_{1t} + \beta_2 X_{2t} + \epsilon_t,$$

where $D_{1t} = 1$ for the second quarter, $D_{2t} = 1$ for the third quarter, and $D_{3t} = 1$ for the fourth quarter. For monthly data, eleven dummy variables would be used (February through December, for example), and then the January seasonality would be captured by the constant term in the regression equation.

It is possible that only one month or quarter has a significant seasonal pattern. However, this is not generally the case in forecasting demand or sales of telephones (for instance): usually *each* month or quarter has a unique seasonal pattern. Dummy variables may be of of limited usefulness if the season changes over time, since the dummy variable approach assumes a *constant* seasonal pattern.

A useful attribute of the X-11 and SABL seasonal adjustment programs (discussed in detail in *The Beginning Forecaster*) is that they may be used with data that do not show a constant seasonal pattern. And most series do seem to exhibit changing seasonality over time. The advantage of using dummy variables in a regression is that seasonality and trend-cycle can be estimated simultaneously, in a one-step process; otherwise, a two-step process would be necessary: (1) seasonal adjustment of data, and (2) performance of the regression. The ease of using seasonal adjustment programs on timeshared computers is minimizing the one-step advantage of the dummy variable approach. In the one-step process, the dummy variable provides an index of average seasonality and this can then be tested for significance by using the *t* test for an individual dummy variable or the incremental *F* test for a group of dummy variables.

Eliminating the Effects of Outliers

To illustrate the use of dummy variables to eliminate the effect of outliers, consider the following telephone application. There was a telephone company strike in April and May of 1968 and disconnections of telephone service because of nonpayment of telephone bills were not reported for those months. This distorted the 1968 annual totals, which were subsequently used in a model for predicting telephone gain. It was decided to incorporate a dummy variable into the model. The dummy variable, in this case, was a variable that can generally be set equal to zero for all observations except the unusual event or outlier. Thus, the values for the dummy variable were equal to zero for all years but 1968, when it had a value of one.

Because the dummy variable equalled zero for all periods except that in which the outlier appeared, the dummy variable explained the outlier perfectly. That is, the predicted value equalled the actual value for the outlier. Use of such a dummy variable tends to reduce the estimated standard deviation of the predicted errors artificially, because what is in fact a period of wide variation has been assigned a predicted error of zero. For this reason, it is recommended that dummy variables be used very sparingly for outlier correction. They tend to result in a model with a higher R-squared statistic than can perhaps be justified.

In the case of outliers caused by nonexistent values, it is usually preferable to estimate a replacement value based on the circumstances that existed at the time the outlier occurred.

You need to be especially cautious when dummy variables and lagged dependent variables occur in the same equation. For example, a dummy variable might be used to correct for a strike that could have a large negative impact on the quantity of a product sold. But it is then necessary to adjust the value for the subsequent period, or the value of the lagged dependent variable will drive next period's predicted value too low. A preferable alternative is to adjust the original series for the strike effect (if lagged dependent variables are included in the model).

You also need to understand that the presence of dummy variables for outlier adjustment will result in an artificially inflated value for the R-squared statistic: the model will appear to explain more of the variation than can be attributed to the independent variables. In some cases, the outliers may be a result of an inadequate demand theory (or supply constraints), and you need to be aware of this.

In some circumstances, robust regression techniques may offer a method for estimating regression coefficients so that the results are not distorted by a few outlying values. The variability in the data will not be understated and the very large residuals will readily indicate that the model, as presently specified, is incapable of explaining the unusual events.

On the other hand, residuals in a model in which dummy variables have been used for outliers suggest that unusual events are perfectly estimated. A robust regression alternative has considerable appeal from a forecasting viewpoint. Since it will not understate the variability in the data, there will be less of a tendency to expect a greater degree of accuracy in the forecast period than can be achieved over the fitted period.

At times it may be necessary to introduce dummy variables because it is almost impossible to estimate a replacement value for a missing or extreme data value: there may be too many unusual events occurring at the same time. For example, a company strike may have coincided with the introduction of wage–price controls or an oil embargo. It would be extremely difficult to determine the demand for a product or service had there been no strike, because too many other variables also changed.

Moreover, if you were to attempt to build a model to predict gasoline consumption, based on data that included the 1973–1974 period, you would likely incur problems. Since the period encompassed a fuel supply shortage, as a result of the Arab oil embargo, actual consumption in that period was a function of supply more than of demand. This situation did not exist in any of the historical data. It might be better to use a robust regression, leaving out the data for 1973–1974, or to use a dummy variable to account for that time period.

LAGGED VARIABLES

Many econometric formulations require the inclusion of lagged independent and dependent variables to incorporate the effects of a variable over time. The impact of a given economic factor may not manifest itself for several time periods in the future, or its effect may be distributed over several time periods. To incorporate these situations in econometric models, econometricians have devised various *distributed-lag schemes* (Dhrymes, 1972; Fisher, 1966; Pindyck and Rubinfeld, 1977).

Distributed Lags

Generally, a *distributed-lag scheme* involving only lagged explanatory variables takes the form

$$Y_t = \alpha + \beta_0 X_t + \beta_1 X_{k-t} + \cdots + \beta_k X_{t-k} + \varepsilon_t,$$

or, alternatively,

$$Y_t = \alpha + \beta(\omega_0 X_t + \omega_1 X_{t-1} + \cdots + \omega_k X_{t-k}) + \varepsilon_t,$$

where the order k is generally unknown. Since there may be quite a few regressor variables, this can lead to a variety of statistical problems, including multicollinearity and the imprecise estimation of lagged coefficients when k is large.

The weights ω_i are interpreted as the proportion of the total effect achieved in a given time interval (this is called the *lag effect*). The coefficient β is interpreted as the economic reaction of Y_t to a sustained unit change in X (this is called the *economic effect*).

The distributed-lag coefficients β_i in the first of the two equations above equal a fixed β multiplied by a variable lag weight ω_i. The long-run response of Y to a unit change in X is the sum of the lag coefficients

$$\sum_{i=0}^{k} \beta_i = \beta \sum_{i=0}^{k} \omega_i.$$

To circumvent some of the estimation problems that arise when an equation contains numerous regressor variables, certain simplifying assumptions must be made in practice. One such assumption made many years ago was that the coefficients should decrease over time. Thus the scheme proposed by Koyck (1954) is an exponentially decreasing sequence given by

$$\beta_j = (1 - \lambda)\lambda^j \qquad \text{for} \qquad 0 < \lambda < 1.$$

A model with such coefficients can be written as

$$Y_t = \alpha + \beta(X_t + \omega X_{t-1} + \omega^2 X_{t-2} + \cdots) + \varepsilon_t.$$

This implies an *infinite lag* with weights declining *geometrically* toward zero.

The model can be simplified by lagging Y_t by one period, multiplying by the weight ω, and subtracting the result from the equation for Y_t (as shown above). Then

$$Y_t - \omega Y_{t-1} = (\alpha - \alpha\omega) + \beta X_t + \beta(\omega X_{t-1} - \omega X_{t-1}) + \cdots + (\varepsilon_t \cdots \omega\varepsilon_{t-1}).$$

Figure 12.2 illustrates the distribution of weights for various values of ω. This implies *a priori* knowledge of the proper weighting pattern.

The model may be further rewritten as

$$Y_t = \alpha(1 - \omega) + \omega Y_{t-1} + \beta X_t + (\varepsilon_t - \omega\varepsilon_{t-1}).$$

This can be fitted by OLS, by letting $\varepsilon_t^* = \varepsilon_t - \omega\varepsilon_{t-1}$ and assuming that the ε_t^*'s are uncorrelated. The simplifying assumption has made the model more tractable at the expense of considerable realism. Analyzing simplified structures may, however, pay off in increased insight into economic relationships and simplified economic interpretations.

Possible weighting schemes

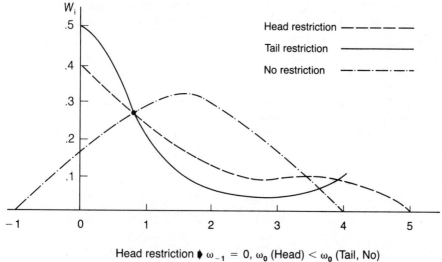

Head restriction $\blacktriangleright$ $\omega_{-1} = 0$, ω_0 (Head) $< \omega_0$ (Tail, No)

Tail restriction $\blacktriangleright$ $\omega_5 = 0$, Lag forced to zero, smooths tail

Figure 12.2 Distribution of weights for a Koyck scheme.

Partial Adjustment and Adaptive Expectations Models

Two other common formulations that also include lagged dependent variables are the *partial adjustment* model (for measuring habit persistence) and the *adaptive expectations* model (for measuring error-caused learning). The rationale behind the partial adjustment model is that inertia, high costs of change, or other factors restrict the immediate response of Y to a given change in X (Y would represent a consumer and X the product). In this case Y_t^* represents the optimum response and is given by

$$Y_t^* = \beta_0 + \beta_1 X_t + \varepsilon_t.$$

Economists suggest that in the current time period the consumer will move only part way from the initial position Y_{t-1} to the desired position Y_t^*, or

$$Y_t - Y_{t-1} = \delta(Y_t^* - Y_{t-1}) \qquad \text{for} \qquad 0 < \delta < 1.$$

This could also be the case when the availability of resources, or the costs of changing the level of production to a higher (or lower) level in response to a change in demand, prevents a manufacturer from making an immediate response.

The final form of the *partial adjustment model* is

$$Y_t = \beta_0\delta + \beta_1\delta X_t + (1 - \delta)Y_{t-1} + v_t, \text{ where } v_t = \delta\varepsilon_t.$$

This model is very similar to the Koyck model.

The assumption that only the current value of X_t is important in determining the optimum value of Y_t is subject to question, and has led to the *adaptive expectations model*. In this model, the observed value of Y_t is related to the expected value of X_t; that is,

$$Y_t = \beta_0 + \beta_1 X_t^* + \varepsilon_t,$$

where X_t^* is the expected value of X_t and is not observable. A further assumption is that the expectations are updated each period by a proportion of the difference between the actual value for X_t and the previous expected value for X_t. Thus

$$X_t^* - X_{t-1}^* = (1 - \lambda)(X_t - X_{t-1}^*) \qquad \text{for } 0 < \lambda < 1.$$

These equations have to be manipulated to eliminate nonobserved values. It can be shown that the final form of the adaptive expectations model is

$$Y_t = \beta_0(1 - \lambda) + \beta_1(1 - \lambda)X_t + \lambda Y_{t-1} + \mu_t,$$

where $\mu_t = \varepsilon_t + \lambda\varepsilon_{t-1}$ represents the error term. This model formulation is also of the Koyck form.

Almon Lags

Another approach that has received a considerable amount of attention in the econometric literature is the Almon lag scheme (see Almon, 1965). Rather than estimating all the β's in the model, the *Almon scheme* postulates that the β's can be approximated by a polynomial of a suitable (but unknown) degree. If the degree of the polynomial is much less than the order k of the model, this simplifies the statistical estimation problem. It is beyond the scope of this exposition to explore the ramifications of the Almon approach. In recent years the seriousness of the difficulties with this approach has somewhat diminished its usefulness in applications.

DUMMIES AND LAGGED VARIABLES CHECKLIST

_____ If quantitative data are not available, are dummy variables appropriate to represent

(a) One-time effects (e.g., strikes, etc.)?
(b) Fixed-duration effects (e.g., war–peace)?
(c) Discontinuities related to changes in qualitative factors?
(d) Seasonal variation?
(e) Differences in slopes or intercepts for different attributes (e.g., sex, race)?

_____ Can a replacement value be used in place of a dummy variable when outliers are present?

_____ Are dummy variables (for outliers treatment) and lagged dependent variables included in the same equation? (If so, carefully examine model predictions for the period following the outlier.)

_____ Does a robust regression offer a better alternative for treating outliers than dummy variables?

_____ Are lagged variables appropriate? Is response to a change in an independent variable distributed over time?

_____ If distributed lags of exogenous variables are called for, have you used Koyck or Almon lags to conserve degrees of freedom (if necessary)?

_____ Are irrelevant variables included in the model?

_____ Are relevant variables excluded from the model?

SUMMARY

The use of dummy variables and lagged variables offers additional flexibility to the econometrician.

- Dummy variables provide the capability to include qualitative factors or effects in the model.
- Lagged variables provide the opportunity to allow responses to some change to have distributions over time.

You need to exercise considerable care when using these variables because of potential forecasting problems and associated statistical problems.

USEFUL READING

ALMON, S. (1965). The Distributed Lag Between Capital Appropriations and Expenditures. *Econometrica* 30, 178–96.

DHRYMES, P. J. (1971). *Distributed Lags: Problems of Estimation and Formulation*. San Francisco, CA: Holden-Day.

FISHER, F. M. (1966). *The Identification Problem*. New York, NY: McGraw-Hill.

KOYCK, L. M. (1954). *Distributed Lags and Investment Analysis*. Amsterdam, Netherlands: North-Holland Publishing Co.

PINDYCK, R. S., and D. L. RUBINFELD (1976). *Econometric Models and Economic Forecasts*. New York, NY: McGraw-Hill.

CHAPTER 13

Dealing with Serial Correlation

This chapter discusses the identification and avoidance of serial (auto) correlation in residuals. The underlying assumptions to be considered are that

- Ordinary least-squares estimation is based on model errors that are uncorrelated.

- Normality implies that model errors are pairwise independent; this may be unrealistic in practice.

Some additional ways of dealing with this problem for time series data are explored in later chapters on ARIMA modeling.

IDENTIFYING SERIAL CORRELATION

In econometric applications serial correlation may arise because of an incorrect specification of the *form* of the relationship for the variables. Serial correlation manifests itself, in forecasting, in a pattern of prolonged overforecasting or under-forecasting, period after period, into the future. We will refer to serial correlation when dealing with residuals or data and will associate autocorrelation with specifications of error in a model.

First-Order Serial Correlation

To begin to think about the occurrence of autocorrelation in forecasting models, consider

$$Y_t = \beta X_t + \varepsilon_t.$$

where ε_t depends on the value of itself one period ago—that is,

$$\varepsilon_t = \rho\varepsilon_{t-1} + \upsilon_t, \qquad \text{when } \upsilon_t \sim N(0,\sigma^2),$$

and υ_t is a normally distributed error term with a mean of zero and a variance of σ^2.

It can be shown that if the true parameter $\rho = 0.8$, the sampling variance of $\hat{\beta}$ will be more than four times the estimated variance given by the ordinary least-squares solution (Johnston, 1972, Chapter 8). Since the estimated variance is understated, a t test could falsely lead to the conclusion that the parameter is significantly different from zero.

Two main consequences of using OLS analysis in models in which the errors are autocorrelated are:

- Sampling variances of the regression coefficients are underestimated and invalid.

- Forecasts have variances that are too large.

When OLS is used for estimation, the calculated acceptance regions or confidence intervals will be narrower than they should be for a specified level of significance. This leads to a false conclusion that estimates of parameters are more precise than they actually are. There will be a tendency to accept a variable as significant when it is not, and this may result in a misspecified model.

The Durbin-Watson Statistic

Because of the serious problems created by autocorrelated errors, it is important to be able to test for their presence. The Durbin-Watson statistic is commonly used to test for *first-order* serial correlation in residuals (Durbin and Watson, 1950, 1951). The ordinary correlogram can also be used as a graphical tool for detecting the presence of first- or higher-order serial correlation.

The Durbin-Watson statistic is calculated from the residuals e_t in a model, and it has the formula

$$d = \frac{\sum\limits_{t=2}^{n} (e_t - e_{t-1})^2}{\sum\limits_{t=1}^{n} e_t^2}.$$

If the residuals $\{e_t, t=1, \ldots, n\}$ are positively correlated, the absolute value of $e_t - e_{t-1}$ will tend to be small relative to the absolute value of e_t. If the residuals are negatively correlated, the absolute value of $e_i - e_{t-1}$ will be large relative to the absolute value of e_t. Therefore d will tend to be small (near 1.0) for positively correlated residuals, large (near 4.0) for negatively correlated residuals, and approximately equal to 2.0 for random residuals.

The sampling distribution of d depends on the values of the independent variable X_t in the sample. Therefore, the test is only able to provide upper (d_u) and lower (d_l) limits for significance testing. One either accepts the null hypothesis of zero autocorrelation or rejects it in favor of first-order positive autocorrelation. If $d<d_l$, the zero autocorrelation hypothesis is rejected in favor of first-order positive auto-correlation. If $d_l<d<d_u$, the test is inconclusive. If $d>4-d_l$, the zero-autocorrelation hypothesis is rejected in favor of first-order negative autocorrelation (see Appendix A, Table 5).

With the advancements that have been made in computer processing, it is often simpler to plot the correlogram of the residuals of a model and to determine any serial correlation patterns (directly discussed under the identification of ARIMA models). Both the Durbin-Watson statistic and the correlogram are *inappropriate if lagged dependent variables* are used as explanatory variables, however.

The Durbin h Statistic

The h statistic can be used to test for serially correlated residuals when a lagged dependent variable appears as an independent variable in a model (Johnston, 1972, Chapter 8). The h statistic is defined by

$$h = r \{n/(1 - n \hat{V}(b_1))\}^{1/2} \quad \text{for} \quad n\hat{V}(b_1) < 1,$$

where $r \simeq 1 - 0.5d$ and $\hat{V}(b_1)$ = estimated variance of b_1 (the coefficient of Y_{t-1}).

The h statistic is a large $(n>30)$ sample statistic used to test for serially correlated residuals in a model. This statistic is tested as a standard normal deviate, and if $h > 1.645$, one rejects the hypothesis that the residuals have zero serial correlation (at the 5-percent level). Since only the estimated variance of b_1 is required to compute h, it does not matter how many independent variables or higher-order lagged dependent variables are included in the model.

ADJUSTING FOR SERIAL CORRELATION

There are several approaches that can be tried to eliminate the problem of serial correlation. These include:

- Modeling the first differences or the percent changes year-over-year in the time series.

- Perform a transformation on the data; the transformation should be based on the assumed nature of the autocorrelated error structure.
- Include an autoregressive term (last period's actual) in the model.
- Build an ARIMA model of the residuals of the regression model.

Taking First Differences

Modeling first differences of the data may solve a serial correlation problem. Let the assumed model be

$$Y_t = \beta_0 + \beta_1 X_t + \varepsilon_t,$$

where the errors are autocorrelated as follows (i.e., $\rho = 1$),

$$\varepsilon_t = \varepsilon_{t-1} + \upsilon_t,$$

and where υ_t is a random error term that is not correlated with ε_t. Replacing $t-1$ for t in the model gives

$$Y_{t-1} = \beta_0 + \beta_1 X_{t-1} + \varepsilon_{t-1}.$$

Subtraction yields

$$Y_t - Y_{t-1} = \beta_1 (X_t - X_{t-1}) + (\varepsilon_t - \varepsilon_{t-1}).$$

Since $\varepsilon_t - \varepsilon_{t-1} = \upsilon_t$, this gives the result

$$Y_t - Y_{t-1} = \beta_1 (X_t - X_{t-1}) + \upsilon_t.$$

If the errors in the original model were autocorrelated, modeling first differences of the variables would eliminate this restricted form of autocorrelation. By definition, the $\{\upsilon_t\}$ are assumed to be independent, identically distributed random variables.

It is often the case that the presence of serial correlation is so great that it becomes obvious from an analysis of the autocorrelogram of the residuals or the residual plot. An example of this problem is illustrated in a model relating growth in business telephones to nonfarm employment. A scatter plot (Figure 13.1) shows a strong relationship, since both nonfarm employment and sales of business telephones are increasing. In this case, both variables increase with time. The residuals of a regression between these variables will be serially correlated. Dummy variables were included in the model to account for seasonality. The regression model explained over 84 percent of the variation in the business telephone data; but the residual plot in Figure 13.2 shows that the residuals are serially correlated. From the correlogram of the residuals in Figure 13.3, a first-order autoregressive pattern is evident.

Figure 13.1 A scatter plot of business telephones against nonfarm employment.

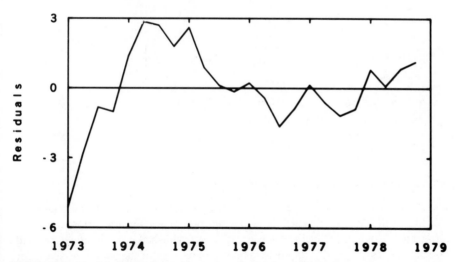

Figure 13.2 Residuals that result from regressing business telephones against nonfarm employment.

A regression between the first differences of the variables has an R-squared value of only 0.41. However, this is because the *changes* in the variables, which are very small relative to the level of the original series, are being modeled. The residuals of this model, shown in Figure 13.4, appear less cyclical. The correlogram of the residuals showed some first-order serial correlation, but of a lesser magnitude than that in Figure 13.3.

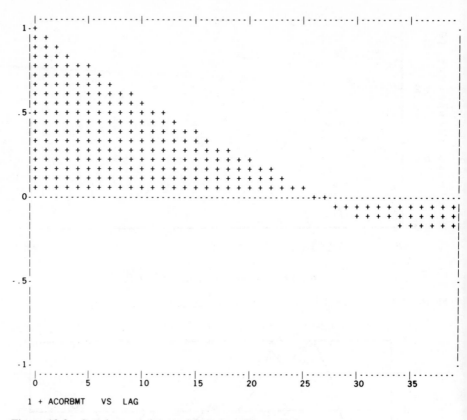

Figure 13.3 Correlogram of the residuals from Figure 13.2.

A First-Order Autoregressive Structure

In a slightly more general situation, it may be possible to assume an autoregressive structure in the error term. Then, by using the appropriate transformation, the transformed model will have uncorrelated errors, and established procedures can be used for analysis. However, this approach requires knowledge of the autoregressive error structure, which may not be realistic in practice.

Consider, for example, a model for a stationary ($|\rho|<1$) error term given by

$$\varepsilon_t = \rho\varepsilon_{t-1} + v_t,$$

in a linear regression model

$$Y_t = \beta_0 + \beta_1 X_t + \varepsilon_t.$$

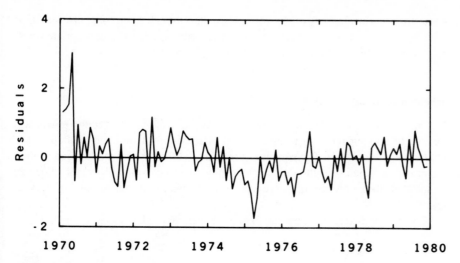

Figure 13.4   Residuals that show less of a cyclical pattern than those in Figure 13.2.

The data are first transformed by the operator $(1 - \rho B)$, where B is the backshift operator, so that

$$BY_t = Y_{t-1}, \quad B^k Y_t = Y_{t-k}.$$

This gives the generalized difference equation

$$(1 - \rho B) Y_t = \beta_0(1 - \rho) + \beta_1(1 - \rho B)X_t + (1 - \rho B)\varepsilon_t.$$

Thus

$$Y_t^* = \beta_0^* + \beta_1 X_t^* + \upsilon_t,$$

where

$$\beta_0^* = \beta_0(1 - \rho),$$

$$Y_t^* = Y_t - \rho Y_{t-1},$$

and

$$X_t^* = X_t - \rho X_{t-1},$$

and where all υ_t are independent, identically distributed errors. Notice that the parameter ρ has to be known for this procedure to work. An *estimate* of ρ could come from a correlogram of the residuals of the model. Care must also be taken with Y_1 and X_1, since their previous values are unavailable. The appropriate transformation for these values are $Y_1^* = (1-\rho^2)^{1/2}Y_1$ and $X_1^* = (1-\rho^2)^{1/2}X_1$.

The Hildreth-Lu Procedure

In the case of first-order autocorrelation, the Hildreth-Lu procedure provides a method of solving for ρ by using a nonlinear maximum-likelihood method due to Hildreth and Lu (1960). The equations are difficult to solve, and the solution is often easier if ρ is varied over the range from minus one to plus one. The best value can be selected from a procedure where ρ is the value that minimizes the sum of squared residuals.

Consider a differencing operation $(1-\rho B)$ on the model

$$Y_t = \beta_0 + \beta_1 X_t + \varepsilon_t.$$

This gives

$$(1-\rho B)Y_t = (1-\rho B)(\beta_0 + \beta_1 X_t + \varepsilon_t).$$

Equivalently,

$$Y_t - \rho Y_{t-1} = \beta_0(1-\rho) + \beta_1(X_t - \rho X_{t-1}) + (\varepsilon_t - \rho\varepsilon_{t-1}).$$

Then, by rewriting,

$$\varepsilon_t - \rho\varepsilon_{t-1} = (Y_t - \rho Y_{t-1}) - \beta_0(1-\rho) - \beta(X_t - \rho X_{t-1}).$$

Then solve for the minimum sum of squared errors, $\Sigma v_t^2 = \Sigma(\varepsilon_t - \rho\varepsilon_{t-1})^2$, using different values of ρ.

The Cochrane-Orcutt Procedure

Another procedure to estimate ρ in the presence of first-order autocorrelation was proposed by Cochrane and Orcutt (1949). This iterative procedure produces successive estimates of ρ until the difference between successive estimates becomes insignificant.

The initial estimate of ρ is derived from the residuals by

$$\hat{\rho} = \frac{\sum\limits_{t=2}^{n} e_t e_{t-1}}{\sum\limits_{t=2}^{n} e_{t-1}^2}.$$

The value $\hat{\rho}$ is substituted into the model

$$Y_t - \hat{\rho} Y_{t-1} = \beta_0(1 - \hat{\rho}) + \beta_1(X_t - \hat{\rho} X_{t-1}) + v_t,$$

and the transformed equation is solved by using OLS. A second estimate of ρ is made in the same manner:

$$\hat{\hat{\rho}} = \frac{\displaystyle\sum_{t=2}^{n} \hat{v}_t \hat{v}_{t-1}}{\displaystyle\sum_{t=2}^{n} \hat{v}_{t-1}^2},$$

where the $\hat{v}_t$'s are the residuals from the OLS fit.

The estimate $\hat{\hat{\rho}}$ is compared to the first estimate $\hat{\rho}$. If these two values are reasonably close, the second estimate is used. If not, another iteration is made.

The Cochrane-Orcutt procedure can also be used to obtain an initial estimate of ρ for use in the Hildreth-Lu procedure. This can minimize the range of ρ that needs to be searched. The Cochrane-Orcutt procedure cannot be used if there are lagged dependent variables in the model.

It should also be noted that the error term in the model can have a more general autoregressive structure. This complicates the problem, because it is now necessary to transform the model with a "generalized difference" of the form

$$(1 - \rho_1 B - \rho_2 B^2 - \cdots - \rho_k B^k).$$

If you feel that the model warrants more general error structures, we recommend that you consider the modeling approach discussed in Part IV.

The previous model was estimated by using the Cochrane-Orcutt (C-O) auto-correlation routine. The final iteration yielded $\hat{\hat{\rho}} = 0.56$. The residual plot appeared to be random and the correlogram of the residuals showed no significant autocor-relations.

Table 13.1 summarizes the statistics from the alternative models. The coefficient of employment in the C-O procedure is 0.032, which is almost identical to the value obtained from the OLS regression performed with the original series. The t statistics for the OLS and C-O models both suggest that β is significantly different from zero; but the value of the t statistic is less in the C-O procedure, as one would expect.

The model with the first differences of the series has a $\hat{\beta} = 0.024$, approximately one standard error less than what the C-O procedure estimates it should be. The C-O model, particularly with regard to serially correlated residuals, is the pre-ferred model.

A Main Gain Model

In Chapter 5, it was shown that the residuals of the main telephone gain model (as a function of housing starts, the FRBI series, and the dummy variables used to

Table 13.1 Summary of statistics in models for business main telephones (BMT) as a function of the nonfarm employment series.

Regression model using	R^2	$\hat{\beta}$ (Employment)	t statistic	$\hat{\rho}$	Durbin-Watson	Estimated standard deviation
BMT	0.84	0.033	10.6	—	0.52	1.77
First differences in BMT	0.41	0.024	3.1	1.00	1.01	1.19
Cochrane-Orcutt adjustment	0.79	0.032	8.7	0.56	1.39	1.12

account for labor strikes) were also autocorrelated. This model was estimated by using the Cochrane-Orcutt (C-O) and Hildreth-Lu (H-L) procedures, and the results are summarized in Table 13.2.

The H-L procedure results in a slightly lower standard deviation for the residuals than the C-O procedure does, indicating that $\hat{\rho}$ is closer to 0.40 than to 0.32.

It is interesting to note that the lost demand attributable to the 1971 strike (D_2) is lessened from $-225{,}270$ (as the OLS regression estimates it to be) to $-148{,}140$ (the H-L estimate). This is close to the value obtained by using the transfer function model for intervention variables (Chapter 23). The coefficient of the housing starts

Table 13.2 Comparison of serial adjustment corrections in a model for main telephone gain (t statistics are shown in brackets below the coefficient estimates).

Parameter	OLS	Cochrane-Orcutt	Hildreth-Lu
R^2	0.65	0.59	0.57
$\hat{\beta}_1$ (HOUS)	568.5 (7.05)	511.9 (5.17)	493.5 (4.62)
$\hat{\beta}_2$ (DFRB)	13,185 (6.32)	13,040 (5.91)	12,940 (5.78)
$\hat{\beta}_3$ (D_1)	$-246{,}720$ (-4.17)	$-237{,}490$ (-4.54)	$-235{,}740$ (-4.64)
$\hat{\beta}_4$ (D_2)	$-225{,}270$ (-3.81)	$-160{,}290$ (-3.08)	$-148{,}140$ (-2.93)
$\hat{\rho}$	—	0.32	0.40
Durbin-Watson	1.33	1.92	2.08
Estimated standard deviation	57,890	54,160	54,100

variable shows a decline from 568.5 (OLS) to 493.5 (H-L). The standard error of the housing starts variable increases from 80.6 (OLS) to 106.9 (H-L) when the serial correlation in the residuals is corrected.

SERIAL-CORRELATION CHECKLIST

_____ Are the residuals of any equation serially correlated?

_____ If serial correlation is present, have you tried:
  - Modeling first differences?
  - Including an autoregressive term?
  - An ARIMA model for the residuals of the initial model?
  - The Hildreth-Lu or Cochrane-Orcutt procedures for first-order autocorrelation?

_____ If a lagged dependent variable is included in the model, has the h statistic been calculated to test the zero autocorrelation hypothesis?

SUMMARY

The presence of serial correlation is a frequent problem in econometric modeling applications. Some techniques to deal with this include:

  - The use of autoregressive terms in the model.
  - Modeling first (or generalized) differences of the variables.
  - Making adjustments based on the Hildreth-Lu or Cochrane-Orcutt procedures.

USEFUL READING

COCHRANE, D., and G. N. ORCUTT (1949). Application of Least Squares to Relationships Containing Autocorrelated Error Terms. *Journal of the American Statistical Association* 44, 32–61.

DURBIN, J., and G. S. WATSON (1950). Testing for Serial Correlation in Least Squares Regression: I. *Biometrika* 37, 409–28.

DURBIN, J., and G. S. WATSON (1951). Testing for Serial Correlation in Least Squares Regression: II. *Biometrika* 38, 159–78.

HILDRETH, G., and J. Y. LU (1960). *Demand Relations with Autocorrelated Disturbances.* Lansing, MI: Michigan State University, Agricultural Experiment Station, Technical Bulletin 276.

JOHNSTON, J. (1972). *Econometric Methods,* 2nd ed., New York, NY: McGraw-Hill.

CHAPTER 14

Additional Specification Issues

This chapter discusses a number of specification issues arising in econometric forecasting applications. These include:

- The identification of heteroscedastic errors by means of residual plots and the Goldfeld-Quandt test.

- Transformations and weighted least squares estimation as solutions for heteroscedasticity.

- The Chow test for determining if a model is structurally stable over time.

- The identification of multicollinearity.

- Ridge regression as an approach for dealing with multicollinearity.

- Errors in the measurement of the independent variables.

HETEROSCEDASTICITY

Errors in the classical linear regression model that do not have constant variance are said to be *heteroscedastic*. In practice, this phenomenon usually manifests itself as an increasing variability in the residuals with increasing values of the independent variable(s).

Heteroscedasticity occurs frequently in cross-sectional analysis. The changes in the dependent and independent variables are likely to be of the same order of magnitude in time series models. However, in cross-sectional studies of sales of firms, for example, the sales of larger firms may be more variable than the sales of smaller firms. In cross-sectional studies of family income and expenditures, the spending of low-income families may be less variable than the spending of high-income families, since high-income families have more income available for "spur-of-the-moment" spending.

The effect of heteroscedasticity is to place more emphasis on variables with greater variance. The problem that heteroscedasticity poses in studies such as these is that while the estimated regression coefficients are unbiased and consistent, they are not efficient and will not become efficient as the sample size increases (see *The Beginning Forecaster*, Chapter 12); this leads to confidence intervals that are narrower than the correct ones. That is, if the variance is understated, the confidence intervals for the estimated regression coefficients are too narrow. Because of this, the confidence intervals will more often than appropriately *not* span zero, which causes one to overstate the significance of each variable in the model.

Heteroscedasticity cannot be detected by looking only at the results of a regression analysis. A plot of the residuals against the predicted values of the dependent variable or of one of the independent variables may indicate whether the residual variance is constant over all observations. Because the residual variance is a function of the distribution of the independent variables as well as the variance of the true error term, statistical tests can be performed to detect the presence of heteroscedasticity.

The Goldfeld-Quandt Test

The Goldfeld-Quandt test for heteroscedasticity (Goldfeld and Quandt, 1965) consists of the following steps:

- The data are sorted in order of ascending value for the independent variable thought to be related to the error variance.

- The middle observations, perhaps as many as 20 percent of them, are omitted and two separate regressions are performed for the two remaining data subsets. For small samples ($n < 30$), no observations need to be eliminated.

- For each regression, the sum of squares of the residuals is calculated.

- Assuming the errors are normally distributed and have zero autocorrelation, the ratio of the larger residual sum of squares to the smaller residual sum of squares is distributed as an F statistic with $(n - d - 2k)/2$ degrees of freedom in both the numerator and the denominator. The number of observations is n, d is the number of omitted values, and k is the number of independent variables.

- The null hypothesis is that there is no significant difference in variance between the two sums. When the computed F statistic is larger than the tabulated F value (Appendix A, Table 4), you reject the assumption that the error term has a constant variance.

If there is an indication of heteroscedasticity, solutions to this problem include overall maximum-likelihood estimation, weighted least squares, and the transformation of variables.

A Weighted Least-Squares Solution

Consider the simple linear regression model

$$Y_t = \beta_0 + \beta_1 X_t + \varepsilon_t$$

ruled by the usual assumptions. The parameters β_0 and β_1 are estimated by minimizing

$$\sum_{t=1}^{n} (Y_t - \beta_0 - \beta_1 X_t)^2.$$

When the error variance σ_t^2 can be assumed known but perhaps not constant, a weighted least-squares analysis is appropriate; the minimization is given by

$$\sum_{t=1}^{n} \frac{1}{\sigma_t^2} (Y_t - \beta_0 - \beta_1 X_t)^2.$$

Thus, the squared deviations are weighted by the known factor $1/\sigma_t^2$ before summing. General mathematical expressions for the weighted least-squares solution may be found in Wonnacott and Wonnacott (1970, Chapter 16). Except for a few simple cases, these expressions are rather complex.

It may be worthwhile to illustrate one situation in which heteroscedasticity can be easily solved. Suppose that the variance of the error term in the model increases in proportion to the square of the independent variable X_t, such that

$$\sigma_t^2 = k^2 X_t^2, \qquad \text{where } t = 1, 2, \ldots, n.$$

By transforming the model into

$$\frac{Y_t}{X_t} = \frac{\beta_0}{X_t} + \beta_1 + \frac{\varepsilon_t}{X_t},$$

it can be seen that the transformed error term has constant variance k^2, and an ordinary least-squares analysis can be applied to this transformed equation. It can be shown that a weighted least-squares analysis would yield the same solution. In general, the expression for σ_t^2 is unknown and may not be simple. Nevertheless, it may be fruitful to try a simple transformation, like the one above, as part of a preliminary exploration prior to specifying the model.

A TEST FOR STRUCTURAL STABILITY

Forecasters are concerned with the *structural stability* of their models over time. If the model is appropriate for the first part of the data, it is important to ask if it is still applicable for more recent data. The *Chow test* (Chow, 1960) is an F test that can be applied in these situations. The procedure is as follows:

- Combine all observations ($n_1 + n_2$) and perform a regression over the entire time period. Calculate the sum of squared residuals (RSS$_0$) with $n_1 + n_2 - k$ degrees of freedom (k = number of parameters to be estimated, including the constant term).

- Perform two separate regressions over time periods T_1 and T_2 (not necessarily equal) and calculate the sum of squared residuals for each model (RSS$_1$ and RSS$_2$). The degrees of freedom are $n_1 - k$ and $n_2 - k$, respectively.

- Add the sums of squared residuals from the separate regressions (RSS$_1$ + RSS$_2$) and subtract this value from the sum of squared residuals of the complete model.

- Compute the F statistic:

$$F = \frac{\{RSS_0 - (RSS_1 + RSS_2)\}/k}{(RSS_1 + RSS_2)/(n_1 + n_2 - 2k)};$$

test this with k and $n_1 + n_2 - 2k$ degrees of freedom.

- If the value of F is significant, you reject the hypothesis that there is no significant difference between the two regressions and conclude that the entire model is structurally unstable.

MULTICOLLINEARITY

Multicollinearity is probably the most widely discussed and most easily misunderstood phenomenon in econometric model building. In its simplest form, *multicollinearity* arises whenever explanatory variables in a regression model are highly correlated. Multicollinearity is difficult to deal with, because it is almost invariably a problem of degree rather than kind. Multicollinearity results in bias, inconsistency, and inefficiency of estimators, which is undesirable, at least from a theoretical viewpoint. Practically, remedies must be found so that model assumptions can be made approximately true without completely destroying the statistical validity of the fit (Belsley et al., 1980).

Some of the practical effects of multicollinearity include imprecise estimation. While the concepts of bias, consistency, and inefficiency have very precise theoretical definitions, their effect on the inferences that are drawn from a model may be difficult to determine in every practical application. Problems of multicollinearity also encourage misspecification of a model if they are not carefully noted. The stability of coefficients can be severely affected, so that different segments or subsets of the data can give rise to vastly different results. This lack of sensitivity precludes the possibility of making sound interpretations of model coefficients.

The presence of multicollinearity can be tested in a variety of ways. Estimated coefficients will lack stability in a sensitivity analysis that makes use of different segments of the data. There will likely be low t values and a high F value in the regression analysis. Simple correlations between variables can be calculated, although these would not be conclusive. For more than two independent variables, partial correlations should be examined as well.

The Farrar-Glauber test (Farrar and Glauber, 1967; Johnston, 1972, Section 5.7) considers the R-squared statistic between each independent variable and the remaining independent variables in a specification. The test is based on the inspection of the F statistics constructed with each R-squared statistic and the appropriate degrees of freedom:

$$F_i = \frac{R_i^2/(k-2)}{(1-R_i^2)/(n-k+1)}, \qquad \text{where } i = 2, \ldots, k,$$

and where R_i^2 is the coefficient of multiple determination between each X_i and the remaining $(k-1)$ explanatory variables, and k is the number of explanatory variables. A significant value for F_i indicates that the variable X_i is not independent.

In time series analysis, collinearity may often be reduced by eliminating trend from each variable. In the example presented in Chapter 3, the product-moment correlations of the *differenced* data are shown below the diagonal in the following matrix. The entries above the diagonal represent a "robust" version of the ordinary correlations (see The Beginning Forecaster, Chapter 17).

		1	2	3
1	GAIN	1	0.32	0.47
2	HOUS	0.72	1	0.36
3	DFRB	0.57	0.42	1

The robust correlations are all somewhat lower, possibly due to outliers or a few influential data values. This can be compared to the matrix of correlations for the original (undifferenced) data:

	1	2	3
1	1	0.07	0.74
2	0.01	1	0.14
3	0.47	0.32	1

While there are some significant correlations among the differenced independent variables, they are probably not large enough to introduce serious multicollinearity problems. The forecaster should become cautious, however, when simple correlations exceed values of from 0.8 to 0.9.

RIDGE REGRESSION

Ridge regression is an analysis for dealing with multicollinearity in independent variables. The central idea in ridge regression is that it is possible to estimate a *biased* point estimator that has a smaller mean squared error than an unbiased ordinary least-squares estimator. This can be visualized in Figure 14.1. Suppose the goal is to estimate the "true" parameter β. The expected value of the OLS estimator $\hat{\beta}$ is β, but in this case, the variance of $\hat{\beta}$ is very large. Assuming the true value of $\beta = 1.0$, it is quite possible to have an OLS estimate of, say, $\hat{\beta} = 1.2$ in a given situation.

As an alternative, a biased estimate $\hat{\beta}^*$ for which the expected value does not equal β may be more desirable if the variance of $\hat{\beta}^*$ is significantly less than that of $\hat{\beta}$. In this case, it may be possible to obtain more stable estimates of β when regressions are performed over subsets of the data or when new observations are included.

A common criteria is to minimize the mean squared error (MSE); that is,

$$\text{MSE} = \text{Variance} + (\text{Bias})^2.$$

The *bias* is the difference between the "true" β and the ridge estimate $\hat{\beta}^*$. Now suppose that the following are found:

True parameter:	$\beta = 1.0$;
OLS estimate:	$\hat{\beta} = 1.2$;
Ridge estimate:	$\hat{\beta}^* = (0.8)\hat{\beta} = 0.96$;

and

$$\text{Var}(\hat{\beta}^*) = 0.64. \quad \text{Var}(\hat{\beta}) = 1.0,$$

Then $\text{MSE}(\hat{\beta}) = \text{Var}(\hat{\beta}) = 1.0$, since $\hat{\beta}$ is unbiased. The bias is given as

The problem is to estimate a parameter β.
Two estimates of β are $\hat{\beta}$ and $\hat{\beta}^*$.

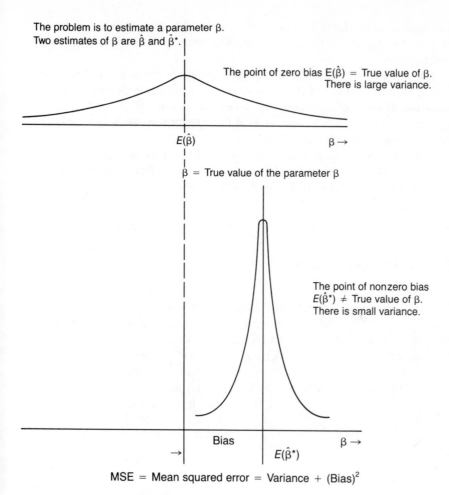

The point of zero bias $E(\hat{\beta})$ = True value of β.
There is large variance.

$E(\hat{\beta})$

$\beta \rightarrow$

β = True value of the parameter β

The point of nonzero bias
$E(\hat{\beta}^*) \neq$ True value of β.
There is small variance.

Bias

$\beta \rightarrow$

$E(\hat{\beta}^*)$

MSE = Mean squared error = Variance + (Bias)^2

Figure 14.1   Variance and bias

$$\text{Bias} = E(\beta - \hat{\beta}^*) = 0.2,$$

where E denotes the expected or average value.
    The mean square error of $\hat{\beta}^*$ is

$$
\begin{aligned}
\text{MSE} &= E(\hat{\beta}^* - \beta)^2 \\
&= (\text{Bias})^2 + \text{Var}(\hat{\beta}^*) \\
&= 0.04 + 0.64 \\
&= 0.68.
\end{aligned}
$$

The ridge estimate (0.96) is closer to β than the OLS estimate (1.2) and the MSE of $\hat{\beta}^*$ is 0.68 versus 1.0 for OLS. It is in these kinds of cases that ridge regression offers promise.

Ridge regression trades a reduction in the mean squared error for an increase in bias. If the increase in bias is not too severe, a more meaningful set of parameter estimates, in terms of signs and magnitudes, may result. Of course, the "reasonableness" of the parameter values is often the agreement between the empirical results and one's *a priori* expectations.

Full treatment of ridge regression is beyond the scope of this book, but the interested reader can refer to Belsley et al. (1980) and to Draper and Smith (1981) for references and a statistical treatment of the subject.

ERRORS IN VARIABLES

The classical regression formulation does not include assumptions about any measurement error in the independent variables. The error term is usually associated with the dependent variable and the regressors are assumed to be fixed. When both the dependent and independent variables are subject to error, estimation methods need to be developed to overcome the limitation of OLS analysis. Among these techniques the *errors-in-variables* method is widely used by econometricians.

If your primary interest is in the estimation and interpretation of statistical parameters (rather than in forecasting), you need to be concerned about the correlation between the error term and the regressors. Consider again the simple linear regression

$$Y_t = \beta_0 + \beta_1 X_t + \varepsilon_{1t},$$

where β_0 and β_1 are parameters to be estimated. If there is also an error term associated with $X_t (= X_t^* + \varepsilon_{2t})$, then the model becomes

$$Y_t = \beta_0 + \beta_1(X_t^* + \varepsilon_{2t}) + \varepsilon_{1t},$$

or

$$Y_t = \beta_0 + \beta_1 X_t^* + (\varepsilon_{1t} + \beta_1\varepsilon_{2t}).$$

The error term $(\varepsilon_{1t} + \beta_1\varepsilon_{2t})$ involves β_1 as well. These complications give rise to problems in estimating coefficients and in determining their statistical properties. A treatment of this subject may be found in Johnston (1972, Section 9.4).

USING SIMULATION TECHNIQUES

Determining Confidence Limits

One way economists and statisticians develop confidence limits for econometric forecasting models is through a method known as *Monte Carlo simulation*. This procedure involves the construction of distributions of possible values for the independent variables.

Consider three independent variables, X_1, X_2, and X_3; a dependent variable Y is a function of these three variables. A distribution of possible values is established for each independent variable. These distributions can be determined subjectively or be the result of some other modeling process; a number of computer programs are available that accept sufficient input data that the mean, standard deviation, median (etc.) of these distributions can be determined, so as to replicate the distributions adequately.

A random sample is taken from each distribution, is inserted into the regression equation, and a value for the dependent variable is thereby determined. The process is repeated as many times as desired, perhaps from 30 to 1000 times, and the result is a full distribution of values for the dependent variable. This distribution can then be the basis for establishing confidence limits. More often than not, the resulting values will appear to be normally distributed. The actual results can be monitored relative to these confidence limits or, more simply, it can be seen if the actual values fall within a reasonably small range of the expected values for the dependent variable. When the independent variables are not truly independent of one another, there is a problem with this approach.

Stability Testing

Simulation testing of econometric models is often referred to as "exercising" the model. If the model is to be used for forecasting, the results of a simulation run can be analyzed to see if the model "blows up" or if the test forecasts appear to be reasonable even when extreme values are substituted for the independent variables. For example, the model may have been fitted over a regression period that was free of a severe recession or high inflation rates. You will want to know whether the model is valid should any of these occurrences happen in the forecast period. By simulating these conditions through distributions of the independent variables, you can determine if the model handles these conditions reasonably well. If it doesn't, you learn the limitations to the use of the model.

Small-Sample Properties

A major analytical problem with econometric systems is that it is difficult to derive explicit statistical properties of various estimating techniques. All properties of such estimators are based on large samples. These are known as *asymptotic* properties.

In order to gain insight into the small-sample properties of estimators, Monte Carlo studies are useful: as stated earlier, various sets of parameters are specified and certain distributional properties of the error terms are postulated. Computerized sampling techniques can be used to make studies of bias and variance in the estimators, since the exact model specifications are known. While the results of such empirical studies are never conclusive, they do point to numerous pitfalls and to likely similarities and differences among estimating techniques.

The most important points considered in these studies are:

- Whether asymptotic differences between OLS and other simultaneous estimators are significant for the small-sample situations that are the most common subjects for study. These comparisons can be expressed in terms of bias, variance, or mean squared errors.

- Whether small-sample properties of estimators are similar to asymptotic-sample properties for the same estimators.

- How estimators are affected by specification errors.

- Differences between forecasts produced by the structural and reduced form equations of the system.

An extensive discussion of this topic with related references from the literature is given in Johnston (1972, Section 13.8). Conclusions of these studies vary, since the methods have to be analyzed on the basis of a set of accepted criteria. Among these, bias and variance are the most widely used. However, other cost/benefit criteria may point to a different approach.

Sensitivity Analysis

It is common practice to vary the values of the independent variables over a range of, say, plus or minus fifteen percent of the basic values in increments of five percent. For each simulation, the forecast values of the dependent variable are printed out. The procedure does not involve as much computer expense as the Monte Carlo simulation does. The analyst can decide to vary one or all of the independent variables.

The purposes of this kind of simulation are:

- You can determine how the model will perform with various values for the independent variables. In other words, how sensitive is the model? Will one or more variables cause wide fluctuations in the forecasts? Are the forecasts reasonable for extreme values of these variables? If the model is multiplicative and/or differenced, this may be the only practical way to determine how sensitive it is to changes.

- Producing various scenarios of the future allows you to estimate some subjective "confidence limits" about the forecast or simply to estimate a range of expected values for the forecast itself. In this manner, the forecaster can talk about the expected variation around the forecast, just as is done with simple regression and ARIMA models.

The difficulty in varying the predictions for the independent variables one at a time is the tremendous number of combinations and permutations that are possible. The forecast user becomes perplexed, since it is not quickly obvious which combinations should be considered seriously. A more desirable alternative is to develop three scenarios—*reasonably pessimistic, best bet,* and *reasonably optimistic.* A consistent set of forecasts can then be made for the independent variables for each scenario. For example, an optimistic scenario (improving economy) would have, as independent variables, higher income, a larger market potential, and lower unemployment.

With the above approach, the forecaster has minimized the time and expense required to express the uncertainty in the forecast and reduced the alternatives to a level that the user can cope with.

A SPECIFICATIONS-PROBLEMS CHECKLIST

_____ Do the residuals from the fitted model show a nonconstant or nonhomogeneous variance?

_____ Has the Chow test been performed to determine if the model is structurally stable over time?

_____ Is multicollinearity of the independent variables evident from

- Coefficients that lack stability when used in different segments of the data?

- Simple correlation with the dependent variable?

- One independent variable correlated with the remaining independent variables (Farrar-Glauber test)?

_____ Does ridge regression provide more reasonable parameter estimates than OLS?

SUMMARY

In this chapter a number of potential modeling pitfalls have been identified. These include

- Specification errors that may manifest themselves in the form of heteroscedastic residuals.
- Structurally unstable models.
- A variety of problems resulting from collinearity in the independent variables.

The plotting and analysis of residual patterns is an essential element of modeling and will help you to identify some of these problems. Coefficients that have wrong signs or wrong magnitudes or that show significant changes when a model is updated or estimated for different segments of the data suggest multicollinearity problems. The solutions proposed will improve model performance in most instances.

USEFUL READING

BELSLEY, D. A., E. KUH, and R. E. WELSCH (1980). *Regression Diagnostics: Identifying Influential Data and Sources of Collinearity*. New York, NY: John Wiley and Sons.

CHOW, G. C. (1960). Tests of Equality Between Sets of Coefficients in Two Linear Regressions. *Econometrics* 28, 591–605.

DRAPER, N. R., and H. SMITH (1981). *Applied Regression Analysis,* 2nd ed. New York, NY: John Wiley and Sons.

FARRAR, D. E., and R. R. GLAUBER (1967). Multicollinearity in Regression Analysis: The Problem Revisited. *Review of Economics and Statistics* 49, 92–107.

GOLDFELD, S. M., and R. E. QUANDT (1965). Some Tests for Homoscedasticity. *Journal of the American Statistical Association* 60, 539–47.

JOHNSTON, J. (1972). *Econometric Methods,* 2nd ed. New York, NY: McGraw-Hill.

WONNACOTT, R. J., and T. H. WONNACOTT (1970). *Econometrics*. New York, NY: John Wiley and Sons.

CHAPTER 15

Specifying a System of Econometric Equations

This chapter presents techniques that are appropriate for systems of equations. The topics treated include:

- Specifying the equations of an econometric system.
- Estimating the system in reduced form by using OLS.
- The identification problem of estimating the system of structural equations.
- Estimating recursive systems by using OLS.
- Estimating simultaneous systems by using a two-stage least squares method.

SPECIFYING AN ECONOMETRIC SYSTEM

In previous chapters, all models were of the single equation form comprising a dependent variable and one or more independent or lagged variables. However, in many cases, one or more independent variables may be *jointly* determined with the dependent variable in an econometric *system of equations*. For example, quantity and price may be simultaneously determined. Price may not be fixed, but may instead be established on the basis of an expectation of the quantity that can be sold to yield the highest profit: if the quantity sold does not meet expectations, price is adjusted to yield the highest profit the marketplace will allow. Therefore, a separate equation is required to account for the behavior of price.

Defining the Variables

In simultaneous equation applications, those variables that are determined *within* the system are called *endogenous* variables. Those variables that are determined *outside* the system of equations are called *exogenous*. Variables related to government

191

expenditures, taxes, and interest rates are often considered as exogenous. This is particularly true when the econometric system of equations is used to predict the demand for a firm's products as contrasted with a model of the entire U.S. economy.

A third kind of variable is called *predetermined* if it is statistically independent of current and future error terms in the model. Endogenous variables cannot be predetermined, though exogenous variables can be; and lagged endogenous variables can also be predetermined if the current error is independent of all past errors. Since its value in the current period is known (i.e., has already occurred), a predetermined variable is treated as an exogenous variable for all practical purposes.

A Classical Example

As an example, consider the classical *consumption function*

$$C_t = \alpha + \beta Y_t \,,$$

which relates consumption C_t to national income Y_t in a simple economic system. The Keynesian *income-determination model* consists of a consumption function,

$$Y_t = C_t + I_t \,,$$

and an income identity,

$$I_t = I \,,$$

where I_t denotes investment (Pindyck and Rubinfeld, 1976, Chapter 10).

The inclusion of the additional equation, even though it is an identity, has complicated the model, since consumption C_t both determines and is determined by income. Investment is assumed to be an exogenous (or predetermined) variable, since its value is determined outside the system. Moreover, I_t and C_t are assumed to be statistically independent. Both Y_t and C_t are regarded as endogenous variables.

More realistically, investment I_t is likely to depend on C_t or Y_t; that is, it too should be an endogenous variable. This can be achieved by the following set of *structural equations*:

$$C_t = \beta_0 + \beta_1(Y_t - T_t) + \varepsilon_{1t} \,,$$
$$I_t = \beta_2 Y_{t-1} + \beta_3 R_t + \varepsilon_{2t} \,,$$

and

$$Y_t = C_t + I_t + G_t \,.$$

This simultaneous equation model includes an additional equation and some *a priori* restrictions—namely, $0 < \beta_1 < 1$, $\beta_2 > 0$, $\beta_3 < 0$. Here G_t denotes government expenditure on goods and services, T_t are taxes on income, and R_t is a government regulator, such as interest rates. In this representation, I_t becomes an endogenous variable; T_t, R_t, and G_t are exogenous variables; and Y_{t-1} is called a *lagged endogenous* variable. There is an equation for each endogenous variable.

Notice that this "classification into endogenous and exogenous is a relative one, depending on the nature and extent of the system being studied and the purpose for which the model is being built" (Johnston, 1972).

The above equations are referred to as structural equations in that they describe the *structure* or behavior of the economy of the nation or the economy of a business firm. Since endogenous variables exist on both the left-hand and right-hand sides of the equation, the independent variables, treated as fixed when applying OLS, are no longer fixed. In the demand model that follows this discussion, it will be seen that the endogenous variables on the right-hand side are often correlated with the error term. In this situation, the OLS estimators can be shown to be inconsistent; OLS is therefore inappropriate for estimating the equations.

IDENTIFICATION PROBLEMS

Identification problems hinge on whether or not the structural parameters can be uniquely determined. Three cases must be recognized:

- *Underidentification*, in which it is impossible to obtain estimates of some or all of the structural parameters.

- *Exact identification*, in which the structural parameters can be uniquely determined from reduced-form coefficients.

- *Overidentification*, in which nonuniqueness of results can occur.

In the last case, there is more than one solution for the structural coefficients, and the solutions are unlikely to be the same. In essence, a *reduced-form* system of equations (a solution of the system for the endogenous variables in terms of the *predetermined*—exogenous and lagged endogenous—variables and residuals) may be appropriate for a variety of different structural equations. This is clearly a problem, since the same data can yield a variety of structural equations. In the case of overidentification, the structural parameters can be solved directly by using the two-stage least squares (2SLS) estimation technique so long as appropriate restrictions are placed on the equations. These will be discussed under estimation procedures.

The most comprehensive (and mathematical) treatment of identification problems is given in Fisher (1966). Other treatments are given in Johnston (1972, Chapter 12) and in Wonnacott and Wonnacott (1970, Chapters 8 and 18), to which the interested reader should refer.

Methods of estimation for econometric systems can be classified according to whether they are applied to each equation separately or to the system as a whole. The single-equation methods include OLS for recursive systems and two-stage least squares (2SLS). The 2SLS is by far the most widely used and will be discussed later in this section. However, full treatment of these methods is beyond the scope of this text, and the reader is therefore referred to any modern text in econometrics for this.

A Recursive System

The flowchart shown in Figure 15.1 represents a *sequential* or *recursive* econometric model for main telephone gain in a geographic area. Each individual regression is performed sequentially, starting with the first, until all the regressions have been run. The predictions from the previous equations are available as forecasts for the exogenous variables in later equations. In this example, the exogenous variables include regional, national, economic, and telephone data. This model would be referred to as an *econometric model* because it contains a set of related equations and it attempts to explain an economic system.

Consider a series of models to help predict the gain in main telephones in a particular region of the United States. Main telephones do not include extension sets. A typical residential customer would have one main telephone and one or more extension telephones. Main telephone gain is defined to be the total telephone gain less removals of telephones as a result of nonpayment of bills. Total telephone gain is considered to be a function of personal income in a region, its unemployment rate, and building permits issued in the region. Removals for nonpayment are considered to be a function of employee hours worked in the nation, main telephones in service (the potential number of telephones that could be terminated), national corporate profits, and the number of removals for nonpayment during the past year.

Each of the variables that determines gross gain is a function of other variables. Personal income is a function of the unemployment rate, national defense purchases, and national personal income. The regional unemployment rate is a function of the national unemployment rate, employee hours worked, and defense purchases. Regional building permits are assumed to be a function of U.S. corporate profits in 1958 dollars, national housing starts, and the interest rate on corporate bonds. The average yield on new issues of high-grade corporate bonds is the particular interest rate used.

The basic philosophy incorporated in this model is that the U.S. economy drives the nation and its various regions; therefore, the economic forecasts for the region are related to the economic forecasts for the United States. The demand for telephone service is then related to these forecasts of economic activity. The regression models define the linkages between the national economy and the region and, in turn, the regional economy and the demand for main telephone service.

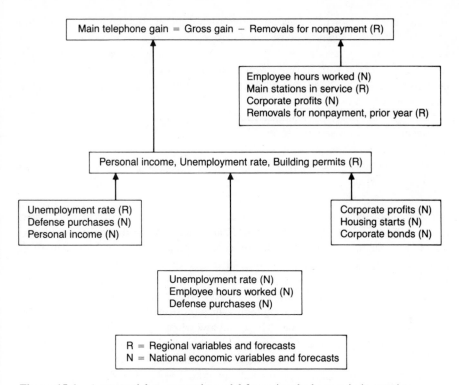

Figure 15.1 A sequential econometric model for main telephone gain in a region.

The Simultaneity Problem

Unlike the example of a recursive system, there are situations in which multi-equation systems do not lend themselves to a satisfactory solution unless the *simultaneous* nature of the economic relations is understood. This involves understanding the mathematical completeness of the system, which requires that the model have as many independent equations as endogenous variables.

The simultaneity problem is illustrated by the way that an independent variable can be correlated with the error term; such correlation violates the OLS assumption.

Suppose that a demand model is built to forecast connections and disconnections of telephones. For connections, a model that includes income, price, disconnections, and an error term is given by

$$\text{Connections}_t = \alpha_0 + \alpha_1(\text{Income})_t + \alpha_2(\text{Price})_t + \alpha_3(\text{Disconnections})_t + \varepsilon_t .$$

The reason for incorporating disconnections is that there are many cases where the customer relocates within a geographical area. The customer's move results in the simultaneous issuance of a disconnection order at the old household location and a connection order at the new household location. Therefore, there would be no connection if there were no disconnection, and there would be no disconnection unless there were a connection. In other words, the move is one transaction as far as the customer is concerned and this transaction results in the simultaneous connection and disconnection of telephone service.

Next, it is hypothesized that disconnections are related to connections, unemployment, price, and an error term:

$$\text{Disconnections}_t = \beta_0 + \beta_1(\text{Connections})_t + \beta_2(\text{Unemployment})_t + \beta_3(\text{Price})_t + v_t .$$

The two preceding equations—one for connections and the other for disconnections—can be looked upon as a system, and an analyst can determine whether or not the independent variables are likely to be correlated with the error term. With all other factors (e.g., income, price) held constant, high values for the error term ε_t will be a result of high values for connections. However, from the disconnections model it can be seen that (unemployment and price held constant) high values of connections mean that there will be high values for disconnections. Therefore, disconnections will be high when ε_t is high and low when ε_t is low. It is in this manner that the exogenous variable in an equation can be correlated with the error term.

The correlation between the independent variable and the error term in an equation can occur even if there is no second equation. The fact that disconnections and connections are simultaneously or jointly determined makes simultaneity a problem: the connections model does not adequately describe the process when OLS is used to estimate the parameters, because of the presence of the disconnections variable. The problem that this kind of correlation presents is that the estimated coefficient for the disconnections variable in the first equation will be too high. The OLS procedure will give too much credit for the variation in connections versus disconnections and not enough credit to the random errors. This means that in making predictions of connections, too much weight will be given to the predictions of disconnections. Therefore, in a system of equations such as those just presented, it is necessary to think through the process involved to determine if any exogenous variables are jointly determined with the endogenous variable.

One way to determine how much *bias* in the forecast results from the simultaneity problem is to see if the same parameters can be estimated by OLS and 2SLS. If both methods yield estimates that are approximately the same, there is little evidence of simultaneity, and one should use OLS. If the parameter estimates differ significantly, then 2SLS should be used to estimate the parameters instead of OLS, because the particular independent variable is correlated with the error.

The method of comparing OLS and 2SLS estimates will also work if the number of equations is small and the number of parameters is small. For large systems with many unknowns, this method may be too costly.

In the example we have used, price, income, and the unemployment rate are considered exogenous: connections and disconnections of telephones do not determine a person's income level or employment status. Factors outside the control of the telecommunications industry determine these variables.

Price could be considered within a company's control, but it is extremely difficult to build a set of equations to describe pricing. Because of regulation in the telephone industry, for example, the price charged is a function of revenues, expenses, investment, government regulations, public utility commission approvals, and many other factors. This means that there is a long delay between the decision that prices need to be changed and the implementation of a new rate schedule. Therefore, price is generally considered exogenous in the telephone industry. In analyzing the agricultural business, where current supply and demand determine price, price would be considered endogenous.

Special techniques, such as the use of instrumental variables, can be used to solve the estimation problem where endogenous variables are correlated with the error term. An *instrumental variable* is one that is very highly related to the independent variable (disconnections, in the telephone example) but not correlated with the error term. It may be that by searching through reams of time series the analyst can find such a variable. However, it turns out that for relatively small models, an easier way is by means of 2SLS. To see how this is so, let us continue to use the telephone example.

A Two-Stage Least Squares Solution

The first step in the two-stage process is to find an instrumental variable D^* that is highly correlated with disconnections but not correlated with the error term. This can be accomplished by regressing the disconnections variable on all exogenous and predetermined variables in *both* equations.

Since the connections variable is also a *right-hand-side exogenous variable,* an instrumental variable C^* is constructed similarly. In the first stage of the two-stage least squares process, each *endogenous* variable that is on the *right-hand side* of the equality sign of any equation is regressed against all the exogenous and predetermined variables in the system (i.e., in all the equations that make up the system). (It should be apparent that 2SLS is not appropriate for systems comprising 100–200 equations: there would be a greater number of exogenous variables than observations, and the first-stage regressions could not be performed.)

The reason for using *all* exogenous and predetermined variables in the first stage is to have only one instrumental variable for each right-hand-side endogenous variable. In the example of disconnections, there are seven potential instrumental variables based on the possible combinations of price, income, and unemployment. The simplest procedure for deciding which one to use is to include all exogenous variables, letting the regression procedure determine the weights to be applied to each exogenous variable. If a particular exogenous variable has no relationship with disconnections, its coefficient will be very low and no harm will be done. The first-stage equation is only used to create instrumental variables, to solve the simultaneity problem. This equation is not used for forecasting. Therefore, primary interest lies in having a very *high R-squared value,* so that the instrumental variable is very much like the dependent variable.

The instrumental variable is a function of exogenous variables. Since the variables are exogenous, each is independent of the error term in a connection or disconnection model. Moreover, a linear combination of exogenous variables is also independent of the error series.

In the second stage of the 2SLS procedure, each equation is estimated by OLS. However, the instrumental variables created from the first stage replace the right-hand-side endogenous variables in the appropriate equations: therefore, it can be seen that the 2SLS procedure incorporates two stages of OLS estimation; the first stage involves the estimation of instrumental variables for each *right-hand-side endogenous variable;* the second stage is the estimation of the parameter values for each equation in which any right-hand-side endogenous variables have been replaced by their first-stage predictions.

If an instrumental variable from the first stage is not highly correlated with an endogenous variable, some other exogenous series will have to be found—if one believes that the problem of error correlation is severe.

The following requirements must be met in order to use the 2SLS methodology:

- There must be an equation for each endogenous variable.

- There must be at least as many exogenous and/or predetermined variables as there are endogenous variables, or there must be a greater number of the former.

- In *any* equation containing right-hand-side endogenous variables, there must be *at least as many exogenous* (and/or predetermined) variables in the *system of equations* that are *not included in that equation* as there are right-hand-side endogenous variables that *are included.*

To illustrate the reason for the last restriction, a simplified example should suffice. Consider again the connections-disconnections model, as follows:

$$\text{Connections}_t = \alpha_0 + \alpha_1(\text{Disconnections})_t + \alpha_2(\text{Price})_t + \varepsilon_t,$$

and

$$\text{Disconnections}_t = \beta_0 + \beta_1(\text{Connections})_t + \beta_2(\text{Price})_t + \nu_t .$$

In this model, the third requirement would mean that, for the connections equation, there must be at least one exogenous/predetermined variable in the system that is not also in the connections equation. Since "price" is common to both equations, this requirement is not met for either the connections or disconnections equation.

The first stage of the 2SLS process for this model involves a regression of disconnections on price, and then a second regression—of connections on price. In the second stage, the connections equation would look as follows:

$$\text{Connections}_t = \alpha_0 + \alpha_1 (\text{Disconnections})_t^* + \alpha_2 (\text{Price})_t + \varepsilon_t .$$

But

$$\text{Disconnections}_t^* = \hat{\delta}_1 + \hat{\delta}_2 (\text{Price})_t$$

from the first stage of 2SLS. Therefore,

$$\text{Connections}_t = \alpha_0 + \alpha_1 (\hat{\delta}_1 + \hat{\delta}_2 \text{ Price}_t) + \alpha_2 (\text{Price})_t + \varepsilon_t .$$

The same variable cannot appear twice on the right-hand side of any equation. Nor can a linear combination of right-hand-side variables appear as an additional independent variable. This is what was described earlier as a problem of identification in econometrics, and it suggests an incorrectly specified model (Fisher, 1966).

In summary, it is possible for an independent variable to be correlated with the errors in an equation. This correlation has the effect of *overstating* the contribution of the independent variable and *understating* the variance of the errors when OLS procedures are used. The 2SLS methodology is an appropriate estimation method when faced with this problem for relatively small models. To establish the need for a simultaneous-equations solution, you can try both OLS and 2SLS estimation and compare the parameter estimates. If the estimates are essentially the same, you should use the simpler OLS estimation. If the parameter estimates differ significantly and the *instrumental variables are good*, use of the 2SLS estimation procedures is recommended.

RELATIONSHIP OF ECONOMETRIC MODELS TO ARIMA TIME SERIES MODELS

It has been noted that many large econometric models have questionable forecasting accuracy (Armstrong, 1978). There is sufficient literature to suggest that simpler, more parsimonious representations often possess better extrapolative properties. While the ARIMA time series models are based purely on historical patterns, they often provide less costly and more accurate forecasts. Thus forecasters must be aware of the pitfalls and advantages of any quantitative forecasting method before applying it in practice. It is often desirable, however, to use both econometric and time series approaches in a forecasting system. The predictions of each can often be subjectively combined into a useful and dependable forecast.

It is shown in Granger and Newbold (1977) that traditional econometric and modern time series (ARIMA) models are closely related. Both approaches can be interpreted as a system in which a number of inputs are entered into a "black box" that transfers the values to an output. When expressed in mathematical terms, the "black box" is a linear system that derives its name from systems engineering. The manner in which the parameters describing the system are estimated and interpreted represents the key difference between the two methodologies.

ECONOMETRIC SYSTEMS CHECKLIST

_____ (1) Are any "independent" variables determined jointly or simultaneously with the dependent variable?

_____ If the answer to 1 is yes, can OLS be applied to the reduced-form equations?

_____ If the answer to 1 is yes, can the equations be arranged so that they can be estimated sequentially or recursively?

_____ If the answer to 1 is yes, has a 2SLS solution been tried? How do the parameter estimates for right-hand-side endogenous variables compare for OLS versus 2SLS? Is a 2SLS solution warranted?

_____ If a 2SLS solution was attempted, are the first-stage instrumental variables adequate (how high is the R-squared value for each instrument)?

_____ If a 2SLS solution was attempted, is there an equation for each endogenous variable?

_____ For 2SLS, are there at least as many exogenous and/or predetermined variables as there are endogenous variables in the system of equations, or more?

_____ For 2SLS, in each equation containing right-hand-side endogenous variables, are there at least as many exogenous or predetermined variables in the system of equations that are not included in that equation as there are right-hand-side endogenous variables that are included in that equation?

SUMMARY

When confronted with econometric systems, the forecaster/analyst has several possible alternatives:

- Use OLS on reduced-form equations.
- Use OLS for recursive systems.
- Use 2SLS (or other specialized techniques) for estimating the structural equations of simultaneous-equations systems.

Since the forecaster is usually less concerned with the implications of the parameters of structural equations and more concerned with accurate forecasts, reduced-form equations may be perfectly satisfactory. Econometricians, in their policy recommendation role, are generally more concerned with what the structural equations imply. They are more likely to pursue 2SLS and other simultaneous-equation estimation methods even when reduced-form equations can be estimated.

USEFUL READING

ARMSTRONG, J. S. (1978). Forecasting with Econometric Methods: Folklore versus Fact. *Journal of Business* 51, 549–64.

FISHER, F. M. (1966). *The Identification Problem*. New York, NY: McGraw-Hill.

GRANGER, C. W. J., and P. NEWBOLD (1977). *Forecasting Economic Time Series*. New York, NY: Academic Press.

JOHNSTON, J. (1972). *Econometric Methods,* 2nd ed. New York, NY: McGraw-Hill.

PINDYCK, R. S., and D. L. RUBINFELD (1976). *Econometric Models and Economic Forecasts*. New York, NY: McGraw-Hill.

CHAPTER 16

Pooling of Cross-Sectional and Time Series Data

The regression models covered up to this point have involved the estimation of model parameters through use of either time series or cross-sectional approaches. In some cases, cross-sectional data are available over time. The analyst has the choice of building time series, cross-sectional models, or pooled models. This chapter addresses:

- The reasons why pooled models are considered.
- Some commonly used pooling methods for estimation.
- The steps in building pooled models.
- Considerations in selecting the pooling method for a forecasting problem.

WHY CONSIDER POOLING?

A *pooled model* includes observations for N cross sections over T time periods. For example, suppose you are interested in per capita electricity consumption for the United States and annual data are available for 50 states for ten years. You can build 50 time series models, ten cross-sectional models, or one pooled model. But what advantage does pooling offer?

There are several reasons why pooled models may be desirable:

- Pooling can increase the reliability of the parameter estimates by increasing the degrees of freedom and decreasing the standard errors of the parameter estimates. In the above example, each of the 50 state models will have only ten time series observations. A pooled model, if appropriate, will have 500 observations.

- Since cross sectional variation is normally substantially greater than time series variation, the estimates for a pooled model may be based on a wider range of variation in a potential independent variable than will exist for time series models.

- Pooling allows for the analysis of data in a unified model that considers both time and cross-sectional variation.

Several pooled models for demand analysis have indicated elasticities that are less in absolute value than those indicated by comparable time series or cross-sectional models. The reason for this, in the case of time series models, may be that data are increasing over time because of a factor not included in the model. This misspecification may not be apparent because of multicollinearity, and the effect may be to overstate the parameter estimates of the model.

In the case of cross sectional models, the reason may be attributable to slow consumer response to changes in income, or to habits of buying that are strong. However, a more basic reason may be the fact that for some data, the cross-sectional model has an inherent misspecification. Suppose that for a given data set, the constant term or slope should be allowed to vary for each state or location. This cannot be accomplished in a purely cross sectional model.

Pooling can be helpful in increasing the reliability of the parameter estimates by greatly increasing the number of observations in the model. An argument can be made that pooling methods should be used even when the model assumptions are not precisely met. You may be willing to accept parameter estimates that have a small bias but much less variance than those of time series or cross-sectional alternatives. However, if the structure of the relationships is substantially different over the cross sections, pooling is inappropriate.

If you examine the residuals of the pooled model related to one cross section (state, area) over time, a serial correlation problem may exist. Similarly, if you examine the residuals across all the cross sections at a point in time, heteroscedasticity may exist. It is also possible that the residuals from one cross section are correlated with the residuals from another cross section. Some relevant journal articles for the interested reader include Balestra and Nerlove (1966), Bass and Wittink (1975), Maddala (1971), Mundlak (1978), Swamy (1970), and Zellner (1962).

METHODS FOR ESTIMATION OF POOLED MODELS

The method used to build a pooled model depends on the underlying assumptions that are involved, and these are based on the characteristics of the data under investigation. Three generic classifications of methods are generally considered:

- Ordinary least squares (OLS).
- Ordinary least squares with dummy variables.
- Generalized least squares (GLS).

Pooling through Use of OLS

The OLS model has the form

$$Y_{it} = \alpha + \beta X_{it} + \varepsilon_{it}, \qquad \text{for } i = 1, 2, \ldots, N \text{ and } t = 1, 2, \ldots, T.$$

N represents the number of cross sections and T denotes the number of time periods. In this case, you can estimate T cross-sectional or N time series models. If both α and β are constant over time for all cross sections, more efficient parameter estimates can be obtained by combining all the data in one large pooled model, with N times T observations (500 in the case of the electricity consumption example).

When it is feasible to assume that the intercepts are fixed (not random) and equal for all cross sections, that the coefficients of the independent variables are fixed and equal for all cross sections, that autocorrelation and heteroscedasticity are *not* present, and there is *no cross correlation* among the residuals of the cross sections, then OLS is the method to be applied. The foregoing represent a rather restrictive set of assumptions that will generally not be satisfied. However, when these conditions exist, there is an opportunity to obtain more efficient (minimum variance) estimators, since the fewest number of parameters are estimated, leaving the largest number of degrees of freedom.

If autocorrelation is present, this can be corrected by transforming the data, by using autocorrelation correction techniques, such as the Cochrane-Orcutt or Hildreth-Lu techniques (Chapter 13). Then, OLS can be used on the transformed data.

When using cross section data, you should generally expect to find heteroscedasticity and adjust for it, when necessary, before pooling. Homoscedasticity is unlikely among cross sections and you should satisfy yourself that the condition truly does exist. In the example that follows, the assumptions of no autocorrelation and presence of homoscedasticity will be examined.

Other pooling methods are considered when conditions exist such that the normal OLS assumptions are not valid. Specifically, aggregation bias may result from combining the cross sections before estimation. The assumption of constant intercepts over the cross sections may not be realistic. In the electricity example, there may be some factors influencing demand that are peculiar to a given geographical region. By constraining the intercepts to be the same for all cross sections, significant bias may be introduced in the estimates of parameters.

The assumption in OLS analysis that there is no correlation between the residuals of the various cross sections may also not be valid. Cross correlation among cross sections can result if there are omitted variables that are common to all equations.

For example, it is frequently found that regions of the country respond in a similar fashion to changes in economic prosperity. These changes cannot always be explained with the independent variables at hand. Since state boundaries are some-what arbitrary relative to regional economic influences, it may be the case that the residuals of individual state models will be correlated. Methods using generalized least squares estimation can be used when this kind of cross-correlation is present.

The ability to generate estimators in situations where traditional OLS cannot generate these is the primary value of analysis of covariance and the generalized least squares method.

The Analysis of Covariance Model

A slightly more general model can result if dummy variables are used to allow for different intercepts for each cross section and/or different time period: such a model is

$$Y_{it} = \alpha + \beta X_{it} + \gamma_2 W_{2t} + \cdots + \gamma_N W_{Nt} + \delta_2 Z_{i2} + \cdots \delta_t Z_{it} + \varepsilon_{it} \, ,$$

where $W_{it} = 1$ for the ith cross-section and zero elsewhere, and $Z_{it} = 1$ for the tth time period and zero elsewhere.

OLS models incorporating dummy variables of this form are referred to as *analysis-of-covariance models*. These models have the same form as those referred to in Chapter 12. When dummy variables are used, the intercept can be viewed as having as many as three components—a "national or regional average," a variation by cross section, and a variation over time. A limitation of the model is that while it allows the intercept coefficients to vary, the cause of the variation is not identified. Throughout the remainder of this chapter, only dummy variables for cross sections will be discussed. Note there are at most $(N - 1)$ dummy variables (e.g., 49 to allow for varying intercepts for the 50 states; the first state effect is captured by the constant α.) Some cross sections may have the same intercept, thereby reducing the number of dummy variables.

The form of the analysis-of-covariance model presented earlier allows for different intercepts for each cross section. A second form of the analysis-of-covariance model allows for different coefficients for one or more independent variables. For example, the price elasticity for consumption of electricity might be the same for all regions, but the income elasticity might be different. (See Chapter 12 for the multiplicative dummy variable model.)

An important aspect of pooling is to determine the constraints that are appropriate for a given study; i.e., are constant or different intercepts appropriate? Can all slope coefficients or only some be considered constant?

As a practical matter, pooling does not yield the "best" estimators in the statistical sense of unbiasedness and minimum variance. A decision is required as to the degree of constraint that will be applied to the intercept and the coefficients of the equation. Since constraint may introduce bias in the estimators, it is important that the constraint be the minimum required, so that the estimators will have minimum mean squared errors. In the example that follows, we will show how to decide on the constraints to be applied, based on the data under study.

Generalized Least Squares Method

Explanations of the generalized least squares (GLS) estimation methods are treated in Johnston (1972, Section 6.3); Kmenta (1971, Chapter 12) and Maddala (1977, Section 14.2). The example that follows shows how cross correlation can be identified. A GLS computer program is used to correct for cross correlation in the residual series.

Another class of model allows for random intercepts and coefficients. The random coefficients and/or random intercepts models assume that an additional error term is required to account for a random variation that is specific to a given cross section. For example, if you were to build 50 models for electricity consumption—one for each state—the assumption of fixed (nonrandom) coefficients and intercepts is reasonable. If you were to select ten states randomly and build a pooled model to estimate electricity consumption for the nation, it might be appropriate to allow for an additional source of variation in the parameters resulting from the sampling process: a selection of ten different states might yield slightly different results. Since forecasters are not usually involved in randomly selecting forecast entities (areas, products, etc.), the random-effects models are not treated in this book.

STEPS IN BUILDING A POOLED MODEL

The considerations involved in building pooled models will be reviewed by extending the example presented in Chapter 10. Recall that the objective was to build a model to estimate the price elasticity of telephone toll service. With only six years of quarterly data available for the state being modeled, the estimated standard deviation of the price parameter was quite large. Since data are available for two additional states for the same time period, it may be possible to build a pooled model that is more efficient (one in which there is smaller parameter-variance).

As with all modeling efforts, this process will begin with an examination of the data. This leads to the specification of an appropriate model. Two alternative models were considered previously. Model 1 had a lagged dependent variable and Model 2 had a telephones-in-service variable in lieu of the lagged variable. With $ln = \log_e$,

Model 1: $\ln Q_t = \alpha + \beta_0 \ln Q_{t-1} + \beta_1 \ln \text{Price} + \beta_2 \ln \text{Income} + \beta_3 D_1 + \beta_4 D_2 + \beta_5 D_3 + \varepsilon_t$.

Model 2: $\ln Q_t = \alpha + \beta_0 \ln \text{Price} + \beta_1 \ln \text{Income} + \beta_2 \ln \text{Telephones} + \beta_3 D_3 + \varepsilon_t$.

In both models D_1, D_2, and D_3 were dummy variables to account for seasonality. (Do not confuse these dummy variables with those that are used to allow for different intercepts for each cross section.) In Model 2, only the fourth quarter seasonality was significantly different from the first quarter (captured by α). Thus, only the dummy variable D_3 was needed.

Recall that there was no autocorrelation for Model 1, but Model 2 did have autocorrelated residuals. The three states for which data are available are labeled A, B, and C. The previous modeling was done for state B alone. In addition to state B, we now want to review the data for states A and C, to determine if a pooled model is appropriate.

The steps to be followed are:

1. *Examine the data and specify an appropriate starting point for time series models for each cross section.*

Figure 16.1 shows a scatter plot of the logarithms of quantity against the logarithms of the number of telephones-in-service for each state. Notice that the range of the

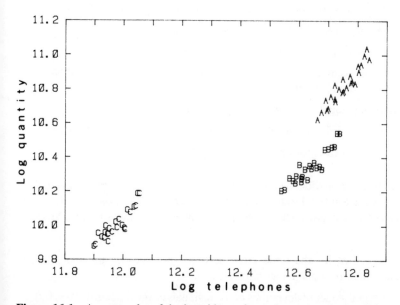

Figure 16.1   A scatter plot of the logarithms of quantity against the logarithms of number of telephones-in-service for each state.

independent variable has been increased with the addition of states A and C. This plot suggests that the same slope may be appropriate for all cross sections, but that the intercepts for each state are different. Plots of the logarithms of quantity against the remaining variables show similar relationships. The only variable for which both the slope and the intercept are constant for all cross sections is the lagged quantity variable. Based on the results so far, the OLS assumption of constant intercepts would be rejected for a simple linear regression model.

2. *Build N separate time series models* (e.g., for cross sections A, B, C).
3. *Develop the assumptions for the tentative pooled model.*
   a. *Analyze the coefficients for constancy across cross sections.*
   b. *Check for potential autocorrelation problems.*

- Are the coefficients equal over all cross sections?

- Are the coefficients equal over some cross sections?

- Are the intercepts equal across all cross sections?

- Are the intercepts equal across some cross sections?

- Are the coefficients equal but the intercepts different for all or some cross sections?

- Are the residuals *within* cross sections autocorrelated? If so,

- Are the autocorrelation parameters equal over *all* cross sections?

- Are the autocorrelation parameters equal over *some* cross sections?

- Did you correct for autocorrelation?

   c. *Check for potential problems of heterosedasticity.*

- Are the residual variances equal across all cross sections (is there homoscedasticity)? If not,

- Are they equal over some cross sections?

- Are they different for all cross sections (is there heteroscedasticity)?

- Did you correct for heteroscedasticity?

Tables 16.1 and 16.2 provide the model statistics and parameter estimates with their standard errors for Models 1 and 2, respectively. Most of the answers to the above questions can be obtained from these tables. First, a review of Table 16.1 for Model 1 indicates:

- The h statistics for the three cross sections (not shown) are insignificant, indicating that residual autocorrelation is not a problem.

- The estimated standard deviations of the residuals of the models (0.023, 0.025, 0.021) do not suggest problems of heteroscedasticity.

Table 16.1 Individual regressions for each cross section
for Model 1. (Standard errors are shown in parentheses.)

Parameter	Cross section		
	A	B	C
Intercept	1.78	4.64	3.96
Log of lagged quantity-coefficient	0.84	0.55	0.65
	(0.09)	(0.15)	(0.13)
Log of price	−0.12	−0.21	−0.24
	(0.09)	(0.08)	(0.08)
Log of income	0.35	0.76	0.94
	(0.18)	(0.20)	(0.24)
D_1	0.06	0.02	0.02
	(0.02)	(0.02)	(0.01)
D_2	0.09	0.02	0.02
	(0.02)	(0.02)	(0.01)
D_3	0.12	0.07	0.07
	(0.01)	(0.02)	(0.01)
R-squared statistic	0.97	0.95	0.95
Residual standard deviation (df = 16)	0.023	0.025	0.021

Note: The Durbin-Watson statistic is not applicable, since there is a
lagged dependent variable in the model.

- The coefficients for cross sections B and C are very similar and are within two standard errors. The coefficients for cross section A appear different for lagged quantity, income, and the three dummies. The standard errors are so large that it is difficult to reject totally the assumption of constant coefficients. An F test will be introduced later to help make this decision.

- The intercepts for cross sections B and C are similar, but they appear different from A.

- Tests for cross correlation of residuals will be taken up later.

Based on the information gathered so far for Model 1, the following assumptions can be made for a tentative pooled model:

- Cross sections B and C are very similar and a pooled model for these cross sections could assume constant slopes and intercepts. If we attempt to include cross section A, at a minimum we should assume different intercepts. The probable difference in slope coefficients is a potential problem when including cross section A in the pooled model.

Table 16.2 Individual regressions for each cross section, for Model 2. (Standard errors are show in parentheses.)

Parameter	Cross section		
	A	B	C
Intercept	−12.56	−1.89	−3.72
Log of telephones-coefficient	1.90	1.00	1.23
	(0.19)	(0.20)	(0.19)
Log of price	−0.21	−0.27	−0.33
	(0.09)	(0.07)	(0.07)
Log of income	0.02	0.63	0.78
	(0.21)	(0.19)	(0.23)
D_3	0.06	0.05	0.05
	(0.01)	(0.01)	(0.01)
R-squared statistic	0.97	0.96	0.96
Residual standard deviation (df = 19)	0.022	0.022	0.020
Durbin-Watson d statistic	1.33	1.65	1.52

- Autocorrelation and heteroscedasticity are not present.
- You may want to pool only cross sections B and C.

A review of Table 16.2 for Model 2 indicates that:

- The Durbin-Watson statistics fall in the indeterminate region in each cross section. It may be desirable to correct for autocorrelation in a pooled model.
- The estimated standard deviations of the residuals of each model do not indicate heteroscedasticity.
- The intercepts are probably different.
- Tests for cross correlation will be considered later.
- The same considerations as for Model 1 are valid regarding pooling across some or all cross sections.

The OLS model with dummy variables for different intercepts, applied with correction for autocorrelation, is a tentative pooling model. For this example, all three states are pooled.

4. *Combine the data from separate cross sections and examine the combined data. If a transformation is now required that was not apparent previously, repeat Steps 1–3 for the transformed data.*

5. *Decide on an appropriate pooling method.* The most desirable condition exists when the coefficients and intercepts are constant and the OLS method can be performed on the combined data. Using the assumptions that seem operative (autoregression, heteroscedasticity, constant/different intercepts, constant/different slopes for some variables, all or some cross sections), select the pooling method.

For Model 1, the tentative pooled model will be an analysis-of-covariance model (different intercepts) with no corrections for autocorrelation or heteroscedasticity. In this example we will pool all three states.

For Model 2, the tentative pooled model will also be an analysis-of-covariance model (different intercepts) with corrections for autocorrelation but not for heteroscedasticity. All three states will be pooled.

The above pooled models with different intercepts can be compared to pooled models with a constant intercept to determine if separate intercepts are required (see Step 7).

6. *Estimate the pooled model.*

7. *Perform statistical (homogeneity) tests to determine if the pooling method is appropriate.* Recall that constraining intercepts or slopes to be constant over the cross sections may introduce bias in the parameter estimates: an F test can be applied to determine the appropriate set of constraints for the data under study. The test determines the increase in explanatory power (i.e., reduction in residual variance) of an equation when intercepts or slopes are left unconstrained:

$$F = \frac{\dfrac{ESS_{\text{more-restrictive model (mrm)}} - ESS_{\text{less-restrictive model (lrm)}}}{df_{\text{mrm}} - df_{\text{lrm}}}}{\dfrac{ESS_{\text{lrm}}}{df_{\text{lrm}}}}.$$

In the example we have been using, the less-restrictive model is one in which separate regressions are made for each cross section and the more-restrictive model is the combined OLS model. Therefore,

$$F = \frac{\dfrac{ESS_{\text{pooled (OLS)}} - ESS_{\text{SR}}}{df_{\text{OLS}} - df_{\text{SR}}}}{\dfrac{ESS_{\text{SR}}}{df_{\text{SR}}}}.$$

If the F statistic is not significant, then there is no evidence that the more restrictive model (all coefficients and intercepts the same) should be rejected. If the mean squared error statistics show significant differences, the more restrictive model assumptions are not valid and OLS is not appropriate. In this way, the F test can be used to help select the appropriate pooling method.

Similarly, the OLS model with dummy variables (analysis-of-covariance or COV model) is less restrictive than the OLS model, since the intercepts may be different for each cross-section. Assume that OLS is not a valid estimation method. You can then use an F test to consider whether the analysis-of-covariance model should be applied.

$$F = \frac{\dfrac{\text{ESS}_{\text{COV}} - \text{ESS}_{\text{SR}}}{\text{df}_{\text{COV}} - \text{df}_{\text{SR}}}}{\dfrac{\text{ESS}_{\text{SR}}}{\text{df}_{\text{SR}}}};$$

ESS = estimated sum of squares, COV = covariance, SR = separate regressions, and df = degrees of freedom. If there is no significant difference in the mean squared errors, the covariance model may be used for pooling.

The F test may also be helpful in determining whether to use OLS or the co-variance method, assuming both techniques are appropriate when compared to separate regressions. If the ratio indicates a significant value for F, the less restrictive covariance model is generally selected, since the more restrictive OLS model has significantly higher residual variance. If you are willing to accept some bias, another test for determining if the unconstrained intercepts differ from one another is to calculate the simultaneous confidence intervals of the difference between all intercepts and see if zero lies within the interval. This will also point to cross sections that may be causing the problem.

For each of Models 1 and 2, we can use the F test to help decide what constraints to apply. The least restrictive model is the separate or individual regressions for each cross section (no pooling). The most-restrictive model is OLS with constant slopes and intercepts. A less-restrictive model is OLS with dummy variables (different intercepts).

Table 16.3 summarizes the F tests for the pooled models. The OLS assumptions are too restrictive in that both F statistics exceed the critical values. However, the bias in Model 1 is much less than that of Model 2. The assumption of different intercepts but constant slopes for all variables is not rejected in either case.

What impact does this result have for this study? Table 16.4 presents the estimated price elasticity and associated standard error for Model 1 for B alone and for the pooled model. Notice the large reduction in the estimated standard deviations of the price coefficient and, therefore, the smaller confidence intervals for the price elasticity.

8. *Analyze residual plots to help verify the appropriate pooling method.*

- A correlogram of the residuals may be used to identify serial correlation in the pooled model. The constraints may introduce bias and autocorrelation.

Table 16.3   A summary of the F tests for the pooled models.

F test	Model 1 (untransformed data)		Model 2 (data transformed for autocorrelation)	
	SSE	df	SSE	df
Sum of individual residuals (SR)	0.02537	48	0.02602	57
Pooled—OLS	0.05163	62	1.1711	67
Pooled—Analysis-of-covariance (COV)	0.03534	60	0.03471	65

Model 1

$$F_{(OLS,SR)} = \frac{(SSE_{OLS} - SSE_{SR})/(df_{OLS} - df_{SR})}{SSE_{SR}/df_{SR}} = \frac{(0.05163 - 0.02537)/(62 - 48)}{0.02537/48}$$

$$= 3.55;$$

$$F_{(14,48)} = 2.25.$$

$$F_{(COV,SR)} = \frac{(SSE_{OLS} - SSE_{SR})/(dr_{COV} - df_{SR})}{SSE_{SR}/df_{SR}} = \frac{(0.03534 - 0.02537)/(60 - 48)}{0.02537/48}$$

$$= 1.57;$$

$$F_{(12,48)} = 2.41.$$

Model 2

$$F_{(OLS,SR)} = \frac{(1.1711 - 0.02602)/(67 - 57)}{0.02602/57} = \frac{0.11451}{0.000456}$$

$$= 251.11;$$

$$F_{(10,57)} = 2.63.$$

$$F_{(COV,SR)} = \frac{(0.03471 - 0.02602)/(65 - 57)}{0.02602/57} = \frac{0.001086}{0.000456}$$

$$= 2.38;$$

$$F_{(8,57)} = 3.02.$$

- Plots of residuals against predictions for each time period or for the total model may be used to identify heteroscedasticity. In this regard, it is convenient to label the residuals from each cross section as shown in Figure 16.2 for Model 2. This plot suggests homoscedasticity (constant residual variance) and shows no problem areas.

Figure 16.3, on the other hand, shows a hypothetical situation where the residuals (while approximately constant in variance) are a problem for individual cross sec-

Table 16.4 Estimates of price elasticities, with their
associated standard errors, for Model 1.

Parameter	Individual regression (State B)		Pooled (COV)	
	Coefficient	Estimated standard deviation	Coefficient	Estimated standard deviation
Constant	4.64	1.46	3.08	0.659
*Intercept B	—	—	−0.06	0.021
*Intercept C	—	—	0.23	0.053
Log of price	−0.21	0.084	−0.20	0.049
Log of income	0.76	0.203	0.61	0.112
Lagged Q	0.55	0.153	0.72	0.064
D_1	0.02	0.017	0.03	0.009
D_2	0.16	0.017	0.04	0.009
D_3	0.72	0.017	0.09	0.009

*Intercept for cross sections B or C equals *constant* plus the coefficients
for the *intercept* B or C.

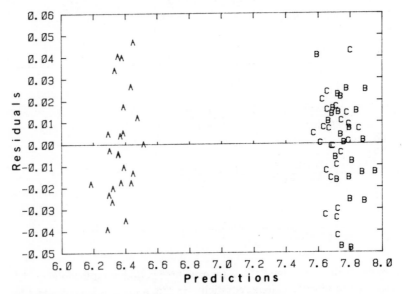

Figure 16.2 Plot of residuals from each cross section.

tions. Note the D residuals are all negative, the B residuals are zero or positive, the
C residuals are generally positive, while the A and E residuals appear reasonable.
In this case, the pooling has gone too far, since the cross sections are obviously

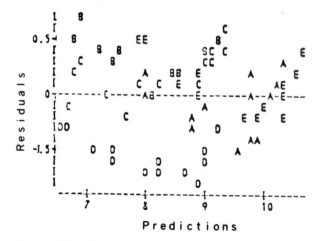

Figure 16.3 Plot of residuals for individual cross sections in a hypothetical situation.

different in some aspect. If an additional variable in a multiple regression model would not solve this problem, pooling across all cross sections should be abandoned.

- The Q-Q plot of the residuals against the normal curve will indicate if the normality assumption is appropriate.

- Cross correlograms indicate whether residuals of one cross section are correlated with the residuals of other cross sections.

The cross correlations for Model 2 between cross sections A and B and A and C were insignificant. The cross correlogram in Figure 16.4 shows significant cross correlation between the residuals of sections B and C at lag zero. This indicates the need for generalized least squares estimation.

Table 16.5 summarizes the results for Model 2 for the individual regression for cross section B corrected for autocorrelation; the pooled model with separate intercepts corrected for autocorrelation (COV); and the pooled model with separate intercepts corrected for autocorrelation and cross correlation (GLS). Notice the significant reduction in the estimated standard deviation of the price coefficient (0.076 to 0.058) with the final pooled model. In this instance, the correction for cross correlation did not improve the efficiency of the price parameter estimate. The greatest improvement in efficiency results when the cross correlation between the residuals of the cross sections is high and the independent variable (price in this case) in each cross section has low correlation with the same variable in other cross sections (see Kmenta, 1971, Chapter 12).

9. *Modify the pooling method as required on the basis of tests and information obtained from the plots.*

10. *Stop when a satisfactory model is obtained.*

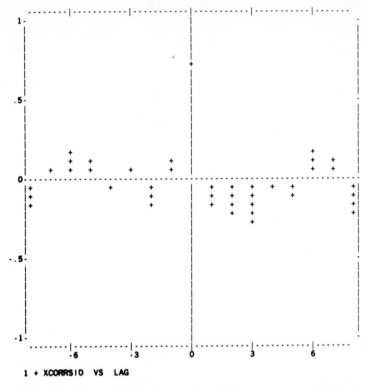

Figure 16.4 A cross correlogram between residuals for cross sections B and C.

Table 16.5 Estimates of price elasticities, with their associated standard errors, for Model 2 (corrected for autocorrelation).

Parameter	Individual Regression		Pooled (COV)		Pooled (GLS)	
	Coefficient	Estimated standard deviation	Coefficient	Estimated standard deviation	Coefficient	Estimated standard deviation
Constant	−1.83	2.593	−6.13	1.73	−7.79	1.792
*Intercept B	—	—	−0.26	0.015	−0.27	0.015
*Intercept C	—	—	−0.59	0.068	0.54	0.068
Log of price	−0.27	0.076	−0.27	0.052	−0.26	0.058
Log of income	0.64	0.202	0.44	0.153	0.23	0.145
Log of telephones	1.00	0.221	1.38	0.146	1.53	0.150
D_3	0.05	0.010	0.05	0.005	0.05	0.005

*Intercept for cross section B or C equals *constant* plus the coefficient for the intercept B or C.

POOLED MODELS CHECKLIST

_____ If both time series and cross-sectional data are available, can a "pooled" model be built to take advantage of the larger data set?

_____ Consider assumptions about each coefficient.

- Is each equal over all cross sections? That is, does $\beta_{i1} = \beta_{i2} = \cdots = \beta_{iN}$?
- If not, is each equal over some cross sections?
- Is it a fixed or random effect?

_____ Consider assumptions about the intercept:

- Is it equal over all cross sections? That is, does $\alpha_1 = \alpha_2 = \cdots = \alpha_N$?
- If not, is it equal over some cross sections?
- Is it a fixed or random effect?

_____ Consider assumptions about autocorrelation.

- Are the error terms within cross sections autocorrelated?
- If so, are the autocorrelation parameters equal over all cross sections?
- If not, are they equal over some cross sections?

_____ Consider assumptions about error variances.

- Are the error variances equal over all cross sections?
- Are they equal over some cross sections?
- Are they different on all cross sections?

_____ Are the error terms correlated over cross sections?

SUMMARY

The subject of pooling time series and cross-sectional data has received increasing attention from forecasters in recent years. Pooling

- Provides the potential for increasing the reliability of the parameter estimates by increasing the sample size and the range of variation of the independent variables.
- With a pooled OLS model, the assumption is that all intercepts and coefficients are the same across all cross sections.
- With the dummy variables for different intercepts (analysis-of-covariance) method, the assumption is that the coefficients are constant but that the inter-

cepts may be different in each cross section and over time. (A multiplicative analysis-of-covariance model also allows for some coefficients to be different for each cross section.)

USEFUL READING

BALESTRA, P., and M. NERLOVE (1966). Pooling Cross Section and Time Series Data in the Estimation of a Dynamic Model: The Demand for Natural Gas, *Econometrica* 34, 585–612.

BASS, F. M., and R. R. WITTINK (1975). Pooling Issues and Methods in Regression Analysis with Examples in Marketing Research. *Journal of Marketing Research* 12, 414–425.

JOHNSTON, J. (1972). *Econometric Methods*. Second Edition. New York, NY: McGraw-Hill.

KMENTA, J. (1971). *Elements of Econometrics*. New York, NY: MacMillan.

MADDALA, G. S. (1977). *Econometrics*. New York, NY: McGraw-Hill.

MADDALA, G. S. (1971). The Use of Variance Components Models in Pooling Cross Section and Time Series Data. *Econometrica* 39, 341–358.

MUNDLAK, Y. (1978). On the Pooling of Time Series and Cross Section Data. *Econometrica* 46, 69–85.

SWAMY, P. A. V. B. (1970). Efficient Inference in a Random Coefficient Regression Model. *Econometrica* 38, 311–323.

ZELLNER, A. (1962). An Efficient Method of Estimating Seemingly Unrelated Regressions and Tests for Aggregation Bias. *Journal of the American Statistical Association* 57, 348–368.

Part 4

The Box-Jenkins
Approach to Forecasting

CHAPTER 17

Filtering Techniques for Forecasting

The fourth part of this book deals with a class of models that can produce forecasts based on a synthesis of historical patterns in data. The construction of these models

- Applies a single, unified theory to the description of a wide range of stationary and nonstationary (seasonal and trending) time series.

- Allows a linear representation of a stationary time series in terms of its own past values and a weighted sum of a current error term and lagged error terms.

- Provides a systematic procedure for forecasting a time series from its own current and past values.

- Allows the capability for incorporating explanatory or leading indicator variables into the equations.

This chapter introduces the concept of stationarity for time series and provides a description of the class of autoregressive moving-average (ARMA) models, which can be used with stationary time series. The next chapter introduces the Box-Jenkins strategy for fitting models to nonstationary as well as stationary time series data, and the remaining seven chapters will treat the specifics of this modeling strategy for a wide variety of other kinds of time series.

FILTERING TECHNIQUES FOR TIME SERIES

Autoregressive moving-average (ARMA) models are a specialized but highly powerful class of *linear* filtering techniques by which a random input is "filtered" so that the output represents the observed or transformed time series. Filtering techniques, widely used in control engineering, have only in recent years found practical use in business forecasting. The general theory of linear filters (Whittle, 1963) is not new, since much of the original work was done by the renowned mathematicians Kolmogorov and Wiener in the 1930's for automatic control problems. The special kind of *linear* filter called the *autoregressive* (AR) model goes back a little further: AR models are generally considered to have been first used by Yule (1927). Another kind of filter, called the *moving-average* (MA) model, was introduced by Slutsky (1937). Autoregressive moving-average (ARMA) theory, in which these models are combined, was developed by Wold (1954).

A general *model-building strategy* for these models can be attributed to Professors G. E. P. Box and G. M. Jenkins; developed during the past three decades, their strategy is the result of their direct experience with forecasting problems in business, economic, and engineering environments. In Box and Jenkins (1976) they present a formally structured class of time series models that are sufficiently flexible to describe many practical situations. Known as *autoregressive integrated moving-average* (ARIMA) models, these are capable of describing various stationary *and* nonstationary (time-dependent) phenomena (unlike ARMA models, which require stationary data) in a statistical rather than in a deterministic manner. Other treatments of the *Box-Jenkins procedure* may be found in Anderson (1976), Granger and Newbold (1977), Mabert (1975), Makridakis and Wheelwright (1978), and Nelson (1973).

WHY USE ARIMA MODELS?

In *The Beginning Forecaster,* it was shown how to analyze a trend/seasonal time series by constructing a two-way table. Although such a table requires the estimation of many coefficients (e.g., monthly and yearly means), the information that is obtained cannot be used effectively for predicting future values. Alternatively, regression analysis offered a means of fitting linear models in which functions of time (e.g., straight lines and polynomials) were independent variables: they often fit well over a limited range, yet often forecast poorly when extrapolated. Moreover, although regression models utilize external (economic and demographic) independent variables, forecasts of these must be added to a comprehensive, final forecast in order to obtain an effective prediction of future values. ARIMA models offer a way of circumventing some of these difficulties.

The ARIMA models have also proved to be excellent *short-term* forecasting models for a wide variety of time series. In a number of studies (see Granger and Newbold, 1977), forecasts from simple ARIMA models have frequently outperformed larger, more complex econometric systems for a number of economic series. Nevertheless, while it is possible to construct ARIMA models with only two years of monthly historical data, the best results are usually obtained when at least five to ten years of data are available—particularly if the series exhibits strong seasonality.

The drawback of ARIMA models is that, because they are univariate, they have very limited explanatory capability. The models are essentially sophisticated extrapolative devices that are of greatest use when it is expected that the underlying factors causing demand for products, services, revenues, etc., will behave in the future much in the same way as in the past. In the short term, this is often a reasonable expectation, however, because these factors tend to change slowly; data tend to show inertia in the short term.

A significant advantage of ARIMA models is that forecasts can be developed in a very short time. More time is spent obtaining and validating the data than in building the models. Therefore, a practitioner can often deliver significant results early in a project for which an ARIMA model is used. The forecaster should always consider ARIMA models as an important forecasting method whenever these models are relevant to the problem being studied.

CREATING A STATIONARY TIME SERIES

The autoregressive moving-average (ARMA) time series models are designed for *stationary* time series; that is, series whose basic statistical properties (e.g., means, variances, and covariances) remain constant over time. Thus, in order to build ARMA models, nonstationarity must first be identified and removed. Nonstationarity typically includes periodic variations and systematic changes in mean (trend) and variance.

The precise definition of stationarity is a complex one. Suffice it to say that for practical purposes, the data used in a random process are said to be *weakly stationary* if the first and second moments of the process are time-independent. This assumption implies, among other things, that the mean and variance of the data are constant and finite.

Another critical assumption in describing stationary time series comes as a result of the chronological order of the data. The difference between conventional regression methods and time series modeling of the sort possible with ARMA models is that independence of the observations cannot be assumed; in fact, in ARMA modeling it is the *mutual dependence* among observations that is of primary interest.

Identifying Nonstationarity

Generally, many kinds of nonstationarity are present in time series data. The simplest is known as *nonstationarity in the level of the mean* and occurs when the level of the mean changes or "drifts" over different segments of the data. A trending time series is a good example.

Nonstationarity can also occur as a result of a seasonal pattern in the data. A seasonal series cannot be considered stationary because the variation in the data is a function of the time of the year.

Another kind of nonstationarity is the result of increasing or decreasing variability of the data with time. Sales data may have a nonlinear trend or show increasing variability in the seasonal peaks and troughs over time.

Still another kind of nonstationarity can occur as a result of a drastic change in the level of some series. Examples of time series that change drastically are price or unemployment data. An unemployment rate is a series that tends to stay at a high or a low level, depending on economic conditions: the change is usually abrupt and time-dependent.

A combination of several or all of the above characteristics will usually be present in nonstationary real data. In reality, most time series are nonstationary.

Removing Nonstationarity through Differencing

In practice, many time series can be made stationary through differencing. Such a nonstationary series is termed *homogeneous*. Fortunately, many of the time series that arise in economics and business are of this type. Let us examine a typical homogeneous time series and see how it can be made stationary.

Consider the seasonally adjusted money supply series shown in Figure 17.1; this is the U.S. Treasury Department M1 Series. It shows nonstationary behavior in its level. This trend pattern is removed by taking differences of order 1 of the data. Figure 17.2 (a) shows the differenced series; it has an increase in level and an increasing variability with time.

Consider taking a second difference of order 1 Figure 17.2(b); this gives rise to a series that now appears to have a constant mean, but continues to show an uneven variability over time. It is unlikely that further differencing is called for.

In most economic data it appears that differences should be taken, at most, twice. The number of times that the original series must be differenced before a stationary series results is termed the *order of homogeneity* of the series. As a first step, it is important to identify the *minimum* amount of differencing required to create a stationary series. "Overdifferencing" can rarely be coompensated for by modeling.

In the case of the money supply data, it is not possible to obtain constancy of variance through differencing. It is better to take logarithms of the data first and then take first differences of the log-transformed data. (It is clear that the logarithms of the money supply data are nonstationary.) By taking first differences of the log-transformed data (growth rates), as shown in Figure 17.3, the resulting series appears stationary.

Another transformation that often works prior to differencing is the square-root transformation. The first differences of the square root of the money supply data did

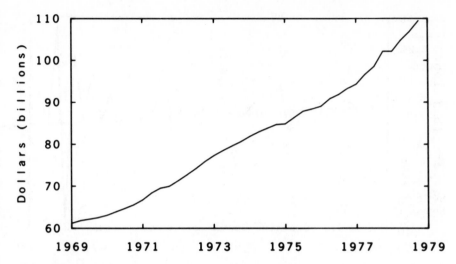

Figure 17.1   A time plot of seasonally adjusted money supply: the U.S. Treasury M1 Series, 1969–79.

Source: Board of governors of the Federal Reserve System.

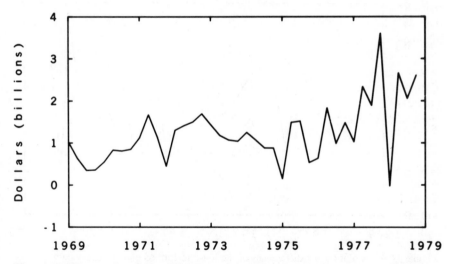

Figure 17.2(a)   A plot of first differences of the money supply data shown in Figure 17.1.

not appear to be very different from the percent changes of the money supply. Thus it would appear to be more natural to work with the percent changes (growth rates) of the money supply series as the stationary series for modeling purposes.

　　There are situations in which differencing alone can change a nonstationary series to a reasonably stationary one. More general situations may require a trans-

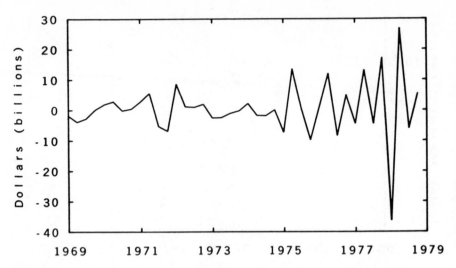

Figure 17.2(b)   A plot of the second differences of the money supply data shown in Fig. 17.1.

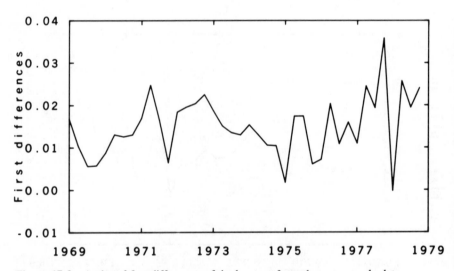

Figure 17.3   A plot of first differences of the log-transformed money supply data.

formation of the Box-Cox type prior to differencing (Granger and Newbold, 1976); this includes taking a square root transformation and a logarithmic transformation. The class of time series models discussed in the remainder of this book assumes that stationarity can be achieved essentially through differencing and transformations.

LINEAR MODELS FOR STATIONARY SERIES

Once a time series has been made stationary, it can be used with a linear filter in order to devise a forecast. Once an appropriate filter has been determined for the series, an *optimal* forecasting procedure follows directly from the theory.

The Linear Filter as a Black Box

The ARMA models are important because they are mathematically tractable; moreover, they are flexible enough to describe many time series. The role they play is equivalent to that of the linear differential equation in the study of deterministic systems, such as are encountered in control theory or physics.

Application of linear models is based on the idea that a time series in which successive values are highly dependent can also be thought of as having come from a process involving a series of independent errors or "shocks" $\{\varepsilon_t\}$. The general form of a (discrete) *linear process* is

$$Y_t = \alpha + \varepsilon_t + \psi_1 \varepsilon_{t-1} + \cdots + \psi_n \varepsilon_{t-n} + \cdots,$$

where α and all ψ are fixed parameters and the $\{\varepsilon_t\}$ is a sequence of identically, independently distributed random errors with zero mean and constant variance. Thus the process is *linear* because Y_t is represented as a linear combination of current and past shocks. It is often referred to as a *black box* or filter because the model relates a random input to an output that is time-dependent. The input is filtered or "damped" by the equation so that what comes out of the equation has the characteristics that are wanted.

A linear process can be visualized as a *black box* in this manner (Figure 17.4): white noise—purely random error $\{\varepsilon_t\}$—is transformed to the observed series $\{Y_t\}$ by the operation of a linear filter; the filtering operation simply takes a weighted sum of previous shocks. The weights are known as ψ coefficients.

The concept of *white noise,* which we explained earlier, is of central importance to time series analysis, just as independent observational error is of central importance to classical statistical analysis. The reason is that the next value ε_t of white noise is *unpredictable* even if all previous values $\varepsilon_{t-1}, \varepsilon_{t-2}, \ldots$, and subsequent values $\varepsilon_{t+1}, \varepsilon_{t+2}, \ldots$, are known.

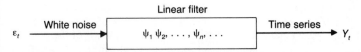

Figure 17.4   A black-box representation of the linear random process.

The ARMA Model as a Linear Filter

In many problems, such as those in which it is required that future values of a series be predicted, it is necessary to construct a *parametric model* for the time series. To be useful, the model should be physically meaningful and involve as few parameters as possible. A powerful parametric model that has been widely used in practice for describing empirical time series is the previously mentioned autoregressive moving-average (ARMA) model:

$$Y_t = \alpha + \emptyset_1 Y_{t-1} + \cdots + \emptyset_p Y_{t-p} + \varepsilon_t - \theta_1 \varepsilon_{t-1} - \cdots - \theta_q \varepsilon_{t-q} ,$$

where p = the highest lag associated with the data, and q = the highest lag associated with the error term.

Consider a time series $Y_t = \varepsilon_t$, $t = 0, \pm 1, \pm 2, \ldots$, where ε_t is independent of all other values $\varepsilon_{t-1}, \varepsilon_{t-2}, \ldots, \varepsilon_{t+1}, \varepsilon_{t+2}, \ldots$. This time series is *purely random*. If all ε_t are normally distributed, as well, the data are said to be *white noise*. The counts of radioactive decay from a long-lived sample and even the changes in the level of stock prices are examples of data that are stationary and can be regarded as white noise.

One simple *linear* random process is the moving average (MA) process. For example, the formula for a *first-order MA process* (MA(1)) is

$$Y_t = \alpha + \varepsilon_t - \theta_1 \varepsilon_{t-1} ,$$

where α and θ are parameters and σ_ε^2 is the variance of all ε_t .

An example of an MA(1) process with $\theta_1 = 0.25$ is given in Figure 17.5. In general, a qth-order MA process has the form

$$Y_t = \alpha + \varepsilon_t - \theta_1 \varepsilon_{t-1} - \theta_2 \varepsilon_{t-2} - \cdots - \theta_q \varepsilon_{t-q} ,$$

where the model is specified by the $q + 2$ parameters $\sigma_\varepsilon^2, \alpha, \theta_1, \ldots, \theta_q$. This model states that the values of Y_t consist of a moving average of the errors ε_t reaching back q periods. The coefficients (parameters) of the error terms are designated by θ's and the minus signs are introduced by convention. It is still assumed that the errors are independent, but the observed values of Y_t are dependent, being a weighted function of prior errors.

Another simple linear random process is the autoregressive (AR) process. The formula for a *first-order AR process* (AR(1), also known as a Markov process), is

$$Y_t = \alpha + \emptyset_1 Y_{t-1} + \varepsilon_t ,$$

where $\sigma_\varepsilon^2, \alpha$, and $\emptyset$, are parameters. This model states that the value Y_t of the process is given by $\emptyset_1$ times the previous value plus an error ε_t . Realizations of an AR(1) process with $\emptyset_1 = +0.9$ and -0.9 are shown in Figures 17.6 and 17.7, respectively.

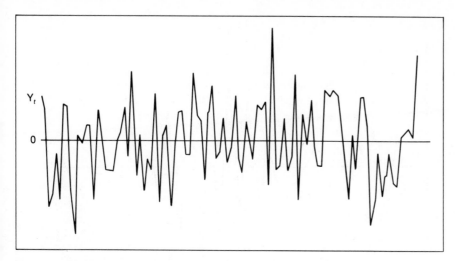

Figure 17.5   Realization of an MA(1) process ($\theta_1 = 0.25$).

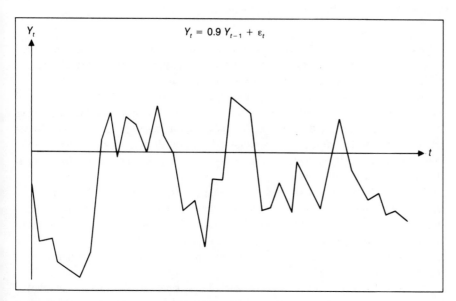

Figure 17.6   Realization of an AR(1) process ($\emptyset_1 = 0.9$).

In general, a pth-order AR process has the form

$$Y_t = \alpha + \emptyset_1 Y_{t-1} + \emptyset_2 Y_{t-2} + \cdots + \emptyset_p Y_{t-p} + \varepsilon_t .$$

The term autoregression that is used to describe such a process arises because an AR(p) model is much like a multiple linear-regression model. The essential

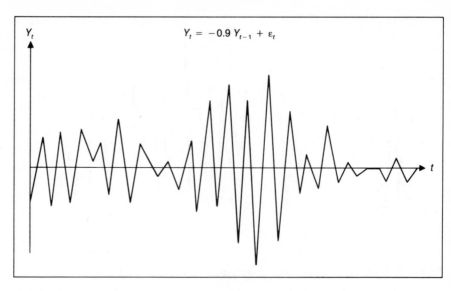

Figure 17.7   Realization of an AR(1) process ($\emptyset_1 = -0.9$).

difference is that Y_t is not regressed on independent variables but rather on lagged values of itself: hence the term *autoregression*. A realization of an AR(2) process, with $\emptyset_1 = +1.0$ and $\emptyset_2 = -0.5$, is shown in Figure 17.8.

It can be shown that any linear process can be written formally as a weighted sum of the current and all past *errors*. If that weighted sum has only a finite number of nonzero error terms, then the process is a moving average process. The linear process can also be expressed as a weighted sum of all past *observations* plus the current error term. If the number of nonzero terms in this expression is finite, then the process is autoregressive.

Thus, an MA process of finite order can be expressed as an AR process of infinite order, and an AR process of finite order can be expressed as an MA process of infinite order. This duality has led to the *principle of parsimony* in the Box-Jenkins methodology (discussed in the next chapter), in which it is advocated that the practitioner employ the smallest possible number of parameters for adequate representation of a model.

It may often be possible to describe a stationary time series with a model involving fewer parameters than either the MA or the AR process has by itself. Such a model will possess qualities of both autoregressive and moving average models: it is called an ARMA process. An ARMA(1,1) process has one prior observation term of lag 1 and one prior error term:

$$Y_t = \alpha + \emptyset_1 Y_{t-1} + \varepsilon_t - \theta_1 \varepsilon_{t-1} .$$

The general ARMA(p,q) process of autoregressive order p and moving-average order q has the form

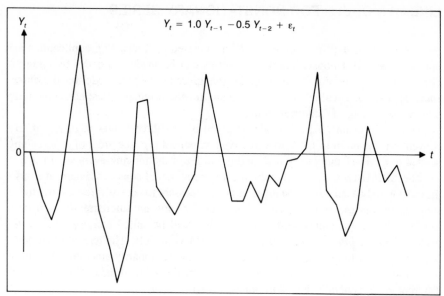

$$Y_t = 1.0\, Y_{t-1} - 0.5\, Y_{t-2} + \varepsilon_t$$

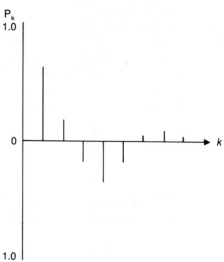

Figure 17.8 Realization and ACS process ($\emptyset_1 = 1.0$, $\emptyset_2 = -0.5$).

$$Y_t = \alpha + \emptyset_1 Y_{t-1} + \emptyset_2 Y_{t-2} + \cdots + \emptyset_p Y_{t-p} + \varepsilon_t \pm \theta_1 \varepsilon_{t-1} - \cdots - \theta_q \varepsilon_{t-q}.$$

In short, the ARMA process is a *linear random* process. It is linear if Y_t is a linear combination of lagged values of Y_t and ε_t. It is random if the errors (also called disturbances or shocks) are introduced into the system in the form of white noise. The random errors ε_t are assumed to be independent of each other and to be identically distributed with a mean of zero and a variance of σ_ε^2.

LINEAR MODELS FOR NONSTATIONARY SERIES

We noted earlier that most time series encountered in forecasting applications are not stationary. If a nonstationary time series can be made stationary by taking d differences (usually of order 0, 1, or 2), this gives an ARMA model for the *differenced* series; this is called an ARIMA model for the original series; hence the term *integrated* to suggest "undifferencing."

The *order* of an ARIMA model is given by the three letters p, d, and q. By convention, the order of the autoregressive component is p, the order of differencing needed to achieve stationarity is d, and the order of the moving-average part is q.

The ARIMA(p,d,q) model is the most general model considered here. It is also the most widely used model. Many times it is necessary to take differences to achieve stationarity, but the resulting series may require only an autoregressive (p) or a moving-average (q) component. These models will be called *autoregressive integrated* (ARI) or *integrated moving-average* (IMA) models. The term "integrated" is used when differencing is performed to achieve stationarity since the stationary series must be summed ("integrated") to recover the original data. The number of differencing is generally 1 or 2 in practice (Figure 17.9).

Many economic forecasting methods use exponentially weighted moving averages and they can be shown to be appropriate for a particular type of nonstationary process. Indeed, the stochastic model through which the exponentially weighted moving average produces an optimal forecast is a member of the class of ARIMA models: a range of ARIMA models exists to provide stationary and nonstationary treatments of the time series met in practice.

SUMMARY

The autoregressive integrated moving-average (ARIMA) models are capable of describing a wide variety of time series for forecasters. They form the framework for

- Expressing various forms of stationary and nonstationary behavior in time series.

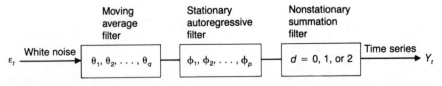

Figure 17.9 A block-diagram representation of an autoregressive integrated moving-average (ARIMA) model.

- A flexible modeling methodology.
- Producing optimal forecasts for a time series from its own current and past values.

USEFUL READING

ANDERSON, O. D. (1976). *Time Series Analysis and Forecasting—The Box-Jenkins Approach*. London, England: Butterworth.

BOX, G. E. P., and G. M. JENKINS (1976). *Time Series Analysis, Forecasting and Control, Revised Edition*. San Francisco, CA: Holden-Day.

GRANGER, C. W. J., and P. NEWBOLD (1977). *Forecasting Economic Time Series*. New York, NY: Academic Press.

MABERT, V. A. (1975). *An Introduction to Short-Term Forecasting Using the Box-Jenkins Methodology*. Production Planning and Control Monograph Series No. 2. Norcross, GA: American Institute of Industrial Engineers.

MAKRIDAKIS, S., and S. C. WHEELWRIGHT (1978). *Forecasting Methods and Applications*. New York, NY: John Wiley and Sons.

NELSON, C. R. (1973). *Applied Time-Series Analysis*. San Francisco, CA: Holden-Day.

SLUTSKY, E. (1937). The Summation of Random Causes as the Source of Cyclic Processes. *Econometrica* 5, 105–46.

WHITTLE, P. (1963). *Prediction and Regulation by Linear Least-Squares Methods*.London, England: English Universities Press.

WOLD, H. O. (1954). *A Study in the Analysis of Stationary Time Series,* 1st edition 1938. Uppsala, Sweden: Almquist and Wicksell.

YULE, G. U. (1927). On a Method of Investigating Periodicities in Disturbed Series, with Special Reference to Wolfer's Sunspot Numbers. *Philosophical Transactions* A 226, 267-98.

CHAPTER 18

The Box-Jenkins Model-Building Strategy

The previous chapter introduced a powerful class of time series models, known as ARIMA models, that have found extensive application in a number of areas.

The selection of ARIMA models or, indeed, of any sufficiently flexible, physically interpretable class of models for describing time series can be achieved through a three-stage iterative procedure. This procedure consists of

- Identification.
- Estimation.
- Diagnostic checking.

A THREE-STAGE PROCEDURE FOR SELECTION OF MODELS

A three-stage strategy for modeling a time series with ARIMA models was developed in the pioneering work of Box and Jenkins (1976).

- *Identification* consists of using the data and any other knowledge that will tentatively indicate if the time series can be described with a moving average model, an autoregressive model, or a mixed autoregressive moving-average model for a selection of the (p,d,q) values of the model.

- *Estimation* consists of using the data to make inferences about the parameters that will be needed for the tentatively identified model, and to estimate values of them.

- *Diagnostic checking* involves the examination of residuals from fitted models, which can result in either

- No indication of model inadequacy, or
- Model inadequacy, together with information on how the series may be better described.

Thus the residuals would be examined for any lack of randomness and, if the residuals are serially correlated, this information would be used to modify the model. The modified model would then be fitted and subjected to diagnostic checking again until an adequate model is obtained.

IDENTIFICATION TOOLS

In identification of an appropriate ARIMA model, the first thing to do is to transform and/or difference the data to produce stationarity, thereby reducing the model to one in the ARMA class. Then the ordinary and partial correlograms for the various patterns of differencing that are found in the adjusted data need to be displayed and compared to a basic catalog of theoretical patterns. These are taken up in Chapter 19.

The Ordinary Correlogram

The basic tools for identification are the autocorrelation function, designated ACF, and the partial autocorrelation function, designated PACF. The *ordinary* and *partial correlograms* are estimates of the ACF and PACF. For example, the ordinary correlogram is determined by correlating the previously adjusted or transformed series with versions of the series in which the units of time have been shifted. This yields a set of numbers, which can be plotted successively, corresponding to each time shift or *lag*. The ordinary or *auto*correlogram is simply a plot of the estimated autocorrelations.

Nonstationarity may be present if the values plotted in the correlogram do not diminish at large lags. When the original series or correlogram exhibits nonstationarity, successive differencing is carried out until the correlogram of the differenced series dies out reasonably rapidly. It is usually sufficient to look at the correlograms of the original series and of its first- and second-order differences.

The correlogram is a powerful tool for deciding whether the process shows pure autoregressive behavior or moving average behavior. The correlogram of a time series can be obtained by computing

$$r_j = c_j/c_0$$

for $j = 0, 1, 2, 3, \ldots, k$, where

$$c_j = \frac{1}{n} \sum_{t=1}^{n-j} [(y_t - \bar{y})(y_{t+j} - \bar{y})] \, ,$$

and

$$\bar{y} = \frac{1}{n} \sum_{t=1}^{n} y_t \, .$$

For large n, r_j is approximately normally distributed with a mean of $-1/n$ and variance of $1/n$. Thus approximate 95-percent confidence limits can be plotted at $\pm 1.96/\sqrt{n}$. Observed values of r_j that fall outside these limits are significantly different from zero at the 5-percent level. If the first 20 values of the correlogram are plotted, then you might expect one significant value even if the r_j's are random.

In practice, the number of r_j's calculated for a correlogram varies with the length of the series and the length of the seasonal cycle. Generally, for a 12-month seasonal series, about 36 to 60 monthly values are appropriate. No more than about $n/4$ to $n/3$ correlations should be calculated for a series of length n. Otherwise, there are not enough terms in the calculation for higher lags for any inferences to be meaningful. It is desirable that $n > 50$.

The Partial Correlogram

The *partial correlogram* complements the ordinary correlogram. Although more difficult to interpret, it measures the strength of the relationship between time periods in a series when dependence on intervening time periods has been removed. If the data in period t are highly related to period $t - k$, then the partial autocorrelogram would result in a large value (or "spike") at lag k.

If you assume that the model for the time series is purely autoregressive, an estimate of the order p can be obtained by successively fitting autoregressive models of orders one, two, three, four, and so on, to the series. The partial autocorrelation coefficient at lag k is an estimate of the *last parameter* ϕ_k in an autoregressive *model* of order k that is fitted to the series. If the order of the model is p, then all partial autocorrelations greater than p should be zero. By noting when the last significant lag in the partial autocorrelogram occurs, you can make an initial estimate of p.

The kth value R_k of a partial correlogram is given by the solution of the set of equations

$$r_j = \emptyset_1 r_{j-1} + \emptyset_2 r_{j-2} + \cdots + \emptyset_k r_{j-k} \, , \qquad \text{for } j = 1, 2, \ldots, k \, ,$$

and with $R_k = \emptyset_k$.

The variance of the kth partial correlogram coefficient, under the hypothesis that the process is of order $p < k$, is approximately $1/n$. If the PACF truncates (that is, if it has a value of zero after a certain number of lags), the process is AR, and the truncation point specifies the order p. A useful simple approximation is that the partial correlogram estimate R_k, $k > p$, is described approximately by a normal random variable with zero mean and variance $1/n$. Hence, if a series of R_k values beyond the pth lie within the 95-percent limits, $\pm 1.96/\sqrt{n}$, it may be reasonable to infer that the process is AR(p).

Some general guidelines for selecting appropriate ARMA models on the basis of ACF and PACF plots and through their interpretations, are covered next.

Interpreting Autocorrelation Functions

Extreme care must be taken in interpreting autocorrelation functions. The interpretation is complex and requires considerable experience. Attention should be directed to the *values* as well as the *patterns* of the autocorrelation coefficients. The ACF's of low-order ARMA models will be used later to help identify the model.

The theoretical autocorrelation function (ACF) of the pure MA(q) process truncates, being zero after lag q, while that for the pure AR(p) process is of infinite extent (Figure 18.1). MA processes are thus characterized by truncation of the ACF while AR processes are characterized by attenuation of the ACF.

In some situations it may not be clear whether the ACF truncates or attenuates. Hence it is useful to supplement the analysis by considering the partial autocorrelation function, PACF. For an AR process, the ACF attenuates and the PACF truncates; conversely, for an MA process, the PACF attenuates and the ACF truncates.

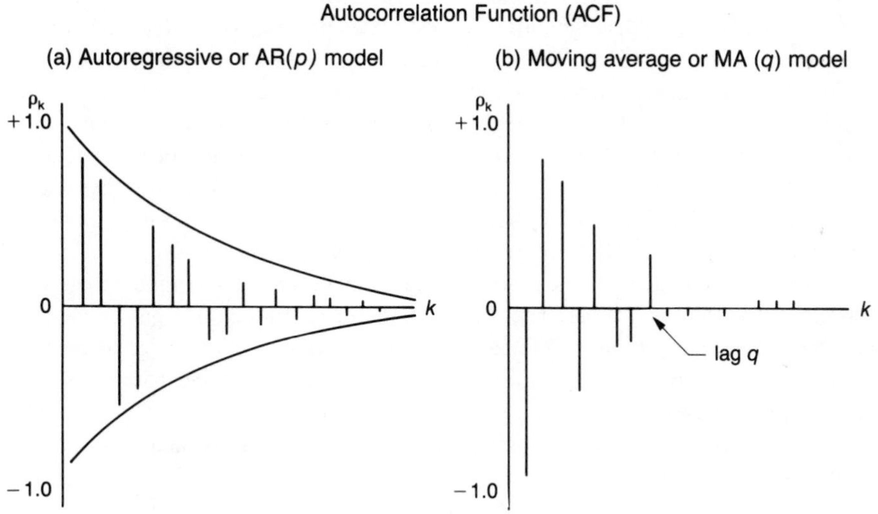

Autocorrelation Function (ACF)

(a) Autoregressive or AR(p) model   (b) Moving average or MA (q) model

Figure 18.1   The autocorrelation function (ACF) of (a) a pure AR(p) model and (b) a pure MA(q) model.

The ACF of an AR process behaves like the PACF of an MA process and vice versa. If both the ACF and the PACF attenuate, then a mixed model is called for. This subject will be treated in more detail in Chapter 19.

PARAMETER ESTIMATION

The second stage of the model-building strategy is the estimation or fitting stage. ARMA models can be fitted by least squares. An iterative nonlinear least-squares procedure is used to obtain the parameter estimates

$$(\hat{\alpha}, \hat{\emptyset}, \hat{\theta},) = (\hat{\alpha}, \hat{\emptyset}_1, \ldots, \hat{\emptyset}_p, \hat{\theta}_1, \ldots, \hat{\theta}_q)$$

of an ARMA(p,q) model. The estimates minimize the sum of squares of errors; that is,

$$S(\alpha, \emptyset, \theta) = \sum_{t=1}^{n} \varepsilon_t^2 ,$$

given the form of the model and the data. Since the procedure is, in general, non-linear (because of the moving-average terms), initial values for the parameters must be estimated so that minimization can be started.

There are a variety of parameter-estimation methods available in the literature (see Box and Jenkins 1976, Chapter 7) to which an interested reader can refer for details. For low-order ARMA models there are software programs that can be used to plot certain data in the neighborhood of estimated parameters, so that the behavior of a model can be visualized. This can provide insight into the sensitivity with which various starting values produce various results, and hence, gives estimates of the precision of the coefficients. For higher-order models few general guidelines exist, except to direct the model builder towards a preference for the use of *parsimonious* models ("the fewer parameters the better"). Parsimony is a practical rather than a theoretical consideration. Experience with many kinds of data series has convinced us that parsimonious models are generally better forecasting models.

Once the parameters are estimated it is also important to determine a matrix of estimated variances and covariances of the coefficients. Based on large-sample theory, the standard deviations of the parameter estimates (usually referred to as standard errors) are obtained from this and used to determine if the parameters are significantly different from zero.

Under the hypothesis that a single parameter $\beta = \beta_0$, approximate tests may be derived from

$$\frac{\hat{\beta} - \beta}{s(\hat{\beta})} \sim N(0,1),$$

where $s(\hat{\beta})$ denotes the estimated standard deviation of $\hat{\beta}$. Usually, the hypothesis to be tested is formulated as: "Is β significantly different from zero?" Also, a 95-percent confidence interval for β is given by

$$\hat{\beta} - 1.96 \, s(\hat{\beta}) < \beta < \hat{\beta} + 1.96 \, s(\hat{\beta}) \ .$$

The variance σ_ε^2 of the error term in the model also needs to be estimated. An estimate of σ_ε^2 is $S(\hat{\alpha}, \hat{\theta}, \hat{\phi},)/n$, where S is the sum-of-squares function minimized for the $p+q+1$ parameters. In practice there appears to be little importance to modifying the denominator n to $n-p-q$, unless n is very small.

It must be remembered that the basic assumption of independence is violated when applying classical regression theory to time series; however, some comfort can be taken in the fact that most results are valid provided there are "enough" observations. In formal terms, the results are asymptotically valid.

DIAGNOSTIC CHECKING

The last stage of the model-building strategy involves diagnostic checking. Analytical techniques are used to detect inadequacies in the model and to suggest model revisions. It may be necessary to introduce new information into the model after this stage and to initiate another three-stage procedure for assessing the new model.

To ensure that the best forecasting model has been obtained it is important to examine the fitted residuals. If the model is ultimately selected for use as a forecasting tool, the performance of the model should also be monitored periodically during its use so that it can be updated when appropriate.

Diagnostic checking involves examining statistical properties of the residuals. Tests should be designed to detect departures from the assumptions on which the model is based and to indicate how further fitting and checking should be performed. When the model is adequate the residual series should be independently and randomly distributed about zero. Any dependence or nonrandomness would indicate the presence of information that could be exploited to improve the accuracy of the model.

There are three ways in which a fitted model can be checked for adequacy of fit.

- The model can be made intentionally more complex by *overfitting;* those parameters whose significance is in question can be tested statistically.

- A more technical test involves the testing of correlogram estimates individually or in an overall *chi-squared test.*

- A third check involves a *cumulative periodogram* test.

These three diagnostic tests will now be treated in greater detail.

Overfitting

When a tentatively identified model is to be enhanced, an additional parameter can be added to the model (this is the method of overfitting) and the hypothesis that the additional parameter is zero can be tested by a t test. Thus an AR(1) model could be tested by overfitting with an AR(2) model or an ARMA(1,1) model, for instance.

The estimate $\hat{\sigma}_\varepsilon^2$ of the square of the residual standard error can also be used for diagnostic checking. A plot of $\hat{\sigma}_\varepsilon^2$ (adjusted for degrees of freedom) against the number of additional parameters should decrease with improved fitting. When overfitting, the improvement, if any, in decreased $\hat{\sigma}_\varepsilon^2$ may be inconsequential.

A Chi-squared Test

Correlograms and partial correlograms are probably the most useful diagnostic tools available. If the residuals are not random these diagrams may suggest residual auto-correlations that can be further modeled.

A chi-squared test can be used to evaluate whether the overall correlogram of the residuals exhibits any systematic error. When the chi-squared statistic exceeds the threshold level, the residual series contains more structure than would be expected for a random series. The test statistic due to Box and Pierce (1970) and modified by Ljung and Box (1978) is given by the formula

$$Q = n(n + 2) \sum_{k=1}^{m} (n - k)^{-1} r_k^2 ,$$

where r_k $(k = 1, \ldots, m)$ are residual autocorrelations, n is the number of observations used to fit the model, and m is usually taken to be 15 or 20. Then Q has an approximate chi-squared distribution with $(m - p - q)$ degrees of freedom. For an ARIMA(p,d,q) model, n is the number of terms in the differenced data. This is a general test of the hypothesis of model adequacy, in which a large observed value of Q points to inadequacy. Even if the statistic Q is not significant a review of the residual time series for unusual values is still appropriate.

It may also appear practical to examine individual values in the correlogram of the residuals relative to a set of confidence limits. Those values that fall outside the limits are examined further. Upon investigation, appropriate MA or AR terms can be included in the model.

Periodogram Analysis

In case of periodic nonrandom effects the cumulative periodogram can be an effective diagnostic tool. The test has its basis in "frequency domain" analysis of time series (Jenkins and Watts, 1968).

For the residual series $\{\varepsilon_t\}$ a *periodogram* $I(f_i)$ is defined by

$$I(f_i) = \frac{2}{n}\left[\left(\sum_{t=1}^{n} \hat{\varepsilon}_t \cos 2\pi f_i t\right)^2 + \left(\sum_{t=1}^{n} \hat{\varepsilon}_t \sin 2\pi f_i t\right)^2\right],$$

where $f_i = i/n$, $i = 0, 1, \ldots, n/2$ is the frequency. Large values of $I(f_i)$ are produced when a pattern with given frequency f_i in the residuals correlates highly with a sine or cosine wave at the same frequency. Then the *normalized cumulative periodogram* is defined by

$$C(f_j) = \frac{1}{ns^2}\sum_{i=1}^{j} I(f_i) ,$$

where s^2 is an estimate of σ_ε^2.

For a white-noise residual series, the plot of $C(f_j)$ against f_j, $j = 0, 1, \ldots, n/2$ would be scattered about a straight line joining the coordinates $(0,0)$ and $(0.5,1)$. Inadequacies in the fit to a model would show up as systematic deviations from this line. A standard statistical test, called the *Kolmogorov-Smirnov test*, uses the maximum vertical deviation of the plot from the straight line as the statistic for determining unsuspected periodicities in the data (Box and Jenkins 1976, Chapter 8).

SUMMARY

This chapter has covered the three-stage ARIMA modeling process; the three stages are

- Identification (or specification) of forecasting models by using data analysis tools (plotting of raw, differenced, and transformed data), correlograms, and partial correlograms, for the purpose of making *tentative guesses* at the order of the parameters in the ARMA model.

- Estimates of parameters for tentative models.

- Diagnostic checking, which is a critical step in looking for model *inadequacies* or for areas where *simplification* can take place.

The Box-Jenkins approach for ARIMA modeling provides the forecaster with a very powerful and flexible tool. It is an excellent method for forecasting a time series from its own current and past values, but it should not be applied blindly and automatically to all forecasting problems. Its complexity requires a fair amount of sophistication and judgment in its use. The results in terms of forecasting accuracy and understanding processes generating data can be significant in the hands of a skilled user.

USEFUL READING

ANDERSON, O. D. (1976). *Time Series Analysis and Forecasting—The Box-Jenkins Approach*. London, England: Butterworth.

BOX, G. E. P., and G. M. JENKINS (1976). *Time Series Analysis: Forecasting and Control, Revised Edition*. San Francisco, CA: Holden-Day.

BOX, G. E. P., and D. A. PIERCE (1970). Distribution of Residual Autocorrelations in Autoregressive Integrated Moving-Average Time-Series Models. *Journal of the American Statistical Association* 65, 1509–26.

JENKINS, G. M., and D. G. WATTS (1968). *Spectral Analysis and Its Applications*. San Francisco, CA: Holden-Day.

LJUNG, G. M., and G. E. P. BOX (1978). On a Measure of Fit in Time Series Models. *Biometrika* 65, 297–303.

NELSON, C. R. (1973). *Applied Time Series Analysis for Managerial Forecasting*. San Francisco, CA: Holden-Day.

CHAPTER **19**

Identifying Regular ARIMA Models

This chapter describes the identification of *regular* ARIMA models. These models are appropriate for

- Nonseasonal time series.
- Seasonally adjusted series.
- Series with strong trend characteristics.
- Series that exhibit a random walk (like changes in stock prices).

Seasonal models are discussed in Chapter 21. In many cases time series are modeled best with a combination of regular and seasonal parameters.

EXPRESSING ARIMA MODELS IN COMPACT FORM

The ARIMA model, in its fullest generality, is cumbersome to write down. It relates a dependent variable to lagged terms of itself and to lagged error terms. Fortunately, there is a convenient notational device for expressing an operation in which a variable is lagged or shifted; it is known as a *backshift operator*. This notation makes the expression and manipulation of a model much simpler and more like an algebraic operation.

The Backshift Operator

The *backshift operator B* is a convenient notational device for expressing ARIMA models in a compact form. It is defined to be B "operating" on the index of Y_t so that BY_t produces Y_{t-1}, which is the value of Y_t shifted back in time by one unit (say

243

one month). Hence $B^2Y_t = B(BY_t) = BY_{t-1} = Y_{t-2}$. The B^2 operation shifts the subscript of Y_t by two time units. Similarly, $B^kY_t = Y_{t-k}$.

In the B notation a first difference is simply

$$Y_t - Y_{t-1} = Y_t - BY_t = (1 - B)Y_t .$$

This looks like a polynomial in B "operating" on Y_t. The AR(1) model becomes

$$(1 - \emptyset_1 B)Y_t = \alpha + \varepsilon_t ,$$

where ε_t is white noise.

Notice that the first difference corresponds to the special AR(1) model in which $\emptyset_1 = 1$. The real advantage of using the B notation will become more evident when you want to write down expressions for multiplicative ARIMA models (Chapter 21). It is also convenient to suppress the constant term α in the model. This can be done by redefining Y_t to include α, so that Y_t henceforth is $(Y_t - \alpha)$. In practice, this is accomplished by modeling $Y_t - \overline{Y}$, where $\overline{Y}$ is the mean of the time series.

The AR(1) model can now be written as $(1 - \emptyset_1 B)Y_t = \varepsilon_t$. Dividing by $(1 - \emptyset_1 B)$ gives Y_t in terms of ε_t :

$$Y_t = \varepsilon_t/(1 - \emptyset_1 B).$$

If we expand the operator, so that

$$1/(1 - \emptyset_1 B) = 1 + \emptyset_1 B + \emptyset_1^2 B^2 + \cdots ,$$

it becomes clear that

$$\begin{aligned} Y_t &= (1 + \emptyset_1 B + \emptyset_1 B^2 + \cdots)\varepsilon_t \\ &= \varepsilon_t + \emptyset_1\varepsilon_{t-1} + \emptyset_1^2\varepsilon_{t-2} + \cdots . \end{aligned}$$

Hence the AR(1) model is equivalent to an MA model of infinite order. Similarly, the MA(1) model $Y_t = (1 - \theta_1 B)\varepsilon_t$ can be regarded as an autoregressive model of infinite order.

The higher-order AR, MA, and ARMA models can be written as special cases of

$$(1 - \emptyset_1 B - \emptyset_2 B^2 - \cdots - \emptyset_p B^p)Y_t = (1 - \theta_1 B - \theta_2 B^2 - \cdots \theta_q B^q)\varepsilon_t .$$

The series Y_t is assumed to be mean-adjusted, so that the α term is suppressed in the above representation. If we write

$$Y_t = \frac{(1 - \theta_1 B - \theta_2 B^2 - \cdots - \theta_q B^q)}{(1 - \emptyset_1 B - \emptyset_2 B^2 - \cdots - \emptyset_p B^p)} \, \varepsilon_t$$

$$= H(B)\varepsilon_t \, ,$$

it is evident that Y_t is represented as the output from a linear filter whose input is a random series ε_t with zero mean and constant variance, and whose filter *transfer function H(B)* is a ratio of two polynomials in the backshift operator B. The purpose of modeling linear models of the ARMA class is to identify and estimate $H(B)$ with as few parameters as possible *(parsimonious representation)*. Once this is done, the representation can be used for forecasting.

Regular ARIMA Models

Let $W_t = (1 - B)Y_t$, so that W_t represents the first difference of Y_t. An ARMA model for W_t is an ARIMA model for Y_t.

Assume that a series can be reduced to stationarity by differencing the series some finite number of times (possibly after removing any deterministic trend). The order of differencing is denoted by d. Then it is assumed that

$$W_t = (1 - B)^d Y_t$$

is stationary.

For a *regular* ARIMA(p,d,q) model, the general form is assumed to be

$$(1 - \emptyset_1 B - \emptyset_2 B^2 - \cdots - \emptyset_p B^p)(1 - B)^d Y_t = (1 - \theta_1 B - \theta_2 B^2 - \cdots - \theta_q B^q)\varepsilon_t \, ,$$

where the ε_t's are white noise—a sequence of identically distributed uncorrelated errors.

It is required that the roots of the two polynomial equations in B—namely,

$$\emptyset(B) = 0 \qquad \text{and} \qquad \theta(B) = 0$$

—all lie outside the unit circle. The first condition ensures the *stationarity* of W_t—that is, the statistical equilibrium about a fixed mean; the second one, known as the *invertibility* requirement, guarantees uniqueness of representation (the weights applied to the past history of W_t to generate forecasts die out).

This notation is frequently simplified to

$$\emptyset_p(B)(1 - B)^d Y_t = \theta_q(B)\varepsilon_t \, ,$$

where the AR(p) terms are given by the polynomial

$$\emptyset(B) = (1 - \emptyset_1 B - \emptyset_2 B_2 - \cdots - \emptyset_p B^p)$$

and the MA(q) terms are

$$\theta_q(B) = (1 - \theta_1 B - \theta_2 B^2 - \cdots - \theta_q B^q) \ .$$

Thus an ARIMA(1,1,1) model takes the form

$$(1 - \emptyset_1 B)(1 - B)Y_t = (1 - \theta_1 B)\varepsilon_t \ .$$

AUTOCORRELATIONS AND PARTIAL AUTOCORRELATIONS

As we indicated in the preceding chapter, the autocorrelation function (ACF) and the partial autocorrelation function (PACF) are the two principal tools used to characterize the structure of a theoretical ARMA model. The ACF and PACF have patterns that are useful for identifying the order and lag structure of an ARMA model.

The Moving Average Process

The moving average model MA(q), of order q, is given by

$$W_t = (1 - \theta_1 B - \theta_2 B^2 - \cdots - \theta_q B^q)\varepsilon_t \ ,$$

where the current value W_t of the time series is assumed to be a linear combination of the current and previous error terms ε_t. In practice, these models are useful for describing events that are affected by random events such as strikes and policy decisions. Many economic and planning series exhibit behavior that can be reasonably described by a model containing MA components.

The theoretical autocorrelation function (ACF) of the pure moving average model, MA(q), *truncates* at lag q. After lag q the values of the ACF are zero. The PACF does not cut off, but *decays* to zero.

The ACF and PACF of an MA(1) model with positive θ are depicted in Figure 19.1(a). There is a single negative spike at the lag 1 in the ACF. There is a decaying pattern in the PACF. The ACF of an MA(1) process with negative θ (Figure 19.1(b))

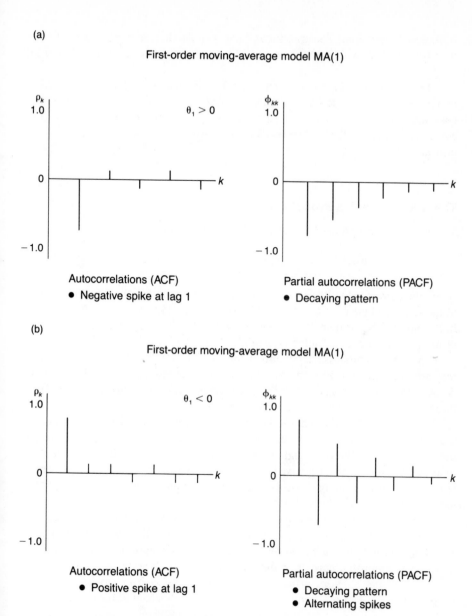

Figure 19.1   The ACF and PACF of an MA(1) process.

shows a single positive spike but the PACF shows a decaying pattern with spikes alternating above and below the zero line.

No restrictions on the size of θ are required for an MA process to be stationary. However, to avoid certain modeling problems an *invertibility* requirement needs to be imposed. This arises because, for an MA(1) model that has a coefficient of either θ or $1/\theta$, the ACF will be the same. An MA(1) model is said to be invertible if $|\theta_1|$ < 1.

The multiplicity of a model is not the only thing that determines its invertibility or noninvertibility. Models are called *noninvertible* because it is impossible to estimate the errors ε_t. Since starting values ε_0, ε_{-1}, . . . are unknown, there will be errors in estimating the early ε_t's. With noninvertible models these errors grow instead of decay.

The Autoregressive Process

The autoregressive process AR(p) of order p is given by

$$(1 - \emptyset_1 B - \emptyset_2 B^2 \cdots - \emptyset_p B^p)W_t = \varepsilon_t ,$$

where the current value W_t is assumed to be a linear combination of previous values of the series and the current error ε_t. In contrast to an MA model, the AR model is like a multiple linear regression model in which the W_t is regressed on past values of itself; hence the term "autoregression."

For an AR(p) process the patterns in the ACF and PACF are reversed from what they are in the moving average process. The values in the ACF diminish to a tail or decay with increasing lags, and in the PACF values (possibly zero) occur from lag 1 through p and then they are zero *(truncate)* thereafter.

The ACF and PACF of an AR(1) process are depicted in Figure 19.2. There is a decaying pattern in the ACF; the decay is exponential if $\emptyset_1 < 1$ (part (b) of the figure). The PACF shows a single positive value at lag 1 if $0 < \emptyset_1 < 1$. The ACF of an AR(1) process with negative $\emptyset_1$ ($-1 < \emptyset_1 < 0$) shows a decaying exponential tail with values alternating above and below the zero line. The corresponding PACF has a single negative value at lag 1.

A somewhat more complicated process that occurs fairly often in practice is the AR(2) process. In this case there are two autoregressive coefficients $\emptyset_1$ and $\emptyset_2$. Figure 19.3 shows the ACF and PACF of an AR(2) model with $\emptyset_1 = 0.3$ and $\emptyset_2 = 0.5$. The values in the ACF diminish according to the formula

$$\rho_k = \emptyset_1 \rho_{k-1} + \emptyset_2 \rho_{k-2} .$$

The PACF shows positive values at lags 1 and 2 only. The PACF is very helpful because it suggests that the process is autoregressive and, more significantly, that it is second-order autoregressive.

If $\emptyset_1 = 1.2$ and $\emptyset_2 = -0.64$, the ACF and PACF have the following patterns, shown in Figure 19.4: the values in the ACF decay in a sinusoidal pattern; the PACF has a positive value at lag 1 and a negative value at lag 2.

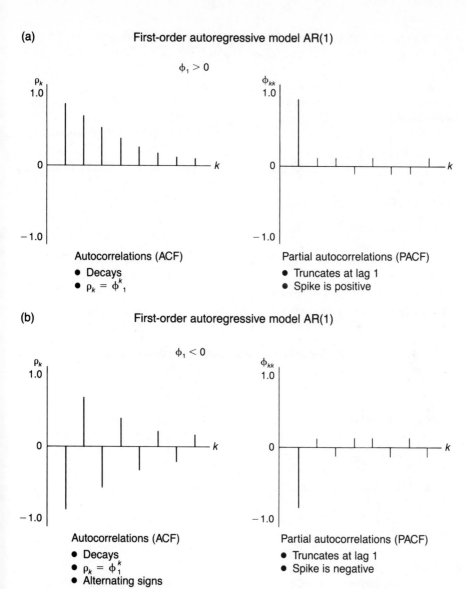

Figure 19.2 The ACF and PACF of an AR(1) process.

There are a number of possible patterns for AR(2) models. The allowable values for $\emptyset_1$ and $\emptyset_2$ in the stationary case are described by the triangular region

$$\emptyset_1 + \emptyset_2 < 1, \qquad \emptyset_2 - \emptyset_1 < 1, \qquad \text{and} \qquad -1 < \emptyset_2 < 1.$$

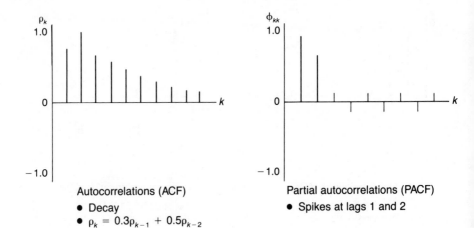

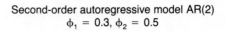

Figure 19.3   The ACF and PACF of an AR(2) process with parameters $\emptyset_1 = 0.3$ and $\emptyset_2 = 0.5$.

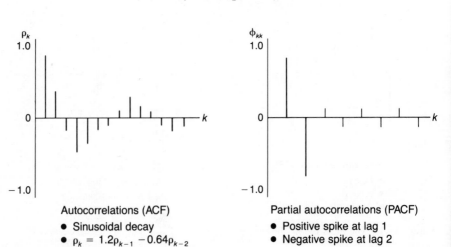

Figure 19.4   The ACF and PACF of an AR(2) process with parameters $\emptyset_1 = 1.2$ and $\emptyset_2 = -0.64$.

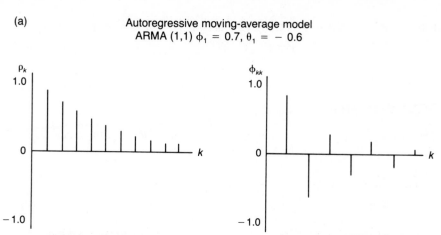

(a)

Autoregressive moving-average model
ARMA (1,1) $\phi_1 = 0.7$, $\theta_1 = -0.6$

Autocorrelations
- Irregular 1st spike
- Decaying pattern after lag 1

$$\rho_k = \phi_1^k$$

Partial autocorrelations
- Decaying pattern
- Alternating spikes

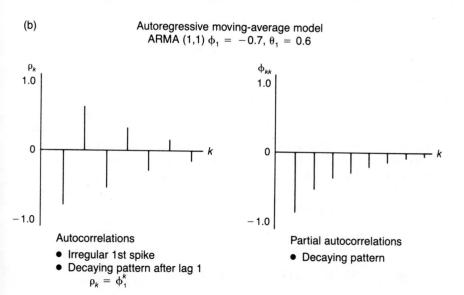

(b)

Autoregressive moving-average model
ARMA (1,1) $\phi_1 = -0.7$, $\theta_1 = 0.6$

Autocorrelations
- Irregular 1st spike
- Decaying pattern after lag 1

$$\rho_k = \phi_1^k$$

Partial autocorrelations
- Decaying pattern

Figure 19.5   The ACF and PACF of an ARMA(1,1) process with parameters $\emptyset_1 = 0.7$ and $\theta_1 = -0.6$.

If $\emptyset_1^2 + 4\emptyset_2 > 0$, the ACF decreases exponentially with increasing lag. If $\emptyset_1^2 + 4\emptyset_2 < 0$, the ACF is a damped cosine wave.

ARMA Processes

The mixed autoregressive moving-average process, ARMA(p,q), contains p AR terms and q MA terms. It is given by

$$(1 - \emptyset_1 B - \emptyset_2 B^2 - \cdots - \emptyset_p B^p)W_t = (1 - \theta_1 B - \theta_2 B^2 - \cdots - \theta_q B^q)\varepsilon_t .$$

This model is useful in that stationary series may often be expressed more parsimoniously (with fewer parameters) in an ARMA model than in the pure AR or MA models.

The mixed ARMA model is a more difficult process to identify. Both p and q must be determined from the sample ACF's and PACF's. The ACF of an ARMA(p,q) process has an *irregular pattern* at lags 1 through q, then the tail diminishes according to the formula

$$\rho_k = \emptyset_1 \rho_{k-1} + \cdots + \emptyset_p \rho_{k-p} , \qquad k > q.$$

The PACF tail also diminishes. So the best way to identify an ARMA process initially is to look for a *decay* or *tail in both the ACF's and PACF's*.

The ACF and PACF of an ARMA(1,1) process with $\emptyset_1 = 0.7$ and $\theta_1 = -0.6$ is shown in Figure 19.5(a). The value at lag 1 in the ACF is high. The remaining values show an exponential decay. The PACF also shows a decay or tail.

The ACF and PACF of the same ARMA(1,1) process, but with the signs of $\emptyset_1$ and θ_1 reversed, would show alternating decaying values in the ACF (Figure 19.5(b)). In the PACF there would be a large negative value at lag 1 followed by an exponential decay in the remaining values.

When Are Correlogram Estimates Significant?

Just as there are for the ordinary correlation coefficient, there are approximate confidence limits for the correlogram that establish which correlogram estimates can reasonably be assumed to be zero. As a rough guide for determining whether or not theoretical autocorrelations are zero beyond lag q, Bartlett (1946) has shown that for a sample of size n the standard deviation of r_k is approximately

$$n^{-1/2}[1 + 2(r_1^2 + r_2^2 + \cdots + r_q^2)]^{1/2} \qquad \text{for } k > q.$$

Quenouille (1949) has shown that, for a pth-order autoregressive model, the standard errors of the partial autocorrelogram estimates $\hat{\phi}_{kk}$ are approximately $n^{-1/2}$ for $k > p$. Assuming normality for moderately large samples, as shown by Anderson (1942), the limits of plus or minus two standard deviations about zero should provide a reasonable guide in assessing whether or not the correlogram estimates are significantly different from zero.

EXAMPLES OF MODEL IDENTIFICATION

So far the theoretical forms of an ACF and PACF have been treated only for ARMA models. To learn how various time series models can be identified by correlograms, refer to Anderson (1976), Box and Jenkins (1976), Granger (1980), Granger and Newbold (1977), Mabert (1975), Makridakis and Wheelwright (1978), Nelson (1973), and Pindyck and Rubinfeld (1976).

Consider now some examples of time series and their associated correlograms. It is helpful to make some observations regarding the displays:

- A time plot is useful to establish what data adjustments and transformations may be needed. The appropriate amount of differencing should be performed to achieve stationarity, and the differenced data should be plotted.

- The ordinary correlogram should be inspected for pure AR or MA structure.

- It is helpful to inspect the partial correlogram to confirm or supplement the information derived from the ordinary correlogram.

- The inspection of the ordinary and partial correlograms together should suggest a preliminary model to be fitted in the first iteration of the model-building process. The most significant patterns in the correlograms should be documented for future reference.

At the early stages of modeling it is generally simpler to attempt to identify only the most obvious patterns. Once the parameter estimates are obtained for a tentative model, the correlogram of the residuals of the model should be inspected for any significant remaining patterns. With experience, it is possible to identify complex patterns relatively quickly. For the beginner, however, a careful consideration of a number of examples is recommended. Table 19.1 provides general guidance for selecting starting models.

The Index of Consumer Sentiment

The first example we will consider is the University of Michigan Survey Research Institute's quarterly index of consumer sentiment. The index of consumer sentiment

Table 19.1   Model identification for nonseasonal time series.

Partial autocorrelation function	Autocorrelation function	
	Decays	Truncates
Truncates	AR	Mixed (ARMA)
Decays	Mixed (ARMA)	MA

is published in the *Business Conditions Digest,* a monthly publication of the U.S. Department of Commerce. The data come from a survey that attempts to ascertain the anticipations and intentions of consumers. While they reflect only the respondent's anticipations (what they expect others to do) or expectations (what they plan to do), and not firm commitments, such information is nevertheless useful as a valuable aid to economic forecasting.

The quarterly index is shown in Figure 19.6; it is not stationary in level, so that a first difference must be taken. Figure 19.7 shows the differenced series, and Figure 19.8 shows the corresponding correlogram and partial correlogram. Both show significant spikes at lag 10, which does not suggest a regular ARIMA model. Possibly the spike is spurious, or a seasonal model is required. A starting model is

$$(1 - B)Y_t = \alpha + \varepsilon_t .$$

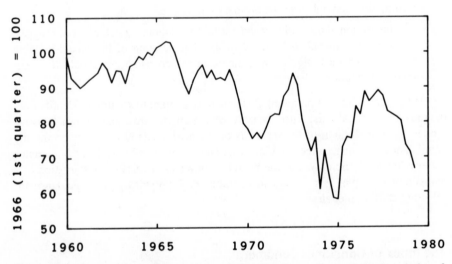

Figure 19.6   A time plot of the University of Michigan Survey Research Institute index of consumer sentiment (quarterly).
Source: University of Michigan, Survey Research Institute.

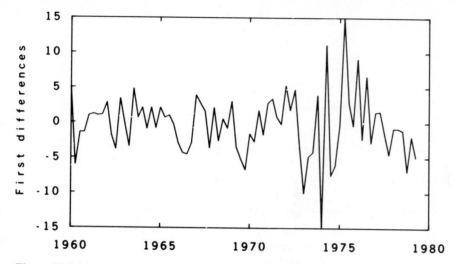

Figure 19.7   A time plot of first differences of the consumer-sentiment index series.

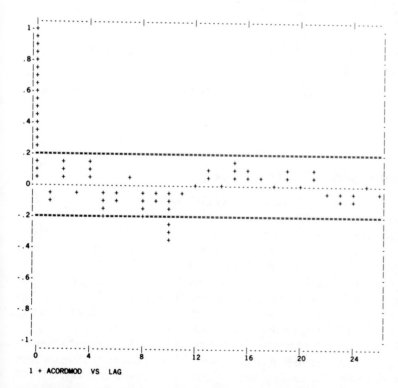

1 + ACORDMOD VS LAG

Figure 19.8(a)   The partial correlogram of first differences of the consumer sentiment index series.

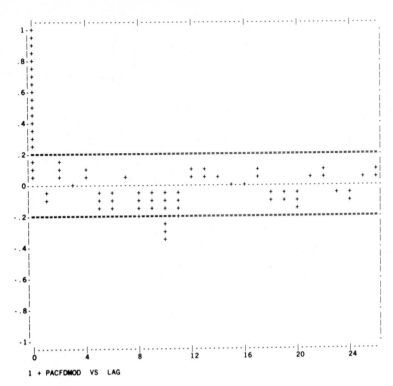

1 + PACFDMOD   VS   LAG

Figure 19.8(b)   The partial correlogram of first differences of the consumer sentiment index series.

Seasonally Adjusted U.S. Money Supply

The seasonally adjusted U.S. money supply (M1) series, shown in Figure 17.1, is also not stationary, and it also requires a first difference. After differencing the series once, the plot still shows a slight trend and increasing variability (Figure 17.2(a)).

Since the original money supply series probably requires a variance-stabilizing transformation, logarithms may indeed give better results. The first differences of the log-transformed data (Figure 17.3) appear stationary. The correlograms (Figure 19.9) suggest the ARIMA(2,1,0) model because of the decay in the ordinary correlogram and the spike at lags 1 and 2 in the partial correlogram. Thus the starting model is an ARIMA(2,1,0) for the log-transformed money supply data.

Another possibility that could be tried on these data is to take a second difference of the original data. The time plot in Figure 17.2(b) depicts a time series with increasing variability. The correlograms in Figure 19.10 suggest an ARIMA(1,2,0) model. This is based on the pattern of alternating decay in the ordinary correlogram and the significant negative spike at lag 1 in the partial correlogram. Figure 19.11

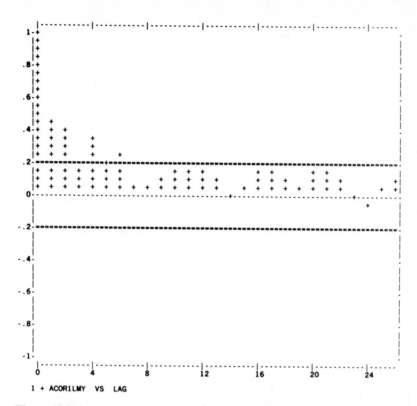

1 + ACOR1LMY VS LAG

Figure 19.9(a)   The ordinary correlogram of first differences of the log-transformed money supply series.

shows the ordinary correlogram of the residuals of this tentative model. The negative spike at lag 2 suggests the addition of a second order moving average parameter to the model. Thus this model could also be contemplated for the money supply data.

Although a forecast test would provide one of the most important criteria for model selection, we will forego making such a test: the first differences of the log-transformed series are closest to stationarity and it is likely that this order of differencing will provide the best results.

The FRB Index of Industrial Production

The FRB Index of Industrial Production is not stationary, since level and slope both change over time. Both the ordinary and partial correlograms of the raw quarterly data would suggest taking first differences.

The first differences of the FRB data do appear stationary. The ordinary correlogram in Figure 19.12(a) shows a significant spike at lag 1. The partial correlogram in Figure 19.12(b) shows alternating decay; these characteristics suggest an

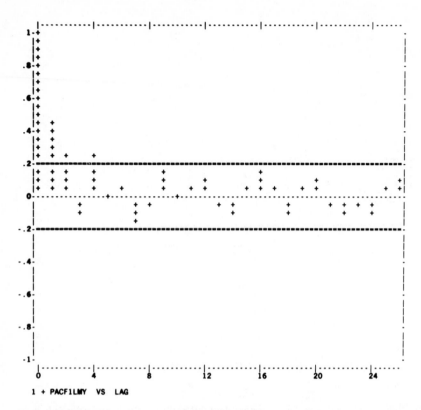

Figure 19.9(b)   The partial correlogram of first differences of the log-transformed money supply series.

MA(1) model of the first differences of the quarterly series. We will return to this model in Chapter 22 as part of the case study for forecasting main telephone gain (see also Levenbach, 1980).

Granger (1980, p. 74) develops a model for the monthly FRB index (seasonally adjusted). His estimated model for the longer time period January, 1948 to October, 1974 is derived from an AR(1) model of the first differences:

$$(1 - B)Y_t = 0.4(1 - B)Y_{t-1} + \varepsilon_t .$$

SUMMARY

The identification of ARIMA models for nonseasonal data is accomplished primarily by analyzing correlograms and partial correlograms. To summarize the rules for identification:

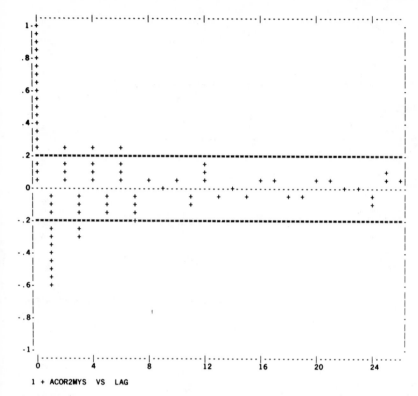

Figure 19.10(a)   The ordinary correlogram of the second differences of the money supply series.

- If the correlogram "cuts off" at some point, say $k = q$, then the appropriate model is MA(q).
- If the partial correlogram cuts off at some point, say $k = p$, then the appropriate model is AR(p).
- If neither diagram cuts off at some point, but does decay gradually to zero, the appropriate model is ARMA(p',q') for some p', q'.

USEFUL READING

ANDERSON, O. D. (1976). *Time Series Analysis and Forecasting: The Box-Jenkins Approach*. London, England: Butterworth.

ANDERSON, R. L. (1942). Distribution of the Serial Correlation Coefficient. *Annals of Mathematical Statistics* 13, 1–13.

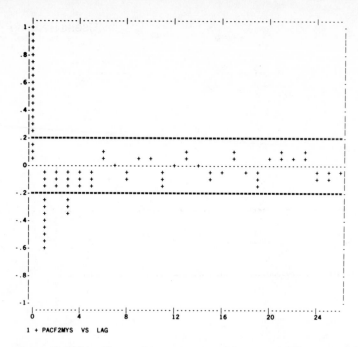

1 + PACF2MYS VS LAG

Figure 19.10(b) The partial correlogram of the second differences of the money supply series.

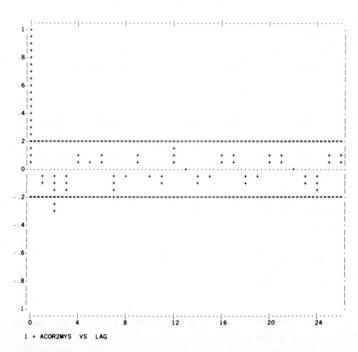

1 + ACOR2MYS VS LAG

Figure 19.11(a) The ordinary correlogram of the residuals of a tentative (1,2,0) model for the seasonally adjusted U.S. money supply series.

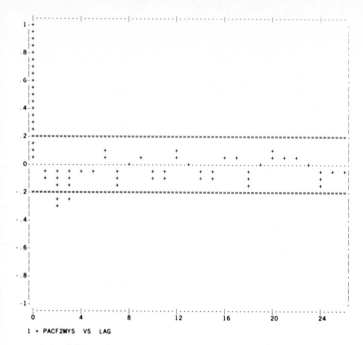

Figure 19.11(b)

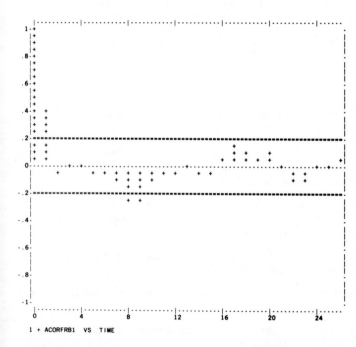

Figure 19.12(a) The ordinary correlogram of first differences of the FRB Index of Industrial Production.

262

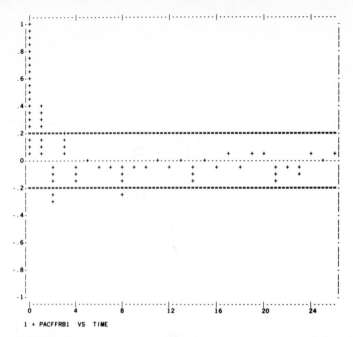

1 + PACFFRB1 VS TIME

Figure 19.12(b) Partial correlogram of first differences of the FRB Index of Industrial Production.

BARTLETT, M. S. (1946). On the Theoretical Specification of Sampling Properties of Autocorrelated Time Series. *Journal of the Royal Statistical Society* B 8, 27–41.

BOX, G. E. P., and G. M. JENKINS (1976). *Time Series Analysis, Forecasting and Control, Revised Edition.* San Francisco, CA: Holden-Day.

GRANGER, C. W. J. (1980). *Forecasting in Business and Economics.* New York, NY: Academic Press.

GRANGER, C. W. J., and P. NEWBOLD (1976). *Forecasting Economic Time Series.* New York, NY: Academic Press.

LEVENBACH, H. (1980). A Comparative Study of Time Series Models for Forecasting Telephone Demand. *In* O. D. Anderson, ed. *Forecasting Public Utilities.* Amsterdam, Netherlands: North-Holland Publishing Co.

MABERT, V. A. (1975). *An Introduction to Short-Term Forecasting Using the Box-Jenkins Methodology.* Production Planning and Control Division Monograph No. 2. Norcross, GA: American Institute of Industrial Engineers.

MAKRIDAKIS, S., and S. C. WHEELWRIGHT (1978). *Forecasting Methods and Applications.* New York, NY: John Wiley and Sons.

NELSON, C. R. (1973). *Applied Time Series Analysis for Managerial Forecasting.* San Francisco, CA: Holden-Day.

PINDYCK, R. S., and D. L. RUBINFELD (1976). *Econometric Models and Economic Forecasts.* New York, NY: McGraw-Hill.

QUENOUILLE, M. H. (1949). Approximate Tests of Correlation in Time Series. *Journal of the Royal Statistical Society* B 11, 68–84.

CHAPTER **20**

Forecast Profiles and Confidence Limits for ARIMA Models

Once a tentative ARIMA model is identified and estimated, forecasts can be computed. This chapter is concerned with

- Generating minimum mean square error (MSE) forecasts from the difference equation of an ARIMA model.
- Developing expressions for prediction variances and confidence limits for forecasts.
- Utilizing forecast errors as a way of monitoring forecasts.

FORECASTS AND FORECAST PROFILES

It is assumed that the number of observations used to fit any model is sufficiently large that errors in estimating the parameters will not seriously affect the forecasts: it is assumed the model is known exactly and that it will remain essentially unchanged for the period for which the forecast is being made. The variance of the forecast error is used to construct confidence limits for forecasts. Confidence limits are used in monitoring forecast in the following ways:

- If a high proportion of the forecast errors begin to fall outside the appropriate limits, the values of the model coefficients may have changed or at least they should be reestimated. However, the new model might produce significantly different forecasts.
- If systematic patterns appear, the structure of the process may also have changed, and additional terms (or transformations) may be required in the model.
- A time plot of the *cumulative sum* of the forecast errors may indicate changes in the structure of the process as a result of changes in external factors.

ℓ-Step-Ahead Forecasts

By expressing an ARIMA model in terms of the backshift operator B and expanding it as a difference equation, a forecast profile can be obtained (Anderson, 1976; Box and Jenkins, 1976; and Nelson, 1973). For example, the AR(1) model

$$(1 - 0.5B)Y_t = 2.0 + \varepsilon_t ,$$

can be expanded and rewritten as

$$Y_t = 0.5Y_{t-1} + 2.0 + \varepsilon_t .$$

Some observed data y_t and fitted errors (residuals) $\hat{\varepsilon}_t$ from the model are given in Table 20.1.

To produce forecasts for future time periods, one first generates a one-step-ahead forecast to obtain an estimate of Y_{t+1}: the subscript t is replaced with $t + 1$ in the equation for Y_t. Since the errors are assumed to have zero mean and constant variance, the best estimate of future errors is zero. Thus, the formula for the one-step-ahead forecast will be

$$\hat{Y}_t(1) = 0.5Y_t + 2$$
$$= 5.0 .$$

The two- and three-step-ahead forecasts are derived through the following steps:

1. Replace t by $t + 2$ in $Y_t = 0.5Y_{t-1} + 2.0 + \varepsilon_t$: then

$$Y_{t+2} = 0.5Y_{t+1} + 2.0 + \varepsilon_{t+2}.$$

2. Use the estimated value of $\hat{Y}_t(1)$ in place of Y_{t+1}.

3. Assume $\hat{\varepsilon}_{t+2} = 0$. Then the two-step-ahead and three-step-ahead forecasts are, respectively,

Table 20.1   Sample data for forecasting.

Time origin	Data y_t	Residuals $\hat{\varepsilon}_t$
$t - 4$	10.0	−3.0
$t - 3$	4.0	2.0
$t - 2$	5.0	1.0
$t - 1$	8.0	0
t	6.0	2.0

$$\hat{Y}_t(2) = 0.5\hat{Y}_t(1) + 2.0 + \hat{\varepsilon}_{t+2}$$
$$= 4.5,$$

and

$$\hat{Y}_t(3) = 0.5\hat{Y}_t(2) + 2.0 + \hat{\varepsilon}_{t+3}$$
$$= 4.3.$$

Notice that the forecast of Y_{t+l} with $\ell \geq 1$ is made from a *time origin*, designated t, for a *lead time*, designated ℓ. This forecast, denoted by $\hat{Y}_t(\ell)$, is said to be a *forecast at origin t, for lead time ℓ*. It is the *minimum mean square error (MSE) forecast* (Box and Jenkins, 1976). When $\hat{Y}_t(\ell)$ is regarded as a function of ℓ for a fixed t, it is referred to as the *forecast function* or *profile* for an origin t.

The forecast error for the lead time ℓ,

$$e_t(\ell) = Y_{t+\ell} - \hat{Y}_t(\ell),$$

has zero mean; thus the forecast is *unbiased*.

It is also important to note that any *linear* function of the forecasts $\hat{Y}_t(l)$ is a minimum MSE forecast of the corresponding linear combination of future values of the series. Hence, for monthly data, the best year-to-date forecast can be obtained by summing the corresponding 1, 2, . . . , *l*-step-ahead forecasts.

A *forecast function* or *profile* can be generated by plotting the *l*-step-ahead forecasts for a fixed time origin, $\ell = 1, 2, 3, \ldots$. It can be seen from Figure 20.1 that for this example there is a geometrical decay to a mean value; the mean equals the constant term divided by $(1 - \emptyset_1)$. In this case, the mean is 4.0. The equation for the *l*-step-ahead forecast $\hat{Y}_t(l)$ is then given as

$$\hat{Y}_t(l) = 4.0 + 0.5^l(Y_t - 4.0).$$

In the same manner as for the AR(1) model, forecasts for the MA(1) model can be developed. Suppose the MA(1) model is given by

$$Y_t = 3.0 + \varepsilon_t - 0.5\varepsilon_{t-1}.$$

Once again t is replaced by $t + 1$ and all future errors are set to zero. Then $y_{t+1} = 4.0$ and all y_{t+i}'s $(i = 2, 3, \ldots)$ are equal to 3.0.

The forecast profile for an MA(1) model is shown in Figure 20.2. It consists of one estimate based on last period's error, followed by the constant mean $(= 3.0)$ for all future periods. This makes sense intuitively because the ACF of the MA(1) model has nonzero correlation only at lag 1, hence a "memory" of only one period.

The forecast profile for an MA(q) can be generalized rather easily. It consists of values at $l = 1, 2, \ldots, q$ that are determined by the past errors, and then equals the mean of the process for periods greater than q.

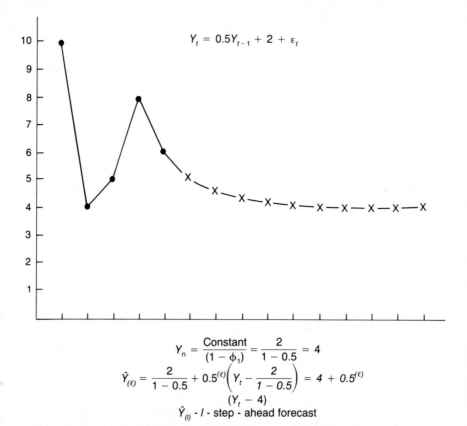

$$Y_n = \frac{\text{Constant}}{(1 - \phi_1)} = \frac{2}{1 - 0.5} = 4$$

$$\hat{Y}_{(\ell)} = \frac{2}{1 - 0.5} + 0.5^{(\ell)}\left(Y_t - \frac{2}{1 - 0.5}\right) = 4 + 0.5^{(\ell)}$$

$$(Y_t - 4)$$

$\hat{Y}_{(l)}$ - l - step - ahead forecast

The forecast profile is an exponential decay to the mean ($=4$) from the last observation.

Figure 20.1 Forecast profile of the AR(1) model.

A Forecast Profile for ARMA(1,q) Models

Consider an ARMA(1,1) model of the form

$$Y_t = 0.5Y_{t-1} + 2.0 + \varepsilon_t - 0.5\varepsilon_{t-1}.$$

By following the same procedure of replacing t by $t + 1$, assuming ε_{t+1}, ε_{t+2}, . . . = 0, and using estimates $\hat{Y}_t(1)$, $\hat{Y}_t(2)$, . . . , one can obtain successive forecasts {6.5, 4.5, . . . , 4.0}.

The forecast profile for this model is shown in Figure 20.3. The forecast for the first period is a function of last period's actual observation and last period's error. All future forecasts are based only on predicted values of Y_t. As in the AR(1) case, it shows a geometrical decay to the mean after the first forecast period.

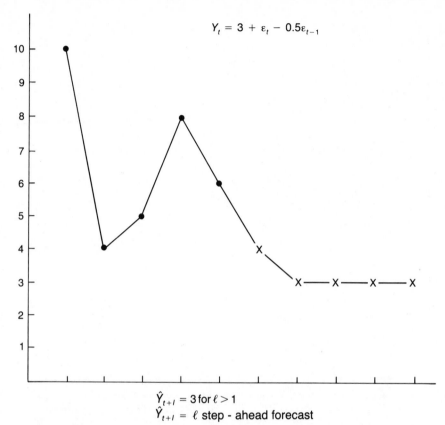

$$\hat{Y}_{t+l} = 3 \text{ for } l > 1$$
$$\hat{Y}_{t+l} = l \text{ step - ahead forecast}$$

The forecast profile is the mean ($=3$) except for the one-step-ahead forecast.
The forecast profile of MA (q) is the mean after q steps ahead. The first q forecasts are a function of prior disturbances.

Figure 20.2   Forecast profile of the MA(1) model.

For the forecast profile for ARMA(1,q) model, the forecasts of the first "q" periods will be a function of observations, estimated values, and errors. After "q" periods ahead, there will be a geometrical decay to the mean.

Forecast Profile for an IMA(1,1) Model

An example of an IMA(1,1) model is

$$Y_t = Y_{t-1} + 3.0 + \varepsilon_t - 0.5\varepsilon_{t-1} .$$

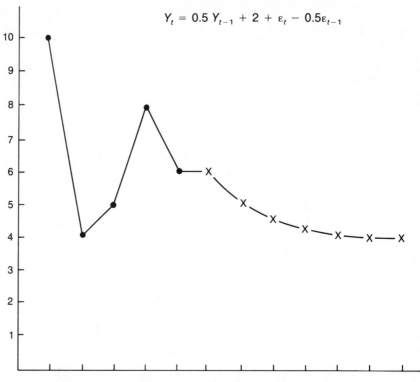

$$Y_t = 0.5\, Y_{t-1} + 2 + \varepsilon_t - 0.5\varepsilon_{t-1}$$

The forecast for $t + 1$ is the function f (Prior observation, Prior error).
The forecasts for $t + 2, t + 3, \ldots$, are the functions f (Prior observations).
The exponential decay to mean is $\dfrac{\text{Constant}}{(1 - \phi_1)} = 4$.

Figure 20.3   Forecast profile of the ARMA(1,1) model.

By following the procedure of replacing t by $t + 1$, assuming $\varepsilon_{t+1}, \varepsilon_{t+2}, \ldots,$ = 0, and using estimates $\hat{Y}_t(1), \hat{Y}_t(2), \ldots$, the forecasts $\{10, 13, 16, 19, \ldots\}$ are obtained.

The forecast profile for this model is shown in Figure 20.4. The forecast profile is a straight line after the one-period-ahead forecast. The slope of the line is equal to the constant α in the model. If $\alpha = 0$, the forecast profile is a horizontal line at the level given by $\hat{Y}_t(1)$.

Forecast Profiles for ARIMA(1,1,1) Models

An example of an ARIMA model is

$$(1 - 0.5B)(1 - B)Y_t = 2.0 + \varepsilon_t - 0.5\varepsilon_{t-1}.$$
When expanded, this becomes

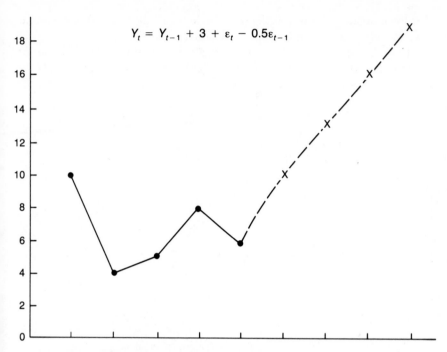

$$Y_t = Y_{t-1} + 3 + \varepsilon_t - 0.5\varepsilon_{t-1}$$

The forecast profile is a straight line after the one-period-ahead forecast. The slope of the line = Constant (3).

Figure 20.4   Forecast profile of the IMA(1,1) model.

$$Y_t = 1.5Y_{t-1} - 0.5Y_{t-2} + 2.0 + \varepsilon_t - 0.5\varepsilon_{t-1} .$$

To generate the forecasts, the following procedure is used:

1. Replace t by $t + 1$.
2. Assume $\varepsilon_{t+1} = 0$. Then

$$\hat{Y}_t(1) = 1.5(6) - 0.5(8) + 2.0 + 0 - 0.5(-2)$$
$$= 8.0 ;$$

$$\hat{Y}_t(2) = 1.5(8) - 0.5(6) + 2.0 + 0 - 0$$
$$= 11.0 ;$$

and

$$\hat{Y}_t(3) = 1.5(11) - 0.5(8) + 2.0 + 0 - 0$$
$$= 14.5 .$$

The forecast profile for this model is shown in Figure 20.5. The forecast profile approaches a straight line with a slope = [Constant/(1 − $\emptyset_1$)].

Modeling first differences, $(1 − B)Y_t$, results in an ARMA(1,1) model for the first difference. After one period the moving average term no longer affects the forecasts. This is simply an AR(1) model of the first differences. As before, the

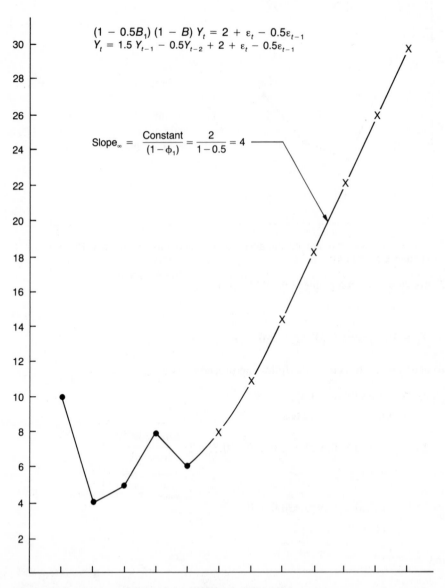

$$(1 − 0.5B_1)\,(1 − B)\,Y_t = 2 + \varepsilon_t − 0.5\varepsilon_{t−1}$$
$$Y_t = 1.5\,Y_{t−1} − 0.5Y_{t−2} + 2 + \varepsilon_t − 0.5\varepsilon_{t−1}$$

$$\text{Slope}_\infty = \frac{\text{Constant}}{(1−\phi_1)} = \frac{2}{1−0.5} = 4$$

Figure 20.5   Forecast profile of the ARIMA(1,1,1) model.

forecast profile for an AR(1) model decays geometrically to a mean value [Constant/$(1 - \emptyset_1)$].

If the first differences approach a constant, the series itself grows by this constant amount each period. The forecast profile of the series becomes a straight line with a constant slope.

Three Kinds of Trend Models

For time series with a linear trend, it is interesting to compare the characteristics of alternative forecasting models. Table 20.2 shows three alternative approaches to forecasting a trending series.

The first model uses "time" as an independent variable. As new observations are added, the forecasts produced by this model would not change, unless the model were to be reestimated. The slope and the intercept of the line of forecasts are constant.

The second model is an expansion of the first:

$$Y_t = Y_{t-1} + \alpha + \varepsilon_t .$$

This model has a slope α. The forecasts from this model are updated as new Y_t's become available. This has the effect of changing the intercept of the line of future forecasts but the slope remains constant.

The third model can be written as

$$Y_t = 2Y_{t-1} - Y_{t-2} + \varepsilon_t .$$

Table 20.2   Three approaches to forecasting a trending time series.

Model type	Equation	Characteristics of updated forecasts	
		Intercept	Slope
Deterministic	$Y_t = \alpha + \beta t + \varepsilon_t$	Constant	Constant
ARIMA with deterministic trend constant	$(1 - B)Y_t = \alpha + \varepsilon_t$	Varies	Constant
ARIMA without deterministic trend constant	$(1 - B)^2 Y_t = \varepsilon_t$	Varies	Varies

Both the slope and the intercept of the lines of updated forecasts will change as new observations become available. The trend can change direction with this model.

Unless there is reason to believe that a deterministic relationship exists (e.g., physical or theoretical reasons), autoregressive models are more adaptive or responsive to recent observations than straight-line models are. The choice of taking first differences with a trend constant or second differences without a trend constant should depend on the data. Since it is not helpful to overdifference a time series, second differences should be used only when first differences do not result in a stationary series.

A Comparison of an ARIMA(0,1,0) Model and a Straight-Line Model

It is of interest to see how the forecasts of an ARIMA(0,1,0) model differ from straight-line models using "time" as the independent variable. Consider a (seasonally adjusted) quarterly series that is generally increasing in trend. Moreover, the series contains business cycles with downturns in 1957–58, 1960–61, 1969–70, and 1974–75.

The correlogram of the data has a gradual decay that suggests a nonstationary series. The original data have a pattern of nonstationarity in trend. Therefore, first differences of the quarterly data were taken; a plot of the first differences shows a constant mean and no signs of increasing variability. The correlogram and partial correlogram of the first differences are shown in Figure 20.6.

There are no significant patterns in either diagram. Therefore an ARIMA(0,1,0) model of the seasonally adjusted quarterly data is appropriate:

$$Y_t - Y_{t-1} = \alpha + \varepsilon_t .$$

The first differences have a mean value that is 1.35 standard deviations from zero, so a deterministic trend constant is included; the fitted values are given by

$$\hat{Y}_t = Y_{t-1} + 332.0 .$$

This is perhaps one of the simplest time series models covered so far. Note that a one-period-ahead forecast would simply be the data value for the current quarter plus 332.0. All information prior to the last observation has no effect on the one-period-ahead forecast. Prior data were used to determine that the trend constant should be 332.0.

To compare the forecast profile of the ARIMA(0,1,0) model with a simple linear regression model with "time" as the independent variable, a straight-line model was fitted to the data:

$$\hat{Y}_t = 7165.2 + 60.8t \qquad t = 1, 2, \ldots .$$

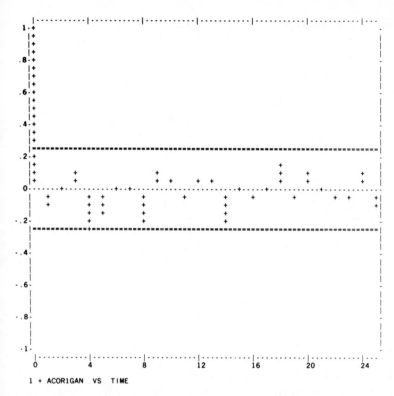

Figure 20.6(a) The ordinary correlogram of first differences of a seasonally adjusted quarterly time series.

You can see that the forecast profile of this model is a straight line with a slope = 60.8. Since "time" is always increasing, the forecasts are always increasing.

To compare the forecasts of the two models over the time periods 1970 and 1971, four regressions were performed, all starting in the first quarter of 1957. The first regression was performed through the fourth quarter of 1969, and then four quarterly forecasts for 1970 were generated. Subsequent regressions and forecasts were also generated by fitting through the second quarter of 1970, and then forecasting through the fourth quarter of 1970; fitting through the fourth quarter of 1970, and then forecasting the four quarters of 1971; and lastly, fitting through the second quarter of 1971, and then forecasting the remaining two quarters of 1971.

The effect of new data on the predictions of the two models can be seen in Figure 20.7, which shows the predictions of the two models for the four quarters of 1970. The predictions that the straight-line regression model gives are closer to the actuals for each of the four quarters when the regression period ends in the fourth quarter of 1969; and when two or more quarters of data are included and the third and fourth

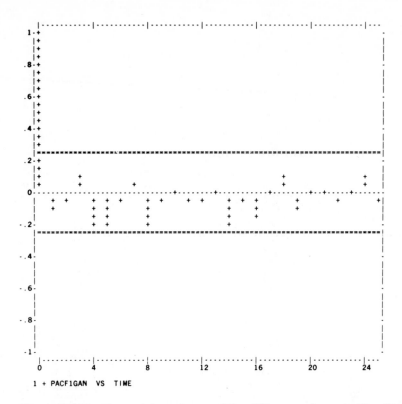

Figure 20.6(b)   The partial correlogram of first differences of a seasonally adjusted quarterly time series.

quarters of 1970 are predicted, once again the straight-line model predictions are close to the actual values.

When more data are added, the predictions made by the models for 1970 are as shown in the following chart:

Period 1970	Model	
	ARIMA	Straight line
Quarter 3	+ 2756	+ 96
Quarter 4	+ 2807	+ 100

Clearly, the ARIMA model responds to the new data by a much greater amount than does the straight-line model. In fact, for this ARIMA model, the forecasts will equal the latest data value plus a constant. Notice that the second quarter of 1970 was unusually high; this distorted the forecasts for the remainder of 1970.

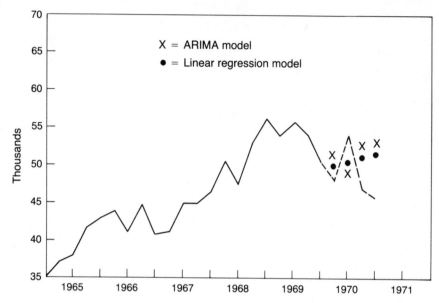

Figure 20.7   Comparison of forecasts generated by the ARIMA(0,1,0) model and the straight-line regression model for 1970.

Next, two more quarters of actuals were added to the regression period and the four quarters of 1971 were predicted. Now the predictions of the straight-line model were once again superior to the ARIMA model.

The last regression with actuals through the second quarter of 1971 was done next. For this time, the plot of the data had started to show a sharp turn upward. The model forecasts showed changes in predictions for 1971 that were as follows:

Period 1971	Model	
	ARIMA	Straight line
Quarter 3	233	116
Quarter 4	11,449	118

Once again, the ARIMA model reacted much more quickly to changing conditions. Figure 20.8 shows that this time the ARIMA model was correct since the data were indeed continuing upward.

The forecast test just described is not done to claim either model as superior. Rather it shows how differently the ARIMA(0,1,0) model and straight-line regression models react to new data. There is almost no reaction in the forecasts of a straight-line model (versus time) when new data are introduced. However, ARIMA

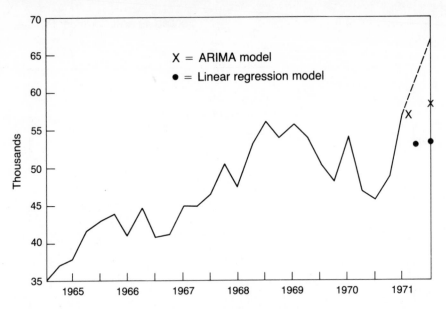

Figure 20.8 Comparison of forecasts generated by the ARIMA(0,1,0) model and the straight-line regression models for the second half of 1971.

models are affected significantly by recent data. Which model should be selected will depend on the circumstances of the application.

If future demand in a given year is expected to be substantially above or below trend (say, for economic reasons), then the forecast of a straight-line model must be modified. A "turning point analysis" (see *The Beginning Forecaster,* Chapter 11) or a trend-cycle curve-fitting approach may be the best way of supplementing judgment about the cycle to a straight-line model (Wecker, 1979).

If it is believed that the current short-term trend in the data will continue, an ARIMA model should be seriously considered. Univariate ARIMA models cannot predict turning points. If the economy is near a turning point, extreme care should be exercised in using the ARIMA models. Regression models with explanatory variables, econometric models, and transfer function models (see Chapters 23 and 24) offer other ways of modifying forecasts to take into account changes in the business cycle.

CONFIDENCE LIMITS FOR ARIMA MODELS

One of the goals of quantitative forecast modeling is to be able to make probability statements about the forecasts. This is normally accomplished by calculating forecast error variances with the assumption that the forecast errors are independent, identically distributed random variables.

To examine confidence limits for ARIMA models, let us first illustrate their use for the MA(2) model.

Confidence Limits for an MA(2) Model

Consider the following model:

$$Y_t = \varepsilon_t - 0.6\varepsilon_{t-1} - 0.4\varepsilon_{t-2} .$$

Future values are

$$Y_{t+1} = \varepsilon_{t+1} - 0.6\varepsilon_t - 0.4\varepsilon_{t-1} ,$$
$$Y_{t+2} = \varepsilon_{t+2} - 0.6\varepsilon_{t+1} - 0.4\varepsilon_t ,$$

and

$$Y_{t+3} = \varepsilon_{t+3} - 0.6\varepsilon_{t+2} - 0.4\varepsilon_{t+1} .$$

As time passes and the actual observations at times $t + 1$, $t + 2$, $t + 3$, . . . become available, there is a discrepancy between the one-step-ahead forecast and the observed value. For the one-step-ahead forecast, the forecast error will be the value of ε_{t+1}. If ε_{t+1} is not zero, the forecast will be off by the value of ε_{t+1}.

For the two-step-ahead forecast, the forecast error will equal

$$\varepsilon_{t+2} - 0.6\varepsilon_{t+1} .$$

The forecast error now is a result of the two unobserved errors $(\varepsilon_{t+2}, \varepsilon_{t+1})$.

For the three-step-ahead forecast, the forecast error is equal to

$$\varepsilon_{t+3} - 0.6\varepsilon_{t+2} - 0.4\varepsilon_{t+1} .$$

For forecasting purposes it is assumed that ε_{t+3}, ε_{t+2}, and ε_{t+1} will be equal to their expected values (zero). In actuality, they will differ from zero by some amount that cannot be known at the time of the forecast. For all forecasts longer than three steps ahead, the forecast error will be the result of three unknown future errors that are assumed to be zero but will actually be somewhat different from zero.

It is of interest to relate the variance of the l-step-ahead forecast to the variance var(Y_t) of the series. For the example of the MA(2) model we have been using,

$$\text{var}(Y_t) = (1 + 0.36 + 0.16)\sigma_\varepsilon^2$$
$$= 1.52\sigma_\varepsilon^2 .$$

Thus the variance of series is 1.52 times the variance of the error term (Box and Jenkins (1976, Chapter 3)). The variance of the one-step-ahead forecast is σ_ε^2—the variance of the error. This is less than the variance of the series—just determined to be $1.52\sigma_\varepsilon^2$. Consequently, you would expect to have less variability in the one-step-ahead forecast than in the series as a whole.

The two-step-ahead forecast error is equal to $\varepsilon_{t+2} - 0.6\varepsilon_{t+1}$. With the assumption that the ε_t's are independent of each other and have a constant variance σ_ε^2, the variance of the two-step-ahead forecast error is equal to $1.36\sigma_\varepsilon^2$. That is, the variance of the two-step-ahead forecast is still less than the variance of the series. Therefore, the model estimates future observations in a smaller range than just the variance of the series. The variance of the three-step-ahead forecast error is $1.52\sigma_\varepsilon^2$—the variance of the series: once three or more periods are forecast with the MA(2) model, the variance of the l-step-ahead forecast equals the variance of the series.

All forecasts beyond the three steps ahead into the future for Y_t will give the mean of Y_t, or zero in this case. To summarize, for the MA(2) model, the best forecast of the distant future is the mean of the process. The variability about that mean (the variance of the forecast) can be estimated from the available data.

This can be visualized in Figure 20.9. The actual observations are connected by a solid line. The dotted line represents the forecasts for one, two, three, and four steps ahead. The confidence limits expand from the one-step-ahead to the three-step-

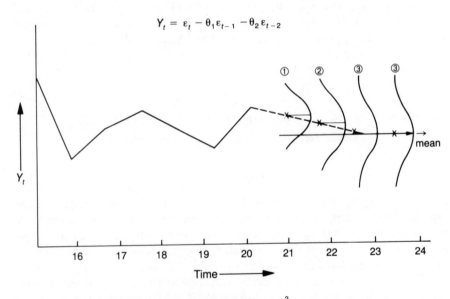

$$Y_t = \varepsilon_t - \theta_1 \varepsilon_{t-1} - \theta_2 \varepsilon_{t-2}$$

① One-step-ahead forecast variance $= \sigma_\varepsilon^2$
② Two-step-ahead forecast variance $= (1 + \theta_1^2)\,\sigma_\varepsilon^2$
③ Three-step-ahead forecast variance $= (1 + \theta_1^2 + \theta_2^2)\,\sigma_\varepsilon^2$

Figure 20.9   Forecasts generated by the MA(2) model with associated confidence limits.

ahead forecast. The variance of the forecast error equals the variance of the series for all forecasts three or more periods into the future.

Let s_ε be an estimate of σ_ε. Then the first three approximate $1 - \eta$ confidence limits are, respectively,

$$\hat{Y}(1) \pm U_{\eta/2} ;$$
$$\hat{Y}(2) \pm U_{\eta/2}(1 + \theta_1^2)^{1/2} ;$$

and

$$\hat{Y}(3) \pm U_{\eta/2} (1 + \theta_1^2 + \theta_2^2)^{1/2} ;$$

in each, $U_{\eta/2}$ is the deviation exceeded by a proportion $\eta/2$ of the standard normal distribution (Appendix A, Table 1). Here $\hat{Y}(3)$ is the mean, which was assumed to be zero. From here on out, the confidence limits (and the forecast) remain the same for the MA(2) model.

The following conclusions are reached for the MA(2) model:

- The confidence limits expand up to the three-step-ahead forecast.

- The confidence limits become constant after the three-step-ahead forecast.

Forecast Error and Forecast Variance for ARIMA Models

Consider the general form of the model given by

$$\emptyset_p(B)(1 - B)^d Y_t = \theta_q(B)\varepsilon_t .$$

This can be expanded as

$$(1 - \emptyset_1 B - \emptyset_2 B^2 - \cdots - \emptyset_{p + d} B^{p + d}) Y_t = (1 - \theta_1 B - \cdots - \theta_q B^q)\varepsilon_t .$$

An observation at time $t + l$ generated by this process can be written as

$$Y_{t+\ell} = \emptyset_1 Y_{t+l-1} + \cdots + \emptyset_{p+d} Y_{t+l-p-d} + \varepsilon_{t+l} - \theta_1 \varepsilon_{t+l-1} - \cdots - \theta_q \varepsilon_{t+l-q} .$$

Alternatively, Y_{t+l} can be written as an *infinite weighted sum* of current and previous errors ε_j :

$$Y_{t+\ell} = \varepsilon_{t+\ell} + \psi_1 \varepsilon_{t+l-1} + \psi_2 \varepsilon_{t+l-2} + \cdots + \psi_l \varepsilon_t + \psi_{l+1} \varepsilon_{t-1} + \cdots .$$

Consider the l-step-ahead forecast, $\hat{Y}_t(l) = \hat{Y}_{t+l}$, which is to be a linear combination of current Y_t, previous observations $Y_{t-1}, Y_{t-2}, \ldots$, and errors $\varepsilon_{t-1}, \varepsilon_{t-2}, \ldots$. It can also be written as a linear combination of current and previous errors $\varepsilon_t, \varepsilon_{t-1}, \ldots$; thus

$$\hat{Y}_t(l) = \psi_l^* \varepsilon_t + \psi_{l+1}^* \varepsilon_{t-1} + \psi_{l+2}^* \varepsilon_{t-2} + \cdots,$$

where the ψ_j^* are to be determined. An important result in ARIMA time series theory is that the mean square error of the forecast is minimized when $\psi_{l+j}^* = \psi_{l+j}$ (Box and Jenkins, 1976).

The variance of the forecast error $e_t^{(\ell)}$ is given by

$$\text{var}(Y_{t+l} - \hat{Y}_t(l)) = (1 + \psi_1^2 + \psi_2^2 + \cdots + \psi_{l-1}^2)\sigma_\varepsilon^2.$$

These estimates of variance are based on the assumption that the ψ_1 are correct. That is, the error in estimating parameters is assumed to be negligible relative to the successive one-step-ahead prediction error.

Consider the simple AR(1) model as an example. The forecast errors are

$$e_t(1) = Y_{t+1} - \hat{Y}_t(1) = \varepsilon_{t+1};$$
$$e_t(2) = Y_{t+2} - \hat{Y}_t(2) = \emptyset_1\varepsilon_{t+1} + \varepsilon_{t+2};$$
$$e_t(l) = Y_{t+l} - \hat{Y}_t(l) = \sum_{j=0}^{l-1} \emptyset_1\varepsilon_{t+l-j}.$$

The variance of the forecast error is

$$\text{var}(e_t(l)) = \sigma_\varepsilon^2 \sum_{j=0}^{l-1} \emptyset_1^{2j}.$$
$$= \sigma_\varepsilon^2(1 - \emptyset_1^{2l})/(1 - \emptyset_1^2).$$

For a stationary AR(1) model, with $-1 < \emptyset_1 < 1$, the variance increases to a constant value $\sigma_\varepsilon^2/(1 - \emptyset_1^2)$ as l tends to infinity. For nonstationary models, on the other hand, the forecast variances will increase without bound with increasing values of l.

An important property of forecast errors, worth noting here, is that while errors in one-step-ahead forecasts are uncorrelated, the errors for forecasts with longer lead times are in general correlated. It is worth remembering in practice that there are two kinds of correlations to be considered:

- The correlation between forecast errors $e_t(l)$ and $e_{t-j}(l)$ made at the *same* lead time l from *different* time origins t and $t - j$.

- The correlation between forecast errors $e_t(l)$ and $e_t(l + j)$, made at *different* lead times from the *same* origin t.

General expressions for these correlations are given in Box and Jenkins (1976).

Confidence Limits for ARIMA Models

The general procedure for calculating confidence limits for ARIMA models follows the same pattern as the derivation of confidence limits for the MA(2) model. First, an expression for the l-step-ahead forecast error $e_t(l) = Y_{t+l} - \hat{Y}_t(l)$ is written down. Next the variance of the forecast error is expressed in terms of σ_ε^2 and the ψ weights, which are computed from the estimated θ's and $\emptyset$'s.

Let s_ε^2 denote an estimate of the variance σ_ε^2. Then approximate $1 - \eta$ confidence limits $Y_t^+(l)$ and $Y_t^-(l)$ for Y_{t+l} are given by

$$Y_{t+l}(\pm) = Y_t(l) \pm U_{\eta/2} \left[1 + \sum_{j=1}^{l-1} \psi_j^2 \right]^{1/2} s_\varepsilon ,$$

where $U_{\eta/2}$ is the deviation exceeded by a proportion $\eta/2$ of the standard normal distribution.

For a simple linear regression model, the confidence limits about the regression line expand in both directions away from the mean value of the independent variable. For a stationary ARIMA model, the concern is not so much with the value of the "independent" variables. The confidence limits for the forecasts of a stationary series will initially expand but will become constant. From this point on, they stay the same for all values into the future.

For a nonstationary ARIMA model, the confidence limits will continue to expand into the future. This is similar to the pattern displayed by the confidence limits of simple linear regression models as the independent variable takes values further away from its mean.

ARIMA FORECASTING CHECKLIST

_____ Has a model been forecast-tested over a sufficient period of time (e.g., several business cycles) to determine its forecasting capabilities?

_____ Is the model parsimonious? Or instead, have you created an overly complex model?

_____ Will a relatively simpler model with perhaps a higher chi-squared statistic perform as well?

_____ Have you performed a forecast test for any alternative models, too?

_____ How rapidly does the model respond to new data?

_____ Does the model predict turning points well (or, at all)?

SUMMARY

This chapter has

- Demonstrated the possibility of generating forecasts from ARIMA models.

- Shown examples of forecast profiles for several simple model types.

- Compared a specific ARIMA model and a simple straight-line regression model to see how responsive each is to new data. It was evident that the ARIMA model's forecasts were greatly influenced by actual current data observations. This was shown to be advantageous during certain time periods but disadvantageous in others.

- Emphasized the importance of understanding models and their limitations so that the forecaster can select the appropriate model for the application at hand.

USEFUL READING

ANDERSON, O. D. (1976). *Time Series Analysis and Forecasting, The Box-Jenkins Approach*. London, England: Butterworth.

BOX, G. E. P., and G. M. JENKINS (1976). *Time Series Analysis, Forecasting and Control, Revised Edition*. San Francisco, CA: Holden-Day.

NELSON, C. R. (1973). *Applied Time Series Analysis for Managerial Forecasting*. San Francisco, CA: Holden-Day.

WECKER, W. E. (1979). Predicting the Turning Points of a Time Series. *Journal of Business* 52, 35–50.

CHAPTER 21

Forecasting Seasonal
Time Series

Seasonal ARIMA models are used for time series in which a seasonal pattern is present. This chapter deals with

- The extension of regular ARIMA models to include the modeling of seasonal patterns.

- The identification of an appropriate seasonal ARIMA model through use of correlograms and partial correlograms.

- The Holt-Winters approach to seasonal forecasting and its relationship to the ARIMA models.

WHY BUILD SEASONAL MODELS?

Many of the economic series that are available in publications and computerized data banks are seasonally adjusted. These data are used in models for forecasting trend-cycle and in cases where seasonal patterns would otherwise mask the information of interest to the forecaster. Regular ARIMA models (discussed in Chapter 19) and econometric regression models are appropriate for these situations. Since seasonal variation is seldom unchanging, seasonal adjustment procedures usually leave some seasonal pattern in the data. Time plots and correlograms, along with the ANOVA (analysis of variance) table and low-resolution displays (discussed in *The Beginning Forecaster*), can provide adequate indication of the relative importance of a seasonal pattern in data.

Often it is desirable to forecast seasonal series that are unadjusted for seasonality. In airline traffic design, for example, average traffic during the months of the year when total traffic is greatest is of central importance. Obviously, monthly traffic data rather than the seasonally adjusted traffic data are required.

Sometimes seasonality shifts with time. For example, changes in school openings and closings during the year can affect a variety of time series, such as energy demand. If the seasonal change persists, models based on unadjusted data rather than seasonally adjusted data are likely to be more flexible and useful. If changing seasonality is expected, it is therefore better to take account of it through the development of a properly specified ARIMA model.

SEASONAL ARIMA MODELING

For time series that contain a seasonal periodic component that repeats every s observations, a supplement to the nonseasonal ARIMA model can be applied. For a seasonal time series with period s, $s = 4$ for quarterly data and 12 for monthly data. In a manner very similar to the regular ARIMA process, *seasonal* ARIMA (P,D,Q) models can be created for seasonal data (to differentiate between the two kinds, it is customary to use capital P, D, and Q when designating the parameters of the seasonal models).

Seasonal Differencing

When strong seasonal fluctuations exist, a seasonal difference of order D may be required to achieve stationarity. This creates a new series:

$$W_t = (1 - B^s)^D Y_t$$
$$= (1 - B^{12}) Y_t \, ,$$

for (as an example) $D = 1$ and seasonal period $s = 12$.

The need for seasonal differencing is evident from a plot of the raw data or from the correlogram. In an example of the monthly main-telephone gain data it is apparent that the data are highly seasonal. The correlogram plotted in Figure 21.1 shows there is a repetitive pattern with spikes at months 12, 24, and 36. In fact, all spikes 12 lags apart have comparable magnitudes; there is no significant decay towards zero.

The period of these data is 12 months. The differences of order 12 of the original data would produce a correlogram that is free of the repetitive spikes. Rather, it would show a pattern that may be described by a regular ARIMA model.

Combining Seasonal and Regular Models

Since models in which seasonal and regular parameters are combined can become unwieldy, the Box and Jenkins modeling strategy recommends that these models

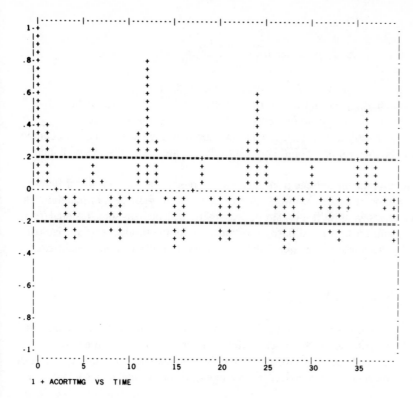

1 + ACORTTMG   VS   TIME

Figure 21.1   A correlogram of monthly main-telephone gain time series.

should be parsimonious. By this is meant that the number of parameters in the model should be kept to a minimum; always try to keep the model simple!

For time series models seasonality is introduced into the model *multiplicatively*. While this is arbitrary, it may be intuitively understood by considering that if a month—say January—is related to the December preceding it and to the previous January, then it is also related to the December thirteen months previous.

It seems reasonable to expect that for monthly data one January would be similar to previous Januarys, and that the remaining monthly subseries would also be related in a similar manner. If the backshift operator B^s is used to denote a relationship between points s time periods apart, a *seasonal ARIMA* model then takes the general form

$$\Phi_P(B^s)(1 - B^s)^D Y_t = \Theta_Q(B^s)a_t,$$

where B^s is the seasonal backshift operator, and where

$$\Phi_P(B^s) = (1 - \emptyset_s B^s - \emptyset_{2s} B^{2s} - \cdots - \emptyset_{Ps} B^{Ps})$$

and

$$\Theta_Q(B^s) = (1 - \Theta_s B^s - \Theta_{2s} B^{2s} - \cdots - \Theta_{Qs} B^{Qs}) \, .$$

Generally $s = 12$ or 4 and P, D, Q have values of 0, 1, or 2.

Notice that a_t need not be an uncorrelated, white-noise process. Indeed, a_t is assumed to be a *nonseasonal* process.

General multiplicative models incorporating both regular and seasonal components can be used for describing a typical twelve-month seasonal series with increasing trend.

To remove the linear trend from such a series requires at least one first difference. This operation is expressed by $(1 - B)^d \, Y_t$, where $d = 0, 1, 2$ in practice. The resultant series would still show strong seasonality with perhaps some residual trend. This may call for taking a difference of order 12 for the residual series. Year-over-year differences often give rise to stationary data (free from the seasonal pattern). Thus

$$(1 - B^{12})(1 - B)Y_t = Y_t - Y_{t-1} - Y_{t-12} + Y_{t-13} \, .$$

Notice that the order in which the differencing operators are applied is immaterial and that the operation of successive differencing is *multiplicative* in nature.

The combined operation suggests that if the residual series resembles a series of random numbers or "white noise," then a model of the original data could have the form

$$Y_t = Y_{t-1} + Y_{t-12} - Y_{t-13} + \varepsilon_t \, .$$

The forecast is then the sum of the values for the previous year's month (Y_{t-12}) and the increase in the values of the previous months ($Y_{t-1} - Y_{t-13}$). This is a special case of an autoregressive model in which past values have been given a constant weight of plus or minus one. It also shows that weights given to Y_{t-1} and Y_{t-13} are equal but have opposite signs.

It occurs relatively frequently in practice that both seasonality and trend are present in data. Hence an AR(13) model of the form

$$Y_t = \alpha + \emptyset_1 Y_{t-1} + \emptyset_2 Y_{t-12} + \emptyset_3 Y_{t-13} + \varepsilon_t$$

can be used to interpret the underlying structure. If the coefficients at lags 1 and 13 are approximately equal and of opposite sign and $\emptyset_2$ equals approximately 1, then the multiplicative model

$$(1 - \emptyset_1 B)(1 - B^{12})Y_t = \alpha + \varepsilon_t$$

is a good starting point for modeling such series parsimoniously.

The Multiplicative ARIMA Model

The regular and seasonal components can be combined into a general multiplicative ARIMA model of the form

$$\emptyset_p(B) \, \Phi_P(B^s)(1 - B)^d(1 - B^s)^D Y_t = \theta_q(B)\Theta_Q(B^s)\varepsilon_t \, ,$$

where the operators are defined as before.

The regular and seasonal autoregressive components, differences, and moving average components are multiplied together in the general model. Fortunately, in most practical examples, most parameters are 0 and the resulting models are often quite simple.

A useful notation to describe the orders of the various components in the multiplicative model is given by

$$(p,d,q) \times (P,D,Q)^s$$

and corresponds to the orders of the regular and seasonal factors, respectively.

By representing a time series in terms of a multiplicative model it is often possible to reduce the number of parameters to be estimated. It also aids in the interpretation of the model structure.

Of course, the forecasting performance of these different representations of seemingly similar model structures should always be evaluated and the models that pass diagnostic checking and have the best performance should be retained.

Consider a monthly seasonal model that has a seasonal difference and also regular and seasonal first-order autoregressive parameters. Thus

$$(p,d,q) \times (P,D,Q)^s = (1,0,0) \times (1,1,0)^{12} \, .$$

This can also be expressed in terms of the backshift operator as

$$(1 - \emptyset_1 B)(1 - \emptyset_{12}B^{12})(1 - B^{12})Y_t = \alpha + \varepsilon_t \, .$$

Expanding the left-hand side gives

$$(1 - \emptyset_{12}B^{12} - \emptyset_1 B + \emptyset_1\emptyset_{12}BB^{12})(1 - B^{12})Y_t = \alpha + \varepsilon_t \, .$$

or

$$(1 - \emptyset_{12}B^{12} - \emptyset_1 B + \emptyset_1\emptyset_{12}B^{13} - B^{12} + \emptyset_{12}B^{24} + \emptyset_1 B^{13} - \emptyset_1\emptyset_{12}B^{25})Y_t = \alpha + \varepsilon_t \, .$$

In terms of the variable Y_t ,

$$Y_t = \alpha + \emptyset_1 Y_{t-1} + (1 + \emptyset_{12})Y_{t-12} - \emptyset_1(1 + \emptyset_{12})Y_{t-13} - \emptyset_{12}Y_{t-24} + \emptyset_1\emptyset_{12}Y_{t-25} + \varepsilon_t.$$

As a multiple linear regression, this AR(25) model has six parameters to be estimated. Expression as a multiplicative model, however, requires only three parameters. A distinct advantage of the multiplicative ARIMA model is that it can be used to represent a wide variety of time series with a minimum number of parameters in an easily interpretable manner. (Parsimony, again!)

Identifying Seasonal ARIMA Models

As with regular ARIMA models, seasonal ARIMA models can be classified as autoregressive and moving average. The seasonal MA process is analogous to the regular MA process. However, the ACF (autocorrelation function) of a pure seasonal MA process has a *single value at the period of the seasonality*. The PACF (partial autocorrelative function) shows a *decaying pattern at multiples of 12* if the seasonality has a 12-month period. This differs from a regular MA process of order 12 where there could be significant values at lags 1 *through* 12 in the ACF. If there is a *single dominant value* at 12, then this indicates a *seasonal* MA process.

In practice, it is possible to find significant spikes at other unusual lags of—say—5 or 9 in the correlogram of a residual series. Generally, these are spurious and may not suggest a secondary seasonal pattern. Their removal through a lag structure generally has little impact on forecasts generated from such a model. It is important, however, to isolate and interpret seasonal lags corresponding to realistic periodicities.

The seasonal AR process is also analogous to the regular AR process. However, the pattern of decay that is evident in the ACF is noticed at multiples of the period. For example, the ACF of a first-order seasonal (monthly) AR process has a decaying pattern in the values at multiples of 12. The PACF has a single value at lag 12. It is worth noting that pure monthly seasonal models look like 12 independent series, so that the ACF and PACF are approximately zero, except at multiples of 12.

The ACF of a particular, simple combined regular and seasonal moving average process is depicted in Figure 21.2. Upon expanding the factors in the model it becomes apparent that the current error as well as errors 1, 12, and 13 periods back in time affect Y_t. The pattern that results in the ACF is a large value at lags 1 and 12 and smaller values at lags 11 and 13.

As a general rule there will be smaller values at lag 12 plus or minus each regular moving average parameter. For example, if the process was regular MA(2), there would be smaller values at lags 10, 11, 13, and 14.

The easiest way to identify the order of combined MA processes is to introduce first a seasonal moving average parameter in the model. The order of the regular MA parameter will then be apparent from the correlogram of the residuals.

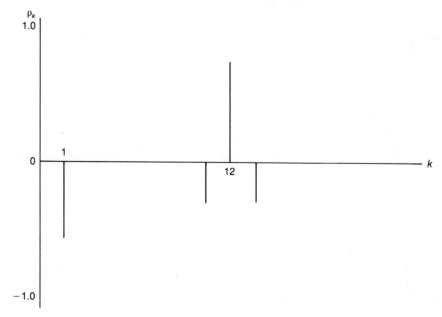

AUTOCORRELATIONS OF COMBINED SEASONAL & REGULAR
MOVING-AVERAGE MODEL

$$Y_t = (1 - \theta_1 B^1)(1 - \theta_{12} B^{12}) \varepsilon_t$$
$$Y_t = \varepsilon_t - \theta_1 \varepsilon_{t-1} - \theta_{12} \varepsilon_{t-12} + \theta_1 \theta_{12} \varepsilon_{t-13}$$

Figure 21.2 An ACF of a combined regular and seasonal MA model.

Another process that occurs in practice is the combination of regular and seasonal autoregressive components. Figure 21.3 shows the ACF of a particular regular AR(1) and a seasonal AR(12) process. A pattern in which the values reach a peak at multiples of 12 is noticeable, as are buildups to and decays from that peak at the other lags.

There are a large variety of patterns that can emerge in the modeling process. The Box-Jenkins approach is so general that it is impossible to catalog all possibilities. So it is essential to follow an iterative procedure in developing successful forecasting models.

AN ARIMA MODEL FOR PREDICTING REVENUES

Let us examine a case study to illustrate the use of ARIMA models. The case study will result in a forecast of telephone toll revenues for a geographic area. The data have been adjusted for rate changes and data recording errors.

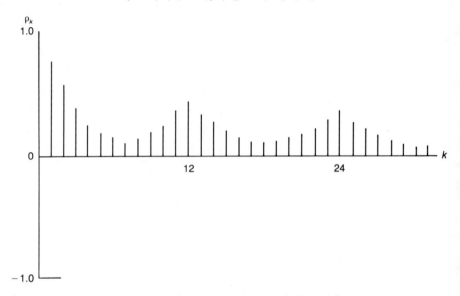

AUTOCORRELATION OF COMBINED SEASONAL & REGULAR
AUTOREGRESSIVE MODEL

$$(1 - \phi_1 B)(1 - \phi_{12} B^{12}) Y_t = \varepsilon_t$$
$$Y_t = \phi_1 Y_{t-1} + \phi_{12} Y_{t-12} - \phi_1 \phi_{12} Y_{t-13} + \varepsilon_t$$

Figure 21.3   An ACF of a combined regular and seasonal AR model.

Preliminary Analysis

Plots of the raw data, of differences of order 1, of 12, and of both 1 and 12 are shown in Figures 21.4, 21.5, 21.6, and 21.7, respectively. An ANOVA decomposition suggests that the variation due to trend, seasonality, and irregularity accounts for 91.6 percent, 7.4 percent, and 1.0 percent, respectively, of the total variation about the mean. Figure 21.4 illustrates the overall trend and also shows increasing dispersion in the seasonal variation. This pattern is reenforced in the first differences of the data (Figure 21.5). The differences of order 12 show a cyclical pattern (Figure 21.6). The combination of differences of order 1 and 12 finally appear somewhat stationary (Figure 21.7). This is designated Series A.

The increasing dispersion in the data was considerably reduced by taking logarithms before differencing. A stationary series (Series B) is obtained by taking differences of order 1 and 12 after making the logarithmic transformation (compare Figures 21.7 and 21.8). For analysis purposes it is sufficient to model the differenced data (of order 1 and 12 together) with and without the logarithmic transformation. For the purposes of exposition models will also be built for the differenced data (of orders 1 and 12, separately). Series C and D data are first differences on the original and transformed data, respectively.

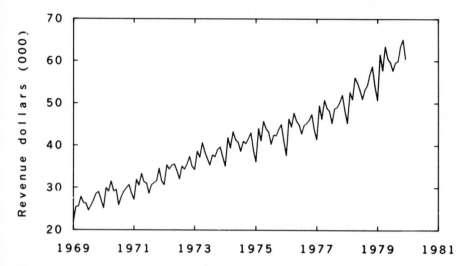

Figure 21.4   A time plot of monthly telephone toll revenues.

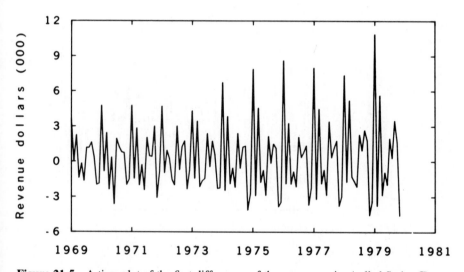

Figure 21.5   A time plot of the first differences of the revenue series (called Series C).

Identification

Figure 21.9 shows the ordinary and partial correlograms of Series A. Figure 21.10 shows the same correlograms for Series B. As expected, the patterns are very similar and suggest the same model. The single, negative spike at lag 1 in the ordinary correlogram suggests a first-order moving-average model. This is confirmed by the

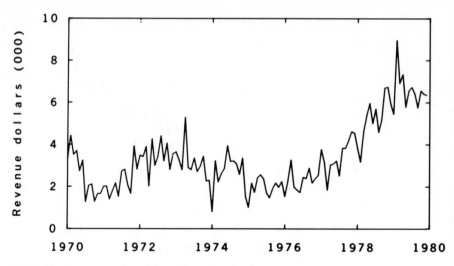

Figure 21.6   A time plot of the differences of order 12 of the revenue series.

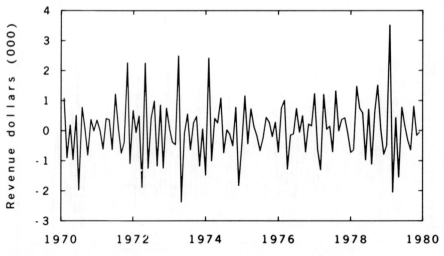

Figure 21.7   A time plot of the first- and twelfth-order differences of the revenue series (called Series A).

decaying pattern in the partial correlogram. The need for moving average parameters frequently occurs with data that have been differenced. Often, a seasonal moving-average parameter is required when a seasonal difference is taken, this is evident by the negative spike at lag 12 in the ordinary correlogram. The $(0,1,1) \times (0,1,1)^{12}$ model was built for both series; when the residuals from these models were examined they showed no significant spikes or patterns in the correlograms.

Figure 21.11 shows the ordinary and partial correlograms of Series C. The corresponding correlograms for Series D are almost identical. The ordinary correl-

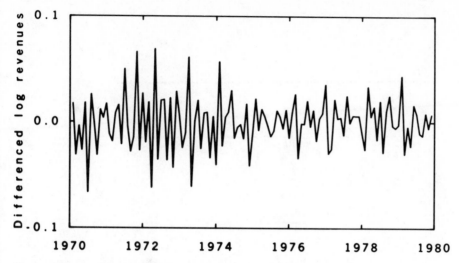

Figure 21.8   A time plot of the first- and twelfth-order differences of the log-transformed revenue series (called Series B).

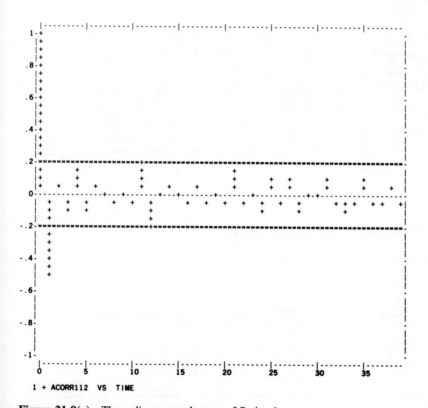

Figure 21.9(a)   The ordinary correlogram of Series A.

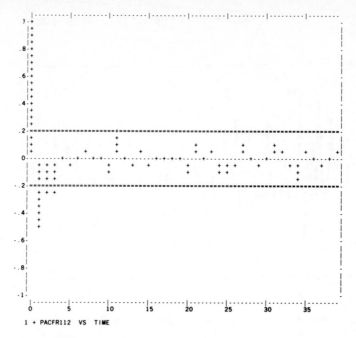

1 + PACFR112 VS TIME

Figure 21.9(b) Partial correlogram of Series A.

ogram suggests a seasonal autoregressive model because of the decaying pattern of spikes at lags 12, 24, and 36 (also 6, 18, 30, and similar patterns of lags separated by 12). This is confirmed by the spike in the partial correlogram at lag 12.

The suggested model (seasonal autoregressive) was fitted to the data and the correlograms of the residuals were plotted (Figure 21.12). The negative spike at lag 1 in the ordinary correlogram and the decaying pattern in the partial correlogram indicates the need to add a first-order regular moving average parameter to the model. After this was done, resultant correlograms of the residuals showed no significant spikes or patterns.

Estimation and Diagnostic Checking

The following models were built:

Series	ARIMA model for revenues or transformed revenues
A	$(0,1,1) \times (0,1,1)^{12}$
B	$(0,1,1) \times (0,1,1)^{12}$
C	$(0,1,1) \times (1,0,0)^{12}$
D	$(0,1,1) \times (1,0,0)^{12}$

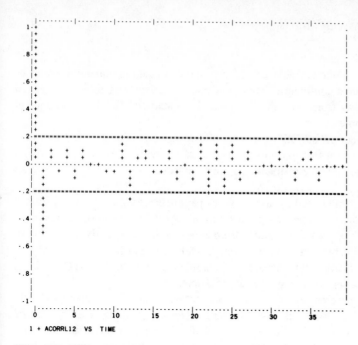

Figure 21.10(a)   The ordinary correlogram of Series B.

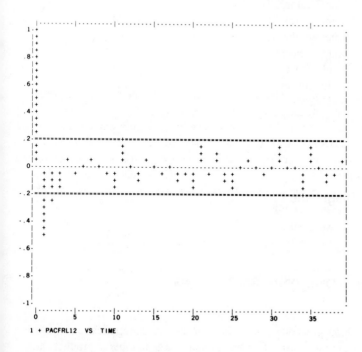

Figure 21.10(b)   A partial correlogram of Series B.

For each of these models, Table 21.1 shows the estimated parameter values, the associated 95-percent confidence limits, how many standard deviations the mean of the residuals is away from zero, the chi-squared test statistic, and the covariance of the parameter estimates.

The chi-squared test statistics for randomness in the residuals are all less than the tabulated values for the appropriate degrees of freedom. Therefore, there is no evidence suggesting that the residuals did not arise from a white noise process in all four cases.

Models A, B, and D need to have only two parameters estimated. The estimated do not span zero, so they appear significant in all cases. The mean of the residuals in Model A are 0.9 standard deviations from zero and thus it appears reasonable to exclude a deterministic trend constant. For Model B, C and D there is clearly no need for a deterministic trend constant. The confidence limits for the DTC in Model C span zero indicating the parameter can be deleted.

In Model C, the estimate of the seasonal autoregressive parameter (SAR-12) exceeds unity, which suggests nonstationarity. However, taking a difference of order 12, instead, would give Model A. In Model D, the estimate of the seasonal autoregressive pattern is less than but close to unity. This may also point the need to take a difference of order 12, instead.

The estimated covariances for the parameter estimates in all models suggest that the parameter estimates are not highly correlated.

Forecast Test Results

Forecast tests were performed for all four models. Four one-year-ahead forecasts (1975–78) were generated. Table 21.2 shows the results of the tests. Model B had the best results (when measured in terms of average absolute percent of forecast error) followed by Models D and A. Model C should not be used since it is non-stationary. The best results were obtained by using the stationary series generated from the combination of differences of orders 1 and 12. Model D needs to be carefully monitored, if used on an ongoing basis, since the estimated seasonal autoregressive parameter is close to unity.

SEASONAL EXPONENTIAL SMOOTHING

Exponential smoothing procedures have their most common application in routine sales forecasting, where forecasts of future sales of a large number of products might be required. The single or linear/quadratic models are inappropriate for seasonal

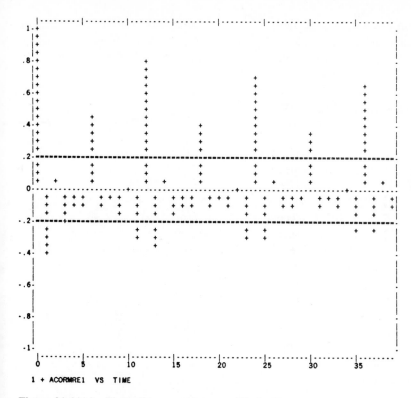

Figure 21.11(a)   The ordinary correlogram of Series C.

data. A three-parameter Winters model is frequently used for the great majority of series that exhibit seasonal patterns. The three parameters are α, β and γ, where:

- α smooths randomness.
- β smooths seasonality.
- γ smooths trend.

Holt (1957) and Winters (1960) developed a low-cost exponential smoothing method to forecast trend-seasonal patterns.

Winters's multiplicative trend-seasonal model is appropriate for time series in which the height of the seasonal pattern is proportional to the average level of the series. The seasonal index I_t is calculated from

$$I_t = \beta \frac{Y_t}{S_t} + (1 - \beta)I_{t-s},$$

where s is the length of the season, and β is a constant between 0 and 1. The current observed seasonal variation is determined from the current value of the series Y_t

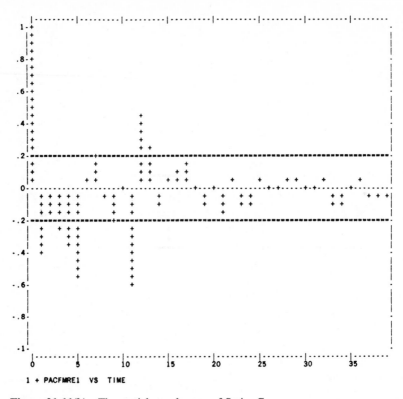

Figure 21.11(b)   The partial correlogram of Series C.

divided by the current single smoothed value for the series S_t. This index is used in the equation

$$S_t = \alpha \frac{Y_t}{I_{t-s}} + (1 - \alpha)(S_{t-1} + T_{t-1}) \,,$$

where α is a smoothing constant between 0 and 1, to adjust the data for randomness A trend term given by T_t is

$$T_t = \gamma(S_t - S_{t-1}) + (1 - \gamma)T_{t-1} \,,$$

where γ is a smoothing constant between 0 and 1.

The trend forecast is $(S_t + mT_t)$ and seasonality is introduced with the multiplicative seasonal factor I_{t-s+m}. The forecast based on Winters's method is computed as

$$S_{t+m} = (S_t + mT_t)I_{t-s+m}, \qquad \text{for } m = 1, 2, \ldots, s \,.$$

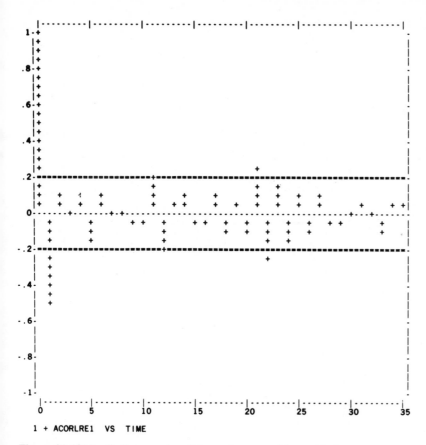

Figure 21.12(a)   Ordinary and partial correlograms of the residuals.

The procedure is developed heuristically and requires three smoothing constants as well as initial values for the smoothed statistics. Several approaches are proposed in the literature (Bowerman and O'Connell, 1979; and Montgomery and Johnson, 1976); however, the smoothing concept is similar to the one performed in the single exponential smoothing method (see *The Beginning Forecaster,* Chapter 8).

The selection of the smoothing parameters α, β, and γ is generally an iterative, *ad hoc* process. The criteria generally used in selection of α, β, and γ is the minimization of the mean square error (MSE), although the mean percent of error (bias) and mean absolute percent of error are also utilized. In an application using airline data (Box and Jenkins, 1976, series G), a variety of Winters-type models were estimated. The impact of inadequate starting values were found in two of the models. For values of α less than 0.2 to 0.3, the initial estimates were very poor and severely distorted comparisons based on average absolute error or MSE. It is interesting to note that over the preceding five years, the highest individual monthly

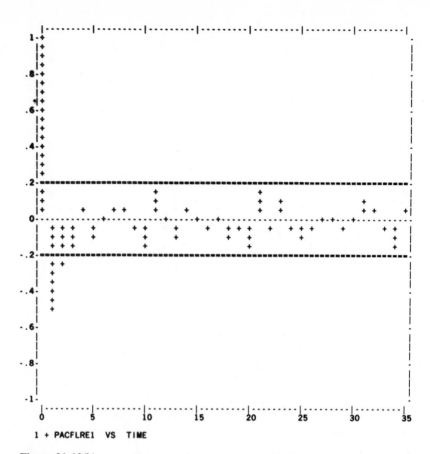

Figure 21.12(b)

error (average) for Model 1 ($\hat{\alpha} = 0.1$, $\hat{\beta} = 0.2$, $\delta = 0.5$) is 2.6 percent and for Model 2 ($\hat{\alpha} = 0.2$, $\hat{\beta} = 0.2$, $\hat{\delta} = 0.5$) is 2.7 percent. Figure 21.13 compares box plots of the monthly errors over the preceding five years for the two models. The relative similarity in model performance, once starting values have stabilized, is not apparent by looking only at the MSE.

By holding two parameters constant and varying the third it became apparent that increasing β improved the monthly percentage error. While the fine-tuning was not completed, Model 3, with $\hat{\alpha} = 0.4$, $\hat{\beta} = 0.5$, and $\hat{\gamma} = 0.4$, produced the best model statistics and minimum monthly errors over the past five years. The smoothing constants in this model are often larger in magnitude than those of the simpler smoothing models. Figure 21.14 shows a box plot of these errors. It is apparent that a seasonal pattern still exists in the errors.

A basic change in the process of generating a time series requires a change in parameter values. An adaptive smoothing modification can be tried, or one can select

Table 21.1 Model statistics for Series A through D.

Model	Parameter	Lower confidence limit	Value	Upper confidence limit	Standard deviation*	Chi-squared statistic	Parameter variance-covariance		
A	RMA-1	0.41	0.57	0.73	0.85	28.5 (48df)	1.0		
	SMA-12	0.02	0.23	0.44			-0.15	1.0	
B	RMA-1	0.38	0.55	0.71	0.64	31.3 (48df)	1.0		
	SMA-12	0.17	0.36	0.55			-0.16	1.0	
C	SAR-12	0.95	1.02	1.08	0.02	26.7 (47df)	1.0		
	RMA-1	0.45	0.61	0.76			-0.26	1.0	
	DTC	-0.04	0.02	0.08			-0.04	-0.01	1.0
D	SAR-12	0.87	0.93	0.99	0.31	29.5 (48df)	1.0		
	RMA-1	0.44	0.59	0.75			-0.03	1.0	

*Number of standard deviations by which the mean of the residuals differs from zero.
DTC = Deterministic Trend Constant

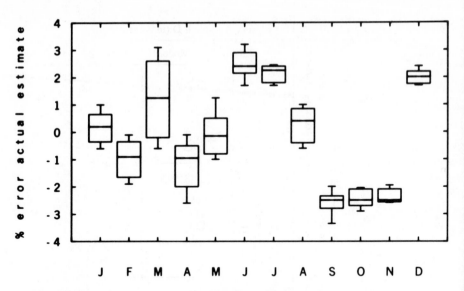

Figure 21.13 (a)   A comparison of box plots for monthly forecast errors over five years for airline Model 1 ($\hat{\alpha} = 0.1$, $\hat{\beta} = 0.2$, $\hat{\delta} = 0.5$).

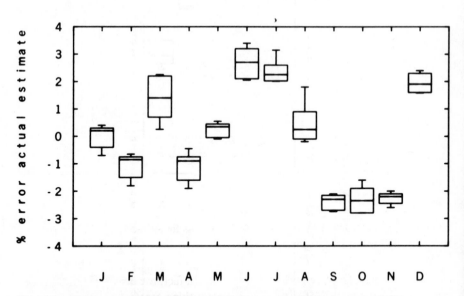

Figure 21.13 (b)   A comparison of box plots for monthly forecast errors over five years for airline Model 2 ($\hat{\alpha} = 0.2$, $\hat{\beta} = 0.2$, $\hat{\delta} = 0.5$).

parameter values in the 0.1-to-0.2 range and allow the forecasts to change gradually. A major drawback of this approach is the large number of possible combinations of parameter values that need to be tested.

Other examples may be found in Bowerman and O'Connell (1979, Chapter 7), Makridakis and Wheelwright (1980, Chapter 4), Montgomery and Johnson (1976),

Table 21.2 Forecast test results for Models A through D (in percent).

	Model A	Model B	Model C	Model D
One-year-ahead results				
1975	−1.31	−1.52	−1.04	−0.94
1976	0.22	−0.21	0.87	0.56
1977	2.21	1.82	2.38	2.20
1978	2.34	1.76	1.71	1.72
Average absolute error	1.52	1.33	1.50	1.36
Two-year-ahead results				
1976	−2.17	−3.24	−0.89	−0.83
1977	2.89	1.76	4.64	3.62
1978	7.35	6.39	7.90	7.36
Average absolute error	4.14	3.80	4.48	3.94

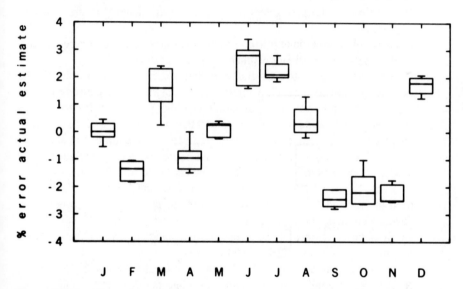

Figure 21.14 A box plot of the monthly forecast errors for airline Model 3 ($\hat{\alpha} = 0.4$, $\hat{\beta} = 0.5$, $\hat{\delta} = 0.4$).

and Sullivan and Claycombe (1977, Chapter 5). In fact, the last of these references provides a listing of FORTRAN programs for making forecasts with a number of techniques including the Winters method of exponential smoothing.

Exponential smoothing methods are intuitively reasonable procedures whose optimality properties can be related to ARIMA processes. Indeed, if the Winters *additive* seasonal predictor is to produce optimal forecasts, it can be shown that the time series must be generated by a five-parameter ARIMA model, whose coefficients are all functions of the three Winters smoothing constants (Granger and Newbold, 1977, Chapter 5).

It is also possible to represent seasonality by combinations of sine and cosine trigometric functions (Harrison, 1965), or even seasonal dummy variables. However, these approaches

- Tend to increase greatly the number of parameters in the model.
- Can induce or leave undisturbed higher order seasonal autocorrelation.
- Imply (in the case of dummy variables) a different error structure than would be adopted when using the Box-Jenkins approach (Thomas and Wallis, 1971).

ARIMA MODELING CHECKLIST

_____ Is the series stationary?

_____ Have all the model parameters (i.e., terms) been included in the model as required?

_____ Examine the correlation matrix. There should *not* be a high degree of correlation between parameter estimates (e.g., over 0.9).

_____ Have the parameter estimates and their standard errors been examined? The confidence interval for each parameter (including the trend constant if there is one) should not span zero, but should be either positive or negative. If the confidence interval does include zero and the interval is basically symmetric about zero, then consideration should be given to eliminating the parameter.

_____ If a regular autoregressive or seasonal autoregressive term is included in the model, the parameter estimates should not be close to 1.0 (e.g., over 0.90 or 0.95). If either is close to 1.0, a regular or seasonal difference should be tried in the model and a forecast test performed to determine which—the differenced model or the autoregressive model—is better.

_____ Do the sum of squares of the errors and the standard error of the residuals become smaller as fits of the model improve?

_____ Does the chi-squared statistic fall below the critical value for the associated degrees of freedom? If so, this indicates white noise. (*Note:* a quick, conservative check for white noise is to see if the model has a chi-squared statistic below the number of degrees of freedom.)

_____ Has a deterministic trend constant been estimated? It is suggested that a deterministic trend constant not be added until the very last. Then if the mean of the residuals of the final model is more than one standard deviation from zero, put a trend constant in the model and test it for significance. (There may be times when one doesn't want the model to put a trend in the forecasts.)

_____ Are there significant patterns in the correlograms of the residuals? Review the ordinary and partial correlograms for any remaining pattern in the residuals. Give primary emphasis to patterns in the correlogram. (*Notes:* (1) The confidence limits on the correlogram and partial correlogram are approximately 95 percent. Hence, 1 spike in 20 would be expected to be outside the confidence limits owing to randomness alone. (2) Decaying spikes in the correlogram, which by visual inspection appear to originate at 1.0, indicate a regular autoregressive process; decaying spikes which originate at a lower level indicate a mixed autoregressive moving-average model.)

_____ Are you watchful of "overdifferencing"? Be alert to the requirement that moving average parameters corresponding to the differencing that has been taken must achieve stationarity. For example, first differences may induce the need for a moving average parameter of order 1. Differencing of order 4 may induce the need for a seasonal moving average parameter in quarterly series. A check of the correlogram helps here. "Overdifferencing" may be apparent if regular and seasonal moving-average parameter estimates are very close to unity.

_____ There are some combinations of parameters that are unlikely to result within the same model. If these combinations are present reevaluate the previous analysis of

- Seasonal differences and seasonal autoregressive parameters.

- Seasonal autoregressive and seasonal moving average parameters.

- Seasonal moving average parameters at other than lags 4 or 12 (or multiples of 4 and 12) for quarterly and monthly series, respectively.

SUMMARY

This chapter has dealt with building forecasting models for seasonal data.

- Regular ARIMA models can be enlarged to include seasonal data by building a multiplicative ARIMA model.

- Seasonal exponential smoothing can be used as an intuitively reasonable, low-cost alternative for short-term forecasting with scarce data.

USEFUL READING

BOWERMAN, B. L., and R. T. O'CONNELL (1979). *Time Series and Forecasting*. North Scituate, MA: Duxbury Press.

BOX, G. E. P., and G. M. JENKINS (1976). *Time Series Analysis, Forecasting and Control, Revised Edition*. San Francisco, CA: Holden-Day.

HARRISON, P. J. (1965). Short-term Forecasting. *Applied Statistics* 14, 102–39.

HOLT, C. C. (1957). *Forecasting Trends and Seasonals by Exponentially Weighted Moving Averages*. Pittsburgh, PA: Carnegie Institute of Technology.

MONTGOMERY, D. C., and L. A. JOHNSON (1976). *Forecasting and Time Series Analysis*. New York, NY; McGraw-Hill.

SULLIVAN, W. G., and W. W. CLAYCOMBE (1977). *Fundamentals of Forecasting*. Reston, VA: Reston Publishing Company.

THOMAS, J. J., and K. F. WALLIS (1971). Seasonal Variation in Regression Analysis. *Journal of the Royal Statistical Society* A. 134, 57–72.

WHEELWRIGHT, S. C., and S. MAKRIDAKIS (1980). *Forecasting Methods for Management, 3rd edition*. New York, NY: John Wiley and Sons.

WINTERS, P. R. (1960). Forecasting Sales by Exponentially Weighted Moving Averages. *Management Science* 6, 324–42.

CHAPTER **22**

Modeling Univariate ARIMA Time Series: Case Studies

This chapter presents a variety of univariate ARIMA models for the series presented throughout the book. In particular,

- A univariate ARIMA model will be derived for the main-telephone gain data as a prelude to developing an explanatory model in terms of the FRB Index of Industrial Production and U.S. housing starts.

- Univariate ARIMA models will be derived for the U.S. housing starts series and the FRB Index of Industrial Production series.

- A univariate ARIMA model will also be derived for nonfarm employment, a series used in the telecommunications forecasting example in *The Beginning Forecaster,* and utilized further in Chapter 24 in a transfer function model for telephone toll revenues.

AN ARIMA MODEL FOR MAIN TELEPHONE GAIN

In Chapter 3 the main-telephone gain series was modeled by using a multiple linear regression approach. In this chapter, an ARIMA model will be built for the same time series. In Chapter 24 a combined transfer function and noise model will be developed relating main telephone gain to U.S. housing starts and the FRB Index of Industrial Production.

Obtaining Stationarity

A time plot of the main gain data, as a quarterly time series from the first quarter of 1959 to the fourth quarter of 1978, is shown in Figure 22.1. The data show **307**

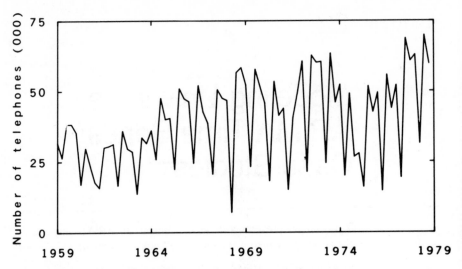

Figure 22.1   A time plot of quarterly main-telephone gain data from the first quarter of 1959 to the fourth quarter of 1978.

increasing volatility in the seasonal pattern over time. An ANOVA decomposition of this series indicated that trend, seasonal, and irregular variation accounted for 7.3 percent, 81.3 percent, and 11.4 percent, respectively of the variation about the mean. The quarters in 1968 and 1971 in which labor strikes occurred are quite evident (low values), although the 1971 strike is not as pronounced as the 1968 strike. However, when comparable quarters of prior and following years are considered, it does appear that the third quarter of 1971 is low.

At this point it is necessary to adjust the original strike data to more reasonable levels. The intervention modeling approach (to be presented in Chapter 23) provides one way of establishing replacement values. To demonstrate the effect that the strike outliers have on the estimation of model parameters, univariate models will be developed for the unadjusted series and a strike-adjusted version of the series. In addition to impacting the parameter estimates, outliers can affect model forecasts dramatically *if* they occur in the most recent data *and* are part of the calculation of the forecasts.

In this example, the same model structure results (in terms of the order of the autoregressive and moving average parameters) when either version of the time series is used. This is not always the case. Even though the order of the models is identical, the model coefficients are different. Since identical model structures apply, the model identification steps will be covered only for the adjusted series.

The ordinary correlogram of the series shown in Figure 22.2 takes a long time to decay, indicating a need for seasonal differencing (differences of order 4).

Figure 22.3(a) shows the ordinary correlogram of the differences of order 4 of the series. It dies out and becomes insignificant relatively quickly, suggesting that the data are stationary.

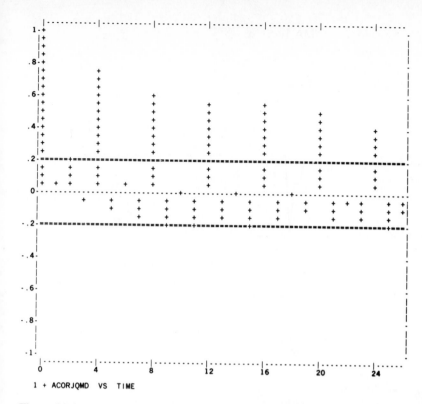

1 + ACORJQMD VS TIME

Figure 22.2 Plot of the ordinary correlogram of the data shown in Figure 22.1

Identification of ARMA Models

The identification of a tentative ARMA model requires the ordinary correlogram and the partial correlogram for the differences of order 4 in the main gain data. The decay in the correlogram (Figure 22.3(a)) suggests a regular autoregressive process. In the partial correlogram (Figure 22.3(b)), the significant positive value at lag 1 confirms an autoregressive model and indicates that the order of the model is equal to 1.

The tentative $(1,0,0) \times (0,1,0)^4$ model was estimated and Figure 22.4 shows the residual ordinary and partial correlograms. The negative spike at lag 4 in the correlogram suggests the need for a seasonal moving average parameter. The pattern of negative spikes at lags 4, 8, and 12 in the partial correlogram confirms the need for a seasonal moving average parameter. The starting model, Model A, in $(p, d, q) \times (P,D,Q)^s$ notation, is $(1,0,0)$ lts $(0,1,1)^4$.

Parameter Estimation and Diagnostic Checking

Table 22.1 shows that the parameter estimates of the two models fitted to the adjusted and unadjusted data are all significant, i.e., the confidence intervals do not span 0.

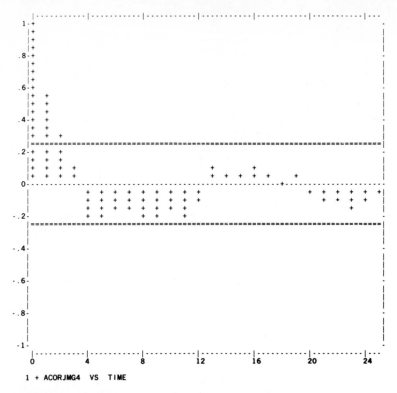

```
        |·········|·········|·········|·········|·········|···
    1·+                                                        ·
        |+
        |+
        |+
    .8·+                                                        ·
        |+
        |+
        |+
    .6·+  +                                                     ·
        |+  +
        |+  +
        |+  +
    .4·+  +                                                     ·
        |+  +
        |+  +  +
        ||==========================================================|
    .2·+  +  +                                                  ·
        |+  +  +
        |+  +  +  +                              +        +
        |+  +  +  +                              +  +  +  +  +     +
     0·········································+········+········+·
        |        +  +  +  +  +  +  +  +                +  +  +  +  +  +|
        |        +  +  +  +  +  +  +  +  +                 +  +  +  +|
        |        +  +  +  +  +  +  +  +  +                       +|
  -.2·|        +  +     +  +  +     +                          ·
        ||==========================================================|
  -.4·|                                                        ·
        |
        |
  -.6·|                                                        ·
        |
        |
  -.8·|                                                        ·
        |
        |
   -1·|                                                        ·
        |·········|·········|·········|·········|·········|···
        0        4        8        12       16       20       24

  1 + ACORJMG4   VS   TIME
```

Figure 22.3(a)   Plots of the ordinary correlogram of the fourth-order differences of the data shown in Figure 22.1.

The estimated parameters in each model are not highly correlated (the estimated variance-covariance matrix is not shown). The mean value of the residuals is 1.4 standard deviations from 0. The analyst has the option of whether or not to include a deterministic trend constant in the model. Omitting a trend constant usually results in a model that is more responsive to new observations. Since the main gain series is volatile, it was decided not to include a trend constant.

The residual correlograms did not show any significant values or patterns. The sample chi-squared statistic ($= 2.67$) is less than the theoretical value ($= 12.59$) for white noise for six degrees of freedom at the 5-percent level (Appendix A, Table 3). There is no evidence to reject the null hypothesis of random residuals. Diagnostic checking has shown that no modifications in the initial models are called for.

Forecast Test Results

Forecast tests were performed for the two models and the results are shown in Table 22.2. Five one-year-ahead forecasts and four two-year-ahead forecasts were generated. In terms of the average absolute percent error, the model based on the adjusted data had slightly more accurate one-year-ahead forecasts but was somewhat inferior

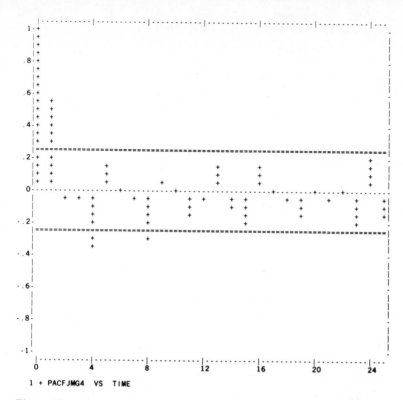

1 + PACFJMG4 VS TIME

Figure 22.3(b) Plots of the partial correlogram of the fourth-order differences of the data shown in Figure 22.1.

for two-year-ahead forecasts. If the outliers occurred in the most current periods, the forecasts from the (unadjusted) model would have been distorted.

In a similar context, Levenbach (1980) studied a number of time series models for main telephone gain by using a methodology due to Priestley (1971). The quarterly series used in that paper differed from the study in this chapter in two respects:

- The timespans used for fitting and evaluating one- and two-year-ahead forecasts were not the same.
- The definition of "strike adjustment" in the data was different.

Consequently, model specifications differ somewhat. However, these differences did not turn out to be significant for forecasting purposes. Clearly, similar model specifications are possible and should not cause too much concern for the practicing forecaster.

AN ARIMA MODEL FOR HOUSING STARTS

The Federal Reserve Board Index of Industrial Production series and U.S. housing starts series were used in multiple linear regression models for quarterly main gain

Table 22.1 Summary of univariate models fitted to main-
telephone gain data.

Model type	Estimated model, and (standard errors)	Residual variance ($\times 10^8$)
Seasonal differences (unadjusted series)	$(1 - 0.66B)(1 - B^4)\text{GAIN} = (1 - 0.58B^4)\varepsilon_t$ (± 0.09) $\qquad\qquad\qquad\qquad (\pm 0.10)$	46.4
Seasonal differences (adjusted series)	$(1 - 0.74B)(1 - B^4)\text{GAIN*} = (1 - 0.51B^4)\varepsilon_t$ (± 0.08) $\qquad\qquad\qquad\qquad (\pm 0.10)$	33.8

Table 22.2 Forecast test results for two models for the
main-telephone gain data.

Model type	Average absolute percent error	
	One year ahead	Two years ahead
Seasonal differences (unadjusted series)	9.7	22.5
Seasonal differences (adjusted series)	9.5	24.0

in Chapter 3. Univariate ARIMA models will be developed now to predict future values for these series. These series are then used in Chapter 24 in a transfer function model for main telephone gain.

The quarterly housing starts series is noticeably seasonal, and this was confirmed by the significant value at lag 4 in the correlogram (Figure 22.4). The series was assumed to be stationary and a model with regular and seasonal autoregressive parameters was tried. The pattern in the ordinary correlogram at low order lags is similar to that described in Figure 21.3. The residual correlogram shown in Figure 22.5 indicates a decaying pattern (autoregressive) at lags 1–3 and a significant negative spike at lag 4. A regular autoregressive parameter (second order was added to account for the remaining autoregressive decay. A seasonal moving average parameter was included to care for the spike at lag 4. At this time the spike at lag 10 was not addressed. The model was estimated and the residual correlogram indicated a significant positive spike at lag 2; the spike at lag 10 was no longer significant. Before attempting to enlarge the model, the estimated parameters were received. The model equation (with standard errors shown below) was

$$(1 - 1.14B + 0.29B^2)(1 - 0.96B^4)\,\text{Housing starts} = (1 - 0.42B^4)\varepsilon_t \ .$$
$$\pm 0.11 \quad \pm 0.11 \qquad \pm 0.02 \qquad\qquad\qquad\qquad \pm 0.14$$

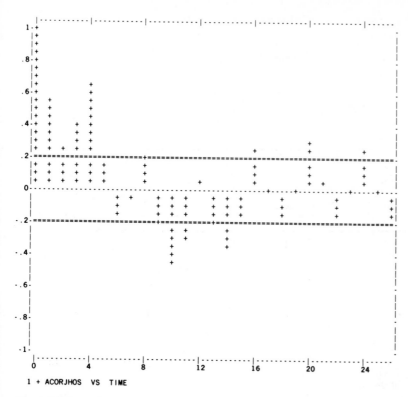

Figure 22.4 A plot of the correlogram of the housing starts data.

The very large value for the seasonal autoregressive parameter (= 0.96) indicated that the model was very close to being nonstationary. This suggests taking seasonal differences prior to modeling.

Figure 22.6 shows the resulting ordinary correlogram and partial correlogram of the differenced series; now the sinusoidal decay in the correlogram (Figure 22.6(a)) indicated an autoregressive process of an order greater than 1. The significant spikes up to lag 3 in the partial correlogram (Figure 22.6(b)) suggested that the order of the process should be 3. If this were to be too high, however, the higher-order parameters would be insignificant when the model is estimated. A negative spike at lag 4 in the residual correlogram and a pattern of decay for negative spikes at lags 4, 8, etc., in the residual partial correlogram indicated the need for a seasonal moving average parameter of order 1. This model was estimated, and the correlogram of the resulting residuals showed no further patterns. However, the second-order autoregessive term was not significant and was deleted in the final model. The residual chi-squared statistic (= 2.35) is less than the tabulated value (= 9.49) for four degrees of freedom.

Since the autoregressive parameters in the previous model suggested a rather complex model, and the second-order term was insignificant, consideration was

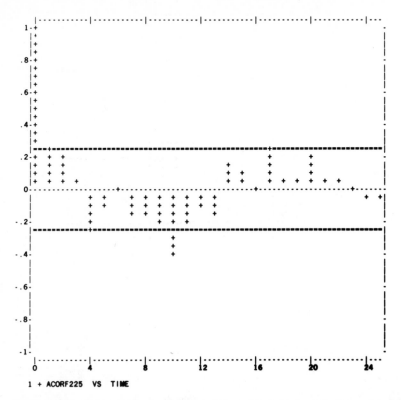

Figure 22.5   A plot of the correlogram of the residuals in the tentative model of the housing starts.

given to reducing the regression period and modeling the more recent observations to see if a simpler model would result. The time period was shortened to cover the 1965–1978 period. In this case, a regular autoregressive model of order 2 was sufficient. This reduced the order of the autoregressive model by one.

Table 22.3 summarizes the housing starts models and Table 22.4 compares the forecast test results.

AN ARIMA MODEL FOR THE FRB INDEX OF INDUSTRIAL PRODUCTION

Figures 22.7 and 22.8 show the quarterly FRB Index of Industrial Production and its correlogram. The series is the sum of a number of components with different seasonalities and is generally reported on a seasonally adjusted basis so there is no significant seasonal pattern in the correlogram. The slow decay in the correlogram

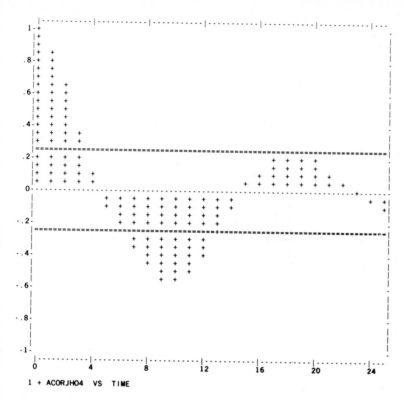

1 + ACORJHO4   VS   TIME

Figure 22.6(a)   Plots of the ordinary correlogram of the differences of order 4 of the housing starts.

indicated the need for taking differences of order 1. Figure 22.9 shows the correlogram and partial correlogram of these first differences of the series. The large positive spike in the correlogram (Figure 22.9(a)) at lag 1 and the alternating decay in the partial correlogram (Figure 22.9(b)) indicated the need for a regular moving average parameter of order 1. This model was estimated, and there was no significant pattern in the correlogram of the residuals.

AN ARIMA MODEL FOR NONFARM EMPLOYMENT

In Chapter 24 a transfer function model will be developed relating telephone toll revenues to nonfarm employment. It will be seen that transfer function modeling begins by building a univariate ARIMA model for the independent (input) series. This model can also be used to generate nonfarm employment forecasts.

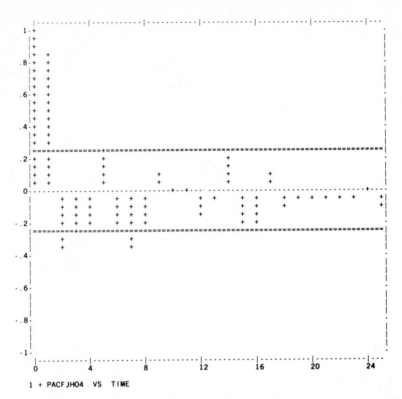

1 + PACF JHO4   VS   TIME

Figure 22.6(b)   Plots of the partial correlogram of the differences of order 4 of the housing starts.

Table 22.3   Summary of univariate models for U.S. housing starts data.

Model type (years)	Estimated model, and (standard errors)	Residual variance ($\times 10^8$)
Seasonal differences (1959–1978)	$(1 - 1.12B + 0.28B^3)(1 - B^4)\text{HOUS} = (1 - 0.76\ B^4)\varepsilon_t$ $(\pm 0.07)\ (\pm 0.07)$ (± 0.10)	1306
Seasonal differences (1965–1978)*	$(1 - 1.21B + 0.32B^2)(1 - B^4)\text{HOUS} = (1 - 0.85B^4)\varepsilon_t$ $(\pm 0.13)\ (\pm 0.13)$ (± 0.05)	1571

*A model of the same order but based on data from 1959–1978 had an estimated residual variance of 1254.

Table 22.4 Forecast test results for two models of U.S. housing starts data.

Model type	Average absolute percent error	
	One year ahead	Two years ahead
$(1 - \phi_1 B - \phi_3 B^3)(1 - B^4)Y_t = (1 - \theta_4 B^4)\varepsilon_t$	21.1	59.0
$(1 - \phi_1 B - \phi_2 B^2)(1 - B^4)Y_t = (1 - \theta_4 B^4)\varepsilon_t$	31.2	81.4

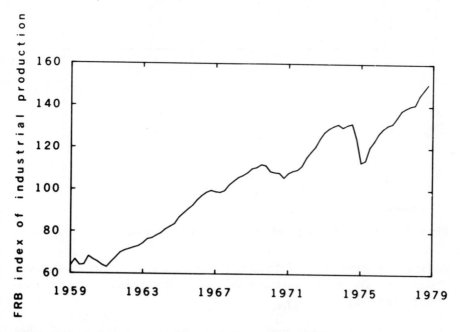

Figure 22.7 A time plot of the quarterly FRB Index of Industrial Production from the first quarter of 1959 to the fourth quarter of 1978. *Source:* Board of Governors of the Federal Reserve System.

Obtaining Stationarity

Figure 22.10 shows the monthly nonfarm employment series and Figure 22.11 shows the ordinary correlogram of the data. The gradual decay in the correlogram indicated the need for regular differences of order 1. The ordinary correlogram of the first-differenced series showed a gradual decay at multiples of lag 12; this indicated the need for taking differences of order 12 as well, to obtain a stationary series. Plots

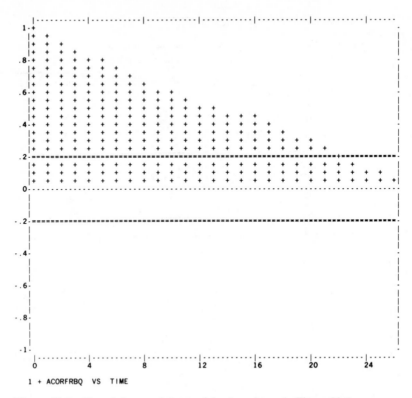

Figure 22.8   Plot of the correlogram of the data shown in Figure 22.7.

of the correlograms for the differenced nonfarm employment data are shown in Figure 22.12. The correlogram in Figure 22.12(a) decays quickly, indicating a stationary series.

Identification

In Figure 22.12(a) the sinusoidal decay in the correlogram at low-order lags suggested a regular autoregressive model of order greater than 1. The negative spikes at lags 12 and 24 in the correlogram (Figure 22.12(a)) indicated a seasonal moving average model of order 2. The spike at lag 2 in the partial correlogram (Figure 22.12(b)) indicated that the order of the autoregressive process should be 2. The decay of negative spikes at lags 12, 24, and 36 in the partial correlogram confirmed that seasonal moving average parameters would be required.

Parameter Estimation and Diagnostic Checking

The final model for nonfarm employment (NFRM) was estimated and took the following form (standard errors are shown beneath the formula for the model):

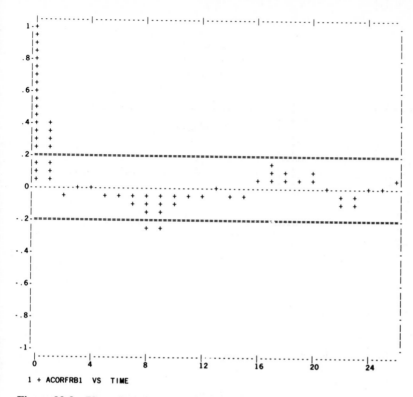

Figure 22.9 Plots of (a) the correlogram and (b) the partial correlogram of the first differences of the FRB Index of Industrial Production.

$$(1 - 1.116B + 0.236B^2)(1 - B)(1 - B^{12})\,\text{NFRM} = (1 - 0.591B^{12} - 0.314B^{24})\varepsilon_t\,.$$
$$\pm\,0.090\quad\pm\,0.090\qquad\qquad\qquad\qquad\pm\,0.085\quad\pm\,0.084$$

In comparison with the standard errors, all the estimates of the model parameters are significant. Moreover, the parameter estimates are not highly correlated. The mean of the residuals is not significantly different from zero, and the chi-squared statistic ($= 2.18$) is less than the theoretical value ($= 31.4$) for 20 degrees of freedom. There were no significant patterns in the residual correlograms.

Forecast Test Results

The model predicts future values with an average absolute one-year-ahead forecast error of 0.77 percent and a two-year-ahead error of 2.79 percent. A similar model using the *logarithms* of nonfarm employment was tested, and the forecast test results were almost identical to those of the untransformed series.

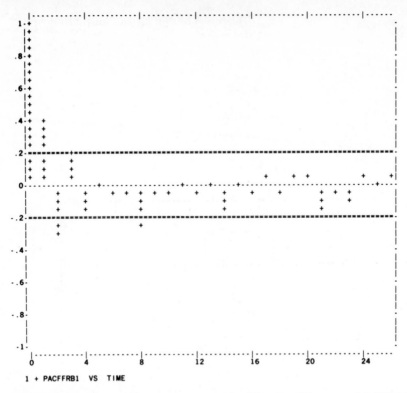

1 + PACFFRB1 VS TIME

Figure 22.9(b) Plot of the partial correlogram of the first differences of the FRB Index of Industrial Production.

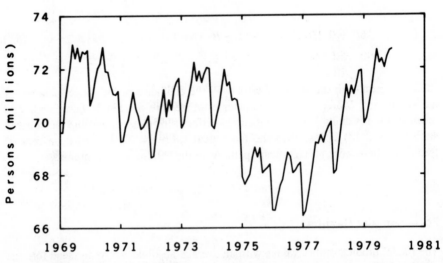

Figure 22.10 A time plot of a monthly nonfarm employment series from January 1969 to December 1979.

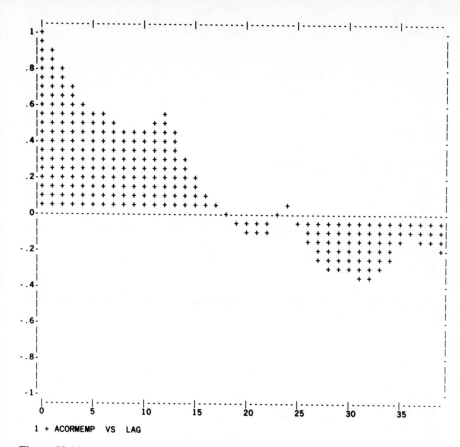

1 + ACORMEMP VS LAG

Figure 22.11 A plot of the correlogram of the data shown in Figure 22.10.

OTHER STUDIES

Numerous studies exist in the literature on univariate ARIMA modeling. In addition to the examples found in the textbooks dealing with the Box-Jenkins modeling strategy, several published studies related to the ones in this chapter include Bhattacharyya (1974), Box et al. (1967), Brubacher and Wilson (1976), Chatfield and Prothero (1973), Jensen (1979), Thompson and Tiao (1971), and Tomasek (1972).

SUMMARY

This chapter provided an illustration of the identification stage for a number of univariate models of three time series.

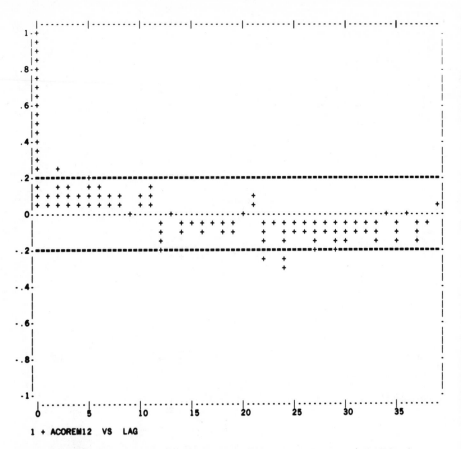

1 + ACOREM12 VS LAG

Figure 22.12(a) Plots of the correlogram of the differences of order 1 and 12 for the nonfarm employment data.

- A univariate model for *quarterly* main-telephone gain to be used further in Chapter 23.

- Univariate ARIMA models for U.S. housing starts and the FRB Index of Industrial Production to be related to main telephone gain.

- A univariate model for a *monthly* nonfarm employment series.

The modeling exercise demonstrated that several tentative models can be entertained for any given series. Diagnostic checks for stationarity, invertibility, significance of coefficients, white noise residuals, and heteroscedasticity may eliminate some of the models from further consideration. Forecasters must use judgment and establish their criteria for selecting the most appropriate model for a given application.

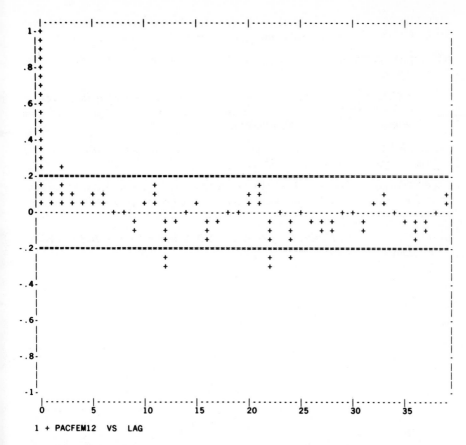

1 + PACFEM12 VS   LAG

Figure 22.12(b)   Plots of the partial correlogram of the differences of order 1 and 12 for the nonfarm employment data.

USEFUL READING

BHATTACHARYYA, M. N. (1974). Forecasting the Demand for Telephones in Australia. *Applied Statistics* 23, 1–10.

BOX, G. E. P., G. M. JENKINS, and D. W. BACON (1967). Models for Forecasting Seasonal and Nonseasonal Time Series. *In* B. Harris, ed. *Advanced Seminar on Spectral Analysis*. New York, NY: John Wiley and Sons.

BRUBACHER, S., and G. T. WILSON (1976). Interpolating Time Series with Application to the Estimation of Holiday Effects on Electricity Demand. *Applied Statistics* 25, 107–16.

CHATFIELD, C., and D. L. PROTHERO (1973). Box-Jenkins Seasonal Forecasting: Problems in a Case Study. *Journal of the Royal Statistical Society* A 136, 295–315.

JENSEN, D. (1979). General Telephone of the Northwest: A Forecasting Case Study. *Journal of Contemporary Business* 8, 19–34.

LEVENBACH, H. (1980). A Comparative Study of Time Series Models for Forecasting Telephone Demand. *In* O. D. Anderson, ed., *Forecasting Public Utilities*. Amsterdam, Netherlands: North-Holland Publishing Co.

PRIESTLEY, M. B. (1971). Fitting Relationships between Time Series. *Bulletin of the International Statistical Institute* 44, 295–324.

THOMPSON, H. E., and G. C. TIAO (1971). An Analysis of Telephone Data: A Case Study of Forecasting Seasonal Time Series. *Bell Journal of Economics and Management Science* 2, 514–41.

TOMASEK, O. (1972). Statistical Forecasting of Telephone Time Series. *ITU Telecommunications Journal* 39, 1–7.

CHAPTER 23

Single Output–Multiple Input Modeling

Chapters 17–22 dealt with the subject of building univariate (single output) time series models by using the Box-Jenkins modeling strategy (Box and Jenkins, 1976). This approach is now extended to single output time series models in which

- One or more related input variables are introduced in a *transfer function relationship* with the output variable to provide an improved description of the process.

- The input variables are used to describe any *additional* information not already contained in the past history of the output variable.

- The time series are *prewhitened* prior to modeling. Prewhitening refers to the process of converting an input variable to a random series.

TRANSFER FUNCTION MODELING

Single-output transfer function models have a single dependent or output variable and one or more independent or input variables (Box and Jenkins, 1976; Jenkins, 1979, and Pack, 1977). Forecasts of the dependent variable are based on its own history and its relation to other variables.

Model Relationships

To understand what a transfer function is, consider a model with one output variable Y_t and two input variables X_{1t} and X_{2t}, as illustrated in Figure 23.1. The transfer function model for this example is:

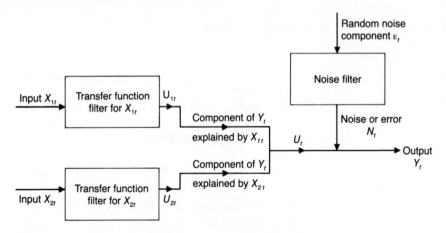

Figure 23.1 A flowchart of a single output–multiple input transfer function model with two inputs X_{1t} and X_{2t}.

$$Y_t = U_{1t} + U_{2t} + N_t \text{,}$$

where

- U_{1t} is that component of Y_t explained exactly by X_{1t}.
- U_{2t} is that component of Y_t explained exactly by X_{2t}.
- N_t is an error or noise term and represents the variation that is not explained by X_{1t} and X_{2t}.

As Figure 23.1 indicates, there is a *filter* for each independent variable: these are termed *transfer function filters*. Consider a nonseasonal transfer function relating Y_t and X_{1t} (the subscript (1) can be left off for convenience) such that

$$Y_t = \omega_0 X_{t-b} \text{,}$$

where ω_0 is a constant to be estimated and b is the *lead time* or *delay* to be estimated from the data. This represents the number of intervals of time before a change in X_t begins to have an effect on Y_t.

Consider next a more complicated transfer function relationship in which there is one lag of the input variable in addition to the ω_0 relationship and one lag of the output variable. Then

$$(1 - \delta_1 B)Y_t = (\omega_0 - \omega_1 B)X_{t-b} \text{,}$$

where

- δ_1 is the autoregressive parameter.

- ω_0 nd ω_1 are "moving average" parameters.
- B is the backshift operator.

Then

$$Y_t - \delta_1 Y_{t-1} = \omega_0 X_{t-b} - \omega_1 X_{t-b-1} \, .$$

In general, the transfer function model has s input lags and r output lags, such that

$$(1 - \delta_1 B - \delta_2 B^2 - \cdots - \delta_r B^r)Y_t = (\omega_0 - \omega_1 B - \omega_2 B^2 - \cdots - \omega_s B^s)X_{t-b},$$

or

$$Y_t = \frac{(\omega_0 - \omega_1 B - \cdots - \omega_s B^s)}{(1 - \delta_1 B - \cdots - \delta_r B^r)} X_{t-b} = \frac{\omega(B)}{\delta(B)} X_{t-b} = v(B)X_t \, ,$$

where

$$v(B) = \frac{\omega(B)}{\delta(B)} B^b \; ;$$

$v(B)$ is a transfer function.

The transfer function $v(B)$ is a polynomial (possibly infinite) in the backshift operator B; values of $v(B)$ are called *impulse response weights* when Y_t is expressed as an infinite sum of X_t. The weights tell which values of X_t have the greatest impact on Y_t, and show how X_t is "reflected in" or *transferring* itself to Y_t. No "feedback" is assumed; that is, X_t determines Y_t but Y_t does not determine X_t.

The Combined Transfer Function–Noise Model

In general, X_t does not perfectly explain Y_t and there is a need to include a noise component, N_t, of the form:

$$(1 - \emptyset_1 B - \emptyset_2 B^2 - \cdots - \emptyset_p B^p)N_t = (1 - \theta_1 B - \theta_2 B^2 - \cdots - \theta_q B^q)\varepsilon_t$$

or

$$N_t = \frac{\theta(B)}{\emptyset(B)} \varepsilon_t \, ,$$

where

- ε_t is the usual white noise process.

- $\theta(B)$ is a moving average polynomial of order q.
- $\emptyset(B)$ is an autoregressive polynomial of order p.

The complete *transfer function–noise model* is

$$Y_t = v(B)X_t + \frac{\theta(B)}{\emptyset(B)}\varepsilon_t \ .$$

Identification

The cross correlogram of the residuals of the prewhitened series is the identification tool used to estimate the order of the transfer function model, where

- $b = $ lead time.
- $s = $ number of input (X_t) lags.
- $r = $ number of output (Y_t) lags.

Since both the transfer function and the noise parts of the model are estimated simultaneously, there is a need to establish an initial noise component. A starting noise model may be assumed to be the univariate model for Y_t. The univariate model is proposed since, if the transfer function equals zero, then $Y_t = N_t$.

The problem of estimating the order of a transfer function model is not as simple as generating a cross correlogram between Y_t and X_t. When two time series are highly autocorrelated (e.g., have strong trend or seasonal patterns), there will be a seemingly high correlation even if both series have no true relationship at all (Bartlett, 1935).

To circumvent this difficulty, one must determine if there is any *additional* information in X_t that is not already captured by the history of Y_t; in part because it accomplishes this, Box and Jenkins (1976) propose *prewhitening* each series. This implies fitting univariate ARIMA models to each series and then cross-correlating the residual series.

Prewhitening can be recommended for two reasons: the first is the spurious correlation described by Bartlett; the second is to establish theoretical cross-correlogram patterns to permit the forecaster to identify b, s, and r.

What if X_t Is White Noise?

Suppose that X_t is white noise and that Y_t is related to X_t as follows:

$$Y_t = \omega_0 X_t - \omega_1 X_{t-1} + N_t \ ,$$

where X_t is independent of N_t.

Then the cross-covariance coefficients between Y_t and X_t are calculated at various lags k as the expected value of $X_t Y_t$, where X_t and Y_t are deviations from their respective means:

$$\gamma_{XY}(k) = E(X_{t-k}Y) = E(\omega_0 X_{t-k}X_t - \omega_1 X_{t-k}X_t + X_{t-k}N_t) \; ;$$

and

$$\gamma_{XY}(0) = \omega_0 \operatorname{var} X_t - \omega_1 E(X_t X_{t-1}) + E(X_t N_t)$$
$$= \omega_0 \sigma_X^2 - 0 + 0 \; .$$

The second term in the second equation is zero since the X_t's are independent by assumption and the third term is zero since X_t and N_t are assumed to be independent of each other.

The *cross-correlation coefficient* γ^k is defined as

$$\rho_{XY}(k) = \frac{\gamma_{XY}(k)}{\sigma_X \sigma_Y} . \text{ for } k = 0, \pm 1, \pm 2, \cdots$$

Then

$$\rho_{XY}(0) = \frac{\omega_0 \sigma_X^2}{\sigma_X \sigma_Y} = \omega_0 \frac{\sigma_X}{\sigma_Y}$$

Next compute

$$\rho_{XY}(1) = \frac{\gamma_{XY}(1)}{\sigma_X \sigma_Y} \; .$$

Since

$$\gamma_{XY}(1) = E(X_{t-1}Y_t) = E(\omega_0 X_{t-1}X_t - \omega_1 X_{t-1}X_{t-1} + X_{t-1}N_t)$$
$$= 0 - \omega_1 \sigma_X^2 + 0 \; ,$$

then

$$\rho_{XY}(1) = \frac{-\omega_1 \sigma_X^2}{\sigma_X \sigma_Y} = \frac{-\omega_1 \sigma_X}{\sigma_Y} \; .$$

All $\rho_{XY}(k)$ for $k > 1$ are equal to zero since there are no longer any subscripts in common.

The conclusion is that if X_t and N_t are independent of each other and if X_t is white noise, then the cross-correlation coefficients are directly proportional to the impulse response weights relating Y_t to X_t.

What if X_t Is Not White Noise?

Since the input series is typically not independently distributed, it is not possible to establish simple theoretical cross-correlogram patterns between Y_t and X_t. This problem can be solved by first prewhitening X_t.

Suppose that X_t and Y_t are stationary series. If

$$\emptyset_x(B)X_t = \theta_x(B)\alpha_t ,$$

then

$$\frac{\emptyset_x(B)}{\theta_x(B)}X_t = \alpha_t .$$

The operation of the univariate ARIMA filter on X_t yields white noise residuals α_t. This is a white noise series for X_t.

Recall that a transfer function–noise model is of the form

$$Y_t = v(B)X_t + N_t .$$

Applying the filter for X_t to the *entire* model yields

$$\frac{\emptyset_x(B)}{\theta_x(B)}Y_t = v(B)\frac{\emptyset_x(B)}{\theta_x(B)}X_t + \frac{\emptyset_x(B)}{\theta_x(B)}N_t .$$

Let

$$\beta_t = \emptyset_x(B)\theta_x^{-1}(B)Y_t$$

and

$$\varepsilon_t = \emptyset_x(B)\theta_x^{-1}(B)N_t .$$

Then

$$\beta_t = v(B)\alpha_t + \varepsilon_t .$$

Multiplying through by α_{t-k}, we obtain

$$\alpha_{t-k} \cdot \beta_t = v(B)\alpha_{t-k} \cdot \alpha_t + \alpha_{t-k} \cdot \varepsilon_t .$$

Taking expectations gives the cross-correlation function, $\gamma_{\alpha\beta}(k)$ for positive k, as

$$\gamma_{\alpha\beta}(k) = E[\alpha_{t-k}\beta_t] = v_k\sigma_\alpha^2 ,$$

assuming α_{t-k} and ε_t are uncorrelated. Then,

$$\rho_{\alpha\beta}(k) = \frac{\gamma_{\alpha\beta}(k)}{\sigma_\alpha\sigma_\beta} = v_k\frac{\sigma_\alpha}{\sigma_\beta} , \qquad k = 0, 1, 2, \cdots$$

The result is that if:

- X_t and Y_t are stationary series
- X_t is prewhitened by its univariate model
- Y_t is prewhitened by the same model used for X_t
- X_t and N_t are uncorrelated,

then the cross-correlation coefficients between the two prewhitened series are directly proportional to the impulse response function relating Y_t to $\overset{*}{X}_t$. Therefore, the cross-correlation function of the two prewhitened series can be used to identify b, s, and r.

The impulse response function provides the tool for identifying b, s, and r. The first significant spike is the value for b. If there is no autoregressive pattern of decay in the sample impulse response function, then $r = 0$. The value for s is one less than the number of significant spikes. A single spike corresponds to $s = 0$ (no lags of the input series). Two spikes indicate that $s = 1$ (one lag of the input series).

When there is an autoregressive decay in the impulse response function, the value of r is the order of the autoregressive process. This can usually be determined by looking at the tail of the impulse response function. The value for s can then be determined by observing the number of spikes that have no fixed pattern preceding the pattern of decay called for by the value of r. The formula $s + 1 - r =$ number of irregular preceding spikes can be used to obtain s once r has been determined.

A regular decaying pattern in the tail of the impulse response function, corresponding to an autoregressive process of order 1, has $r = 1$. If the pattern is preceded by no values that are irregular, then $s = 0$. A decaying pattern preceded by one irregular spike has $r = 1$ and $s = 1$. A decaying pattern preceded by two irregular spikes has $r = 1$ and $s = 2$.

A sinusoidal decaying pattern, corresponding to an autoregressive process of order 2, has $r = 2$. With no preceding irregular spikes, $s + 1 - 2 = 0$ or $s = 1$. A sinusoidal decaying pattern with a preceding irregular spike has $r = 2$ and $s = 2$. A sinusoidal decaying process with two preceding irregular spikes has $r = 2$ and $s = 3$. (When $s < r$, no preceding irregular values occur, so it is possible that s may be less than the value given by the formula.)

The distinctions between s and r for practical applications àre very difficult to make. As a consequence, one must try several combinations of s and r and see which gives the best results.

Diagnostic Checking

In diagnostic checking, the adequacy of both the transfer function and noise models are evaluated by:

- Plotting the residuals to identify possible outliers.

- Plotting the autocorrelogram of the residuals to identify nonrandom patterns suggesting the need to modify either the transfer function or noise components (or both).

- Plotting the cross correlogram of the residuals of the transfer function–noise model and the prewhitened input series. Significant spikes or patterns suggest that the transfer function component is inadequate. An assumption of the method is that residuals are independent of the input series.

- Insignificant parameters in either transfer function or noise components are eliminated to have parsimonious models. By introducing the transfer function component, the appropriate noise component may now be simpler.

- The correlation matrix of parameter estimates is inspected for evidence of high correlation. When high correlation exists, parameters should be considered for elimination from the model, or more parsimonious parameterization should be considered (e.g., perhaps fewer moving average parameters could replace numerous autoregressive parameters or vice versa).

INTERVENTION MODELS

Intervention models are transfer function models where the input series are dummy variables (Box and Jenkins, 1976; Jenkins, 1979). This is similar to the use of dummy variables that was described in Chapter 12.

Uses of Intervention Models

Intervention models are applicable to measure

- The effect of strikes, unusual weather, and other one-time effects. In this case the dummy equals 1 for the unusual event and 0 elsewhere.

- The effect of a change in policy. Here the dummy variable equals 0 before the policy change and 1 after the change.

- The effect of the presence or absence of advertising on sales volume. The dummy equals 1 during an advertising campaign and 0 at other times.

As has been emphasized throughout the book, outliers and unusual values can significantly impact the estimates of the parameters in a model. In the case of strikes, the intervention model provides a vehicle to prevent the unusual data from distorting the relationship between output and input series. In the case of advertising, where it is generally difficult to quantify advertising in terms of an independent variable (the dollars expended do not necessarily equate with effectiveness), the transfer function model can show the incremental volume associated with the advertising program.

Identification

It is not possible to identify the orders of intervention models by the prewhitening process used for transfer function models. Instead, the transfer function that is selected is chosen to fit the observed or theoretical response to the dummy variable.

Figure 23.2 shows both a *pulse input* and a *step input* as possible dummy variables. If the observed or theoretical response takes the form of any of the figures in the middle or right-hand column, the appropriate transfer function is the one indicated in the left-hand column.

The responses are similar to the forecast profiles for ARIMA models presented in Chapter 20. The orientation is different, however. For example, an observed response that looks like a step function can result from a transfer function of type a (ω_0) applied to a step input or a transfer function of type d applied to a pulse input.

The forecaster is generally aware of the appropriate input series; for example, a pulse input is equivalent to an outlier and a step input is equivalent to a change in policy. The appropriate input is a function of environmental conditions surrounding the forecast problem. Therefore, depending on the type of dummy variable selected, if the forecaster observes a response that is similar to one of the panels 1–8 in the figure, the transfer function indicated in the left-hand column would be appropriate.

To see why this is the case, assume a transfer function of type b and an input series of the pulse type. Then

$$Y_t = (\omega_0 - \omega_1 B)\varepsilon_t = \omega_0\varepsilon_t - \omega_1\varepsilon_{t-1}$$

Table 23.1 shows the response to a pulse input at time period 2; $\varepsilon_t = 1$ for $t = 2$ and 0 for all other times. The response is ω_0 in time period 2 and $-\omega_1$ in time period 3. The response is 0 for all other time periods.

Table 23.2 shows the response to a step input beginning at time period 2. The response is 0 up to time 2, ω_0 at $t = 2$, and $\omega_0 - \omega_1$ for $t \geq 3$.

We have already indicated that by reversing the procedure and starting with the observed response and a pulse or step input, you can determine what the transfer function should be, and this is also true for cases not shown in Figure 23.2.

334

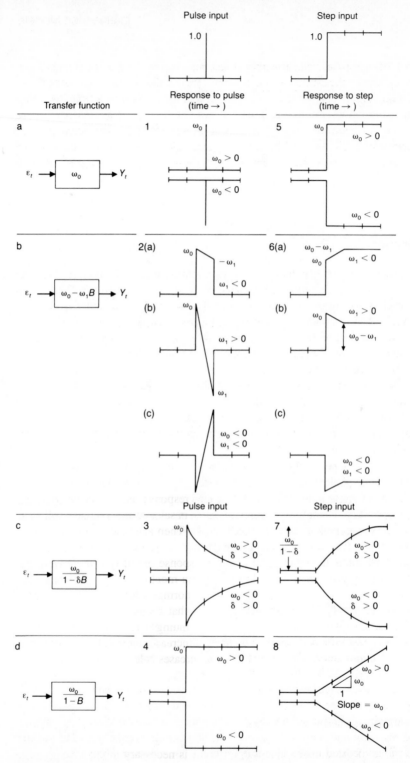

Figure 23.2 Pulse and step impulse responses for four common transfer functions.

Table 23.1 The response to a pulse input at time period 2.

Time t	Input ε_t	Response $\omega_0\varepsilon_t - \omega_1\varepsilon_{t-1}$
0	0	$0 - 0 = 0$
1	0	$0 - 0 = 0$
2	1	$\omega_0 - 0 = \omega_0$
3	0	$0 - \omega_1 = -\omega_1$
4	0	$0 - 0 = 0$
5	0	$0 - 0 = 0$

Table 23.2 The response to a step input beginning at time period 2.

Time t	Input ε_t	Response $\omega_0\varepsilon_t - \omega_1\varepsilon_{t-1}$
0	0	$0 - 0 = 0$
1	0	$0 - 0 = 0$
2	1	$\omega_0 - 0 = \omega_0$
3	1	$\omega_0 - \omega_1 = \omega_0 - \omega_1$
4	1	$\omega_0 - \omega_1 = \omega_0 - \omega_1$
5	1	$\omega_0 - \omega_1 = \omega_0 - \omega_1$

Panel 2(b) of Figure 23.2 shows the kind of response that might be observed when a strike occurs. There is an immediate falloff in sales of an amount ω_0 followed by a recovery of amount ω_1 after the strike ends. Then sales return to the normal level.

Panel 3 of Figure 23.2 shows the kind of response that might result from the introduction of a promotional campaign in a given month. There is an immediate increase in sales followed by a gradual return to normal sales levels once the promotion is ended. Panel 7 of Figure 23.2 shows what might happen as a result of introducing an advertising program and then continuing it for many months at the same level. As customer awareness builds, sales increase until a saturation level is reached when continued advertising no longer increases sales.

An Intervention Model with Noise

As in the case of the transfer function model, it is necessary to introduce a noise term into the intervention model. However, in this instance, the noise term explains

(a)

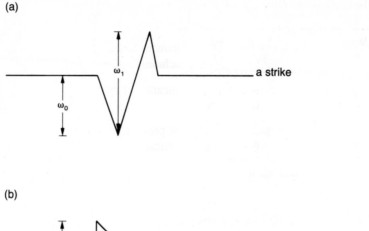

a strike

(b)

A promotional campaign ($\delta > 0$)

(c)

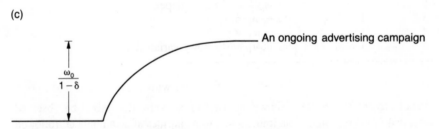

An ongoing advertising campaign

Figure 23.3   Examples of dynamic effects described by a pulse input (in (a) and (b)) and a step input (in (c)).

what would happen if no unusual event had occurred. The model that explains this is univariate for a series that is

- Fitted over that portion of the data containing no unusual events, or
- Fitted over the entire time span by using robust estimation to reduce the effect of unusual events in estimating the model.

To account for trading day or holiday variation or the occurrence of strikes of differing intensity, or the introduction of different kinds of advertising in different time periods, it is often necessary to introduce several transfer functions of the type shown in Figure 23.3 to account for the various intervention effects.

SUMMARY

The chapter has emphasized the development of transfer function and intervention models. The transfer function modeling process includes:

- The analysis of the time series and the identification of appropriate univariate models for the output and input series.

- The use of the cross correlogram (relating the prewhitened input series to the output series that is prewhitened with the identical filter) to identify:

  - The lead time b.

  - The number of input lags s.

  - The number of output lags r.

- The use of the univariate model for the output series as an initial noise model.

- The estimation and diagnostic checking (residual analysis) stages to improve on the initial model.

The intervention modeling approach takes into account effects such as unusual one-time effects (e.g., strikes), changes in policy, and advertising effects, which are often not measurable in terms of an acceptable independent variable. The process includes:

- The selection of impulse and/or step dummy variables as may be appropriate for the kind of phenomena under study.

- The analysis of the data and residuals, together with a chart like that in Figure 23.2, to identify the initial transfer functions.

- The initial noise model is the univariate model based on the history *prior to* the intervention effect. The use of robust estimation methods with *all the available data* is also possible. Remember, robust methods fit the bulk of the data and this is what is desired in determining the noise term of an intervention model.

- The usual diagnostic checking steps to improve on the initial model.

USEFUL READING

BARTLETT, M. S. (1935). Some Aspects of the Time-Correlation Problem in Regard to Tests of Significance. *Journal of the Royal Statistical Society* B 98, 536–43.

BOX, G. E. P. and G. M. JENKINS (1976). *Time Series Analysis, Forecasting and Control, Revised Edition*. San Francisco, CA: Holden-Day.

JENKINS, G. M. (1979). *Practical Experiences with Modelling and Forecasting Time Series*. Jersey, Channel Islands: GJ&P (Overseas) Ltd.

PACK, D. (1977). Revealing Time Series Interrelationships. *Decision Sciences* 8, 377–402.

CHAPTER 24

Applications of Transfer Function and Intervention Models

This chapter presents three applications of the transfer function method. The examples involve:

- Predicting toll revenues based on their relationship to nonfarm employment.

- Using an intervention model to estimate the effect of strikes in main-telephone gain data.

- Predicting a main-telephone gain series on the basis of its own history and its relationships with the U.S. housing starts series and the FRB Index of Industrial Production.

As discussed in Chapter 23, a transfer function model relates an output series, such as toll revenues, to an input series that influences it. The influence leads in one direction—no feedback is assumed; i.e., the output series does not influence the input series.

THE TRANSFER FUNCTION PROCEDURE

The transfer function procedure is iterative and involves the following steps:

- Obtain stationary series for X_t and Y_t.

- Prewhiten X_t by using its univariate ARIMA model to obtain a white noise residual series.

- Prewhiten Y_t by the *same model* used for X_t.

- Examine the cross correlogram relating the residuals of the prewhitened X_t and prewhitened Y_t series to tentatively identify the transfer function components:

- b = the lead time or delay.
- s = the number of lags of the input series.
- r = the number of lags of the output series.
- Identify a preliminary noise component.
- Estimate the tentatively specified transfer function–noise model.
- Examine the sample residual function and cross-correlation function between the residuals of the composite model and the residuals of the prewhitened X_t series to determine the adequacy of the tentatively specified transfer function and noise components.
- Examine the variance-covariance matrix of the parameter estimates to uncover opportunities for a more parsimonious model specification.
- Modify the transfer function or noise components, where applicable, and continue the process until an acceptable model is obtained.

A TRANSFER FUNCTION MODEL RELATING TOLL REVENUES TO NONFARM EMPLOYMENT

The prediction of telephone toll revenues based upon the level of employment in a given market area is a forecasting situation where the absence of feedback is a reasonable consideration. A transfer function model will be developed for this example, using the procedure outlined in the preceding section.

From prior analysis of these monthly series (Chapter 22), we know that regular and seasonal differences of the logarithms of the observations yield stationary series. We also know the univariate model for the nonfarm employment series.

In Chapter 22 a model with regular autoregressive parameters of order 2 and seasonal moving-average parameters of order 2 on the differences of order 1 and 12 of the logarithms of nonfarm employment (NFRM) was developed. It is given by

$$(1 - 0.124B - 0.232B^2)(1 - B)(1 - B^{12}) \, ln \, \text{NFRM} = (1 - 0.615B^{12} - 0.292B^{24})\varepsilon_t .$$

Figure 24.1 shows the cross correlations between the input and output series that have been prewhitened by the univariate nonfarm employment model. The single significant correlation at lag 0 suggests that $b = 0$, $s = 0$, and $r = 0$.

The transfer function–noise model is

$$Y_t = \omega_0 X_t + N_t .$$

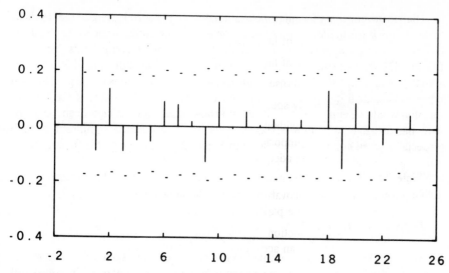

Figure 24.1   Cross correlogram between the prewhitened nonfarm employment and telephone-toll revenue series.

An initial noise component N_t can be identified by using the parameters of the univariate model for the toll revenue (REV) series. Another approach to identifying an initial noise component is to use the impulse response function from the prewhitening stage to calculate a residual series. Then, the correlogram of the residual series can be used to identify a tentative noise model.

A univariate model for the output series has the form:

$$(1 - B)(1 - B^{12}) \, ln \, \text{REV} = (1 - 0.506B)(1 - 0.882B^{12})\varepsilon_t \, .$$

Combining equations yields

$$ln \, \text{REV} = \omega_0 \, ln \, \text{NFRM} + \frac{(1 - 0.506B)(1 - 0.882B^{12})}{(1 - B)(1 - B^{12})}\varepsilon_t \, .$$

Multiplying through by the differencing operators results in the model

$$(1 - B)(1 - B^{12}) \, ln \, \text{REV} = \omega_0(1 - B)(1 - B^{12}) \, ln \, \text{NFRM} +$$
$$(1 - 0.506B)(1 - 0.882B^{12})\varepsilon_t \, .$$

This transfer function–noise model was then estimated and all parameters were found to be significant. There was no significant correlation found between these parameter estimates.

Diagnostic checking includes the examination of the correlogram of the residuals to determine if the model is appropriate. Patterns in the correlogram of the residuals can result from inappropriate transfer functions or noise components (or both). Figure 24.2 shows there were no significant patterns, thus indicating that the noise component is adequate. An examination of the cross correlations of the prewhitened input and residual output series was then undertaken to determine if the transfer function would be appropriate. Figure 24.3 shows there were no significant patterns, with the possible exception of lag 2. This was not deemed significant enough (especially when compared with lag 0 in Figure 24.2) to change the transfer function.

Table 24.1 shows the estimated parameters of the model. For interpretation, divide through by the common differencing operators and obtain:

$$ln \text{ REV} = 0.85 ln \text{ NFRM} + N_t \, .$$

Since the logarithmic transformation has been applied to both series, the model is readily interpreted. A 1-percent increase in nonfarm employment increases toll revenues by 0.85 percent. This is an example of a model with no memory. A change in X_t does not change any *future* values of Y_t.

The estimated standard deviation (the square root of the residual variance) is 0.018. Since the logarithmic transformation was applied to the output series, multiplying the residual standard deviation by 100 yields a value of 1.8 percent. This indicates that the residual standard deviation is approximately 1.8 percent of the level of the series. With this model, approximately two-thirds of the one-step-ahead

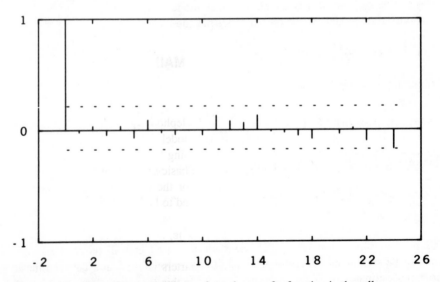

Figure 24.2 The residual correlogram from the transfer function in the toll revenue–nonfarm employment model.

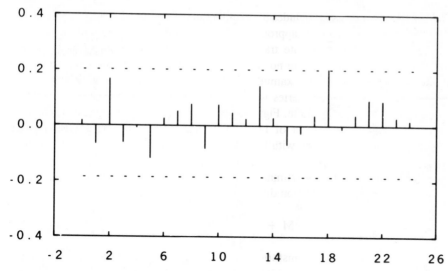

Figure 24.3   Cross correlogram between the prewhitened input and residual output series.

forecast errors can be expected to be within 1.8 percent of the predicted value for revenues.

It is interesting to note that transfer function estimation takes into account moving average processes in the residuals that cannot be addressed with linear regression methods.

AN INTERVENTION MODEL FOR THE MAIN TELEPHONE GAIN

Recall from Chapter 22 that the original main-telephone gain series was impacted by strikes in 1968 and 1971. An intervention model can be built to estimate the impact of these strikes in terms of lost gain by using "strike" dummy variables.

Since the strikes (unusual events) perturb the basic series, the intervention modeling process begins with a univariate model for the quarterly main gain series QTMG. In Chapter 22, this model was determined to be:

$$(1 - 0.66B)(1 - B^4)QTMG = (1 - 0.58B^4)\varepsilon_t .$$

The observed data suggest that the strike quarters were unusually low but that the following quarters were not unusually high. In this case, the strike effect can be estimated with an intervention model of the form

Table 24.1 Summary of parameter estimates for the revenue model.

Model type	Estimated model	Residual variance and (percent standard deviation)
Univariate	$(1 - B)(1 - B^{12})ln$ REV $= (1 - 0.51B)(1 - 0.88B^{12})\varepsilon_t$ $\pm 0.08 \qquad \pm 0.03$	0.0003544 (1.9 percent)
Transfer function	$(1 - B)(1 - B^{12})ln$ REV $= 0.85(1 - B)(1 - B^{12})ln$ NFRM ± 0.24 $+ (1 - 0.72B)(1 - 0.38B^{12})\varepsilon_t$ $\pm 0.07 \qquad \pm 0.09$	0.0003304 (1.8 percent)

$$Y_t = \omega_0 X_t + N_t \, ,$$

where N_t is the univariate model for main telephone gain; namely,

$$N_t = \frac{(1 - \theta B^4)}{(1 - B^4)(1 - \phi B)} \varepsilon_t \, .$$

Combining the above equations yields

$$Y_t = \omega_0 X_t + \frac{(1 - \theta B^4)}{(1 - B^4)(1 - \phi B)} \varepsilon_t \, .$$

Multiplying through by the differencing operator results in

$$(1 - B^4)Y_t = \omega_0(1 - B^4)X_t + \frac{(1 - \theta B^4)}{(1 - \phi B)} \varepsilon_t \, .$$

This is the model that is estimated by using the transfer function estimation program. The coefficient ω_0 provides an estimate of the strike effect.

This suggests the use of impulse dummy variables and a transfer function of $b = 0, s = 0, r = 0$. In other words, the strike effect should be coincident with the strike ($b = 0$) and should be of the form of a single negative spike ($s = 0$, $r = 0$). An alternative transfer function of $b = 0, s = 1, r = 0$ could be tried if one believed there was a recovery effect in the quarter following the strike.

Table 24.2 shows the results of the intervention model where D_1 is a dummy variable with a value of 1 for the 1968 strike quarter and 0 elsewhere and D_2 is a dummy variable with a value of 1 for the 1971 strike quarter and 0 elsewhere. The

Table 24.2 Estimated transfer function models for main telephone gain.

Model type	Estimated model, and (standard errors)	Residual variance ($\times 10^8$)
Intervention	$(1 - B^4)\text{GAIN}_t = -164,970(1 - B^4)D_1 - 145,400(1 - B^4)D_2 + (1 - 0.71B)^{-1}(1 - 0.49B^4)\varepsilon_t$ $(\pm 41,587) \quad\quad (\pm 41,720) \quad\quad (\pm 0.09) \quad (\pm 0.11)$	34.89
Transfer function	$(1 - B^4)\text{GAIN}_t^* = 470.8(1 - B^4)\text{HOUS}_{t-1} + (1 - 0.62B)^{-1}(1 - 0.45B^4)\varepsilon_t$ $(\pm 133.0) \quad\quad (\pm 0.10) \quad (\pm 0.11)$	29.36
Transfer function	$(1 - B^4)\text{GAIN}_t^* = (14,600 + 3095B - 4793B^2)(1 - B)(1 - B^4)\text{FRBI}_t + (1 - 0.70B)^{-1}(1 - 0.33B^4)\varepsilon_t$ $(\pm 2503)\,(\pm 2312)\,(\pm 2430) \quad\quad (\pm 0.09) \quad (\pm 0.13)$	23.25
Transfer function	$(1 - B^4)\text{GAIN}_t^* = 11,800(1 - B)(1 - B^4)\text{FRBI}_t + 354.0(1 - B^4)\text{HOUS}_{t-1} + (1 - 0.62B)^{-1}(1 - 0.34B^4)\varepsilon_t$ $(\pm 2356) \quad\quad (\pm 117.0) \quad\quad (\pm 0.10) \quad (\pm 0.12)$	21.77

impact of the strikes was a loss in main station demand of approximately 165,000 in 1968 and 145,000 in 1971.

The residual correlogram showed no significant patterns. The cross correlogram of the residuals of main telephone gain and D_1 showed no patterns. The cross correlogram between the residuals and D_2 showed a significant correlation at lag 2. Since the strike did not have a carryover effect two quarters after its conclusion (based on knowledge of the backlog of orders), a more complicated intervention model was rejected.

A MULTIPLE INPUT TRANSFER FUNCTION MODEL

A transfer function–noise model for main telephone gain includes the U.S. housing starts series (QHS) and the FRB Index of Industrial Production series (FRBI). The main gain series (QTMG) has been adjusted for the 1968 and 1971 strikes by using the values obtained from the intervention model in the previous section.

A *multiple input transfer function–noise model* begins with the identification of individual transfer function–noise models between the output series (QTMG) and each input series separately. The univariate models for the input series (QHS, FRBI) are the models developed in Chapter 22.

Figure 24.4 shows a plot of the cross correlations between the residual main gain series and the housing starts series prewhitened by the univariate housing starts model $(2,0,0) \times (0,1,1)^4$. The single significant correlation (spike) at lag 1 suggests that $b = 1$, $s = 0$, and $r = 0$. The transfer function model is

$$Y_t = \omega_0 X_{t-1} + N_t ,$$

where N_t is the univariate model for main telephone gain. The results for the model are shown in Table 24.2. The residual correlogram and residual cross correlogram had no significant patterns.

Next, a transfer function–noise model was developed for the QTMG and FRBI series. The univariate model for the FRBI series indicated that only a first difference was required to obtain a stationary series. However, in this example, we wanted to replicate the prior OLS model (Chapter 3) where the seasonal differences of main telephone gain were related to the seasonal differences of housing starts and the regular and seasonal differences of the FRBI series. Therefore a univariate ARIMA model was developed for the regular and seasonal differences of the FRBI series.

A seasonal moving average process was introduced by taking seasonal differences; then the deterministic trend constant was no longer required. The univariate model (beneath it are the standard errors) is

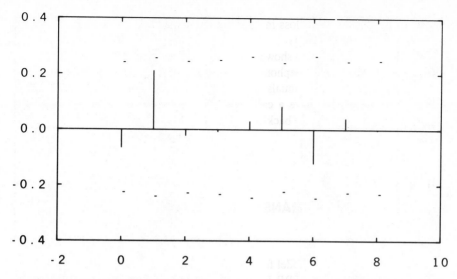

Figure 24.4 Cross correlogram between residual main-telephone gain series and residual (prewhitened) housing starts series.

$$(1 - B)(1 - B^4)\text{FRBI} = (1 + 0.71B)(1 - 0.90B^4)\varepsilon_t .$$
$$\pm\, 0.09 \qquad \pm\, 0.03$$

The high value for the seasonal moving average parameter confirmed that seasonal differencing would not be required if the goal was simply to obtain a stationary series.

The first significant cross correlation occurs at lag 0, indicating that $b = 0$. The remaining values are less than two standard errors. A tentative transfer function of $b = 0$, $s = 0$, $r = 0$ was investigated:

$$(1 - B^4)\text{QTMG}_t = \omega_0(1 - B)(1 - B^4)\text{FRBI}_t .$$

The univariate model for QTMG was chosen as an initial noise component:

$$(1 - \emptyset B)N_t = (1 - \theta B^4)\varepsilon_t .$$

This selection was consistent with the correlogram of the residuals of the noise series that is obtained by using the impulse response weights of the prewhitened series.

The model was estimated; the parameters were significant, and the cross correlogram of the residuals of the model and the residuals of the prewhitened FRBI series showed the value at lag 2 appeared significant (Figure 24.5). Owing to limitations in the software program, it was necessary to include lags 0, 1, and 2.

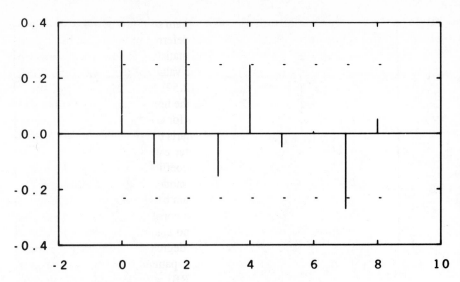

Figure 24.5 Cross-correlogram between the residual main-telephone gain series and residual (prewhitened) FRB Index of Industrial Production.

Table 24.3 Comparison of OLS and transfer function–noise models for the (strike-adjusted) main-telephone gain series.

Regression	$\hat{\beta}_1$ QHS_{t-1}	$\hat{\beta}_2$ $FRBI_t$	$\hat{\rho}$	Seasonal moving average	D-W statistic	Estimated residual standard deviation
OLS	436.4 (5.31)	12200 (5.62)	—	—	0.88	58,310
Cochrane-Orcutt	377.7 (3.59)	12490 (6.11)	0.51	—	1.84	48,510
Hildreth-Lu	354.9 (3.15)	12570 (6.24)	0.60	—	2.02	48,350
Transfer Function	354.0 (3.02)	11880 (5.04)	0.62 (6.20)	0.34 (2.83)	ACF—Not significant	46,660
Hildreth-Lu (Lag on housing starts = 2)	397.2 (3.86)	13271 (6.64)	0.50	—	2.04	48,020

The model was reestimated and the cross correlogram still showed a significant value at lag 0. Further effort on this latter model was deferred until the combined model was investigated. Table 24.2 shows the model statistics. The correlogram of the residuals of the model showed no pattern and the value for the chi-squared statistic ($= 4.96$) is less than the theoretical value ($= 12.59$) for 6 degrees of freedom.

The final multiple input model included both the housing starts (QHS) and FRBI series and is also shown in Table 24.2. The values for ω_1 and ω_2 in the FRBI transfer function became insignificant when the housing starts transfer function was included in the multiple input model. All of the parameter estimates were significant and there were no significant correlations among the coefficient estimates. (When there is a significant correlation, a more parsimonious model should be considered.)

The residual correlogram, the cross-correlogram between the residual series and the prewhitened housing starts series, and the cross-correlogram between the residual series and the prewhitened FRBI series showed no significant patterns or relationships. The chi-squared statistic ($= 1.72$) on 6 degrees of freedom indicated no significant correlations in the residual series. The pattern that was apparent in the residual cross-correlation between QTMG and FRBI is no longer significant when housing starts are included in the model.

Table 24.3 compares the OLS and transfer function–noise models for the strike-adjusted main gain series. The transfer function methodology indicated that housing starts should be lagged 1 instead of 2 quarters. Moreover, the error structure indicated by the transfer function–noise model included a first-order autoregressive and a seasonal moving average relationship. For comparison purposes a one-quarter lag was introduced in the OLS model and serial correlation correction techniques were tried. The Cochrane-Orcutt procedure yielded $\hat{\rho} = 0.51$ and the Hildreth-Lu procedure estimated $\hat{\rho} = 0.60$, which compares to the value of 0.62 from the transfer function model. A search for a finer grid with the Hildreth-Lu procedure might have resulted in the same value for $\hat{\rho}$; but the seasonal moving-average noise parameter is not available with typical regression programs.

The lower part of Table 24.3 shows the statistics for the main gain model with a two-quarter lag on the housing starts series. The residual standard deviation for this model exceeds that of the transfer function model but is slightly less than the comparable model with a one-quarter lag.

Figure 24.6 shows a plot of the forecasts and 50-percent confidence limits on the forecasts. The forecasts for housing starts and the FRBI series were generated by the respective univariate ARIMA models. Where available, forecasts from other sources may be utilized for the input series.

If the input series are highly correlated with each other, it may be necessary to take a more indirect approach to the estimation of the model. In a manner similar to the regression-by-stages approach discussed in Chapter 4, you could build a transfer function model relating input series X_{2t} to input series X_{1t}. The residuals of this model will be independent of X_{1t} and can be denoted X_{2t}^*. Then a multiple input model can be developed relating the output series to X_{1t} and X_{2t}^*.

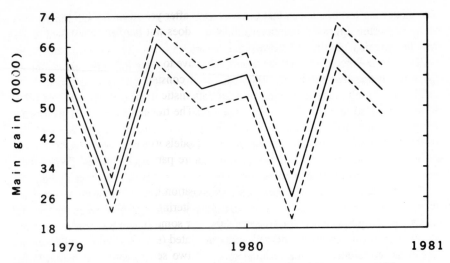

Figure 24.6 Plot of the main gain forecasts and the associated 50-percent confidence limits.

OTHER STUDIES

The use of transfer function techniques in fitting equations to macroeconomic data is illustrated in Wall et al. (1975). Some examples of intervention analysis include Box and Tiao (1975), Glass et al. (1975), and Jenkins (1979). Liu (1980) has a transfer function model for describing calendar effects in a time series model. Damsleth and Sollie (1980) developed a transfer function model for main telephone demand in Norway that incorporates a price change as an intervention.

COMPARISON WITH THE ECONOMETRIC APPROACH

The filtering approach to building multivariate time series modeling is different from the econometric approach in a variety of ways. Unlike most econometric methods, the ARIMA method does not attempt to decompose a time series into seasonal, trend-cycle, irregular, and trading day components. Seasonality in the ARIMA method is part of the overall ARIMA model that is used. The seasonal pattern may change in many different ways and this effect will be captured by the appropriate ARIMA model. In econometric modeling, one frequently seasonally adjusts the time series before modeling. In other instances, dummy variables can be used to account for a seasonal pattern that is stable year after year. Jenkins (1979) believes that the filtering approach is more flexible and does not involve manipulating the data and possibly introducing other problems.

The ARIMA models tend to be more responsive to random changes in trend. By differencing the time series, *deterministic* trend constants are often not required. The forecaster may choose to leave out a deterministic trend constant to ensure that there is no deterministic trend in the forecasts. The trend is then more adaptive or responsive to stochastic events.

The use of distributed lags in econometric models was presented in Chapter 12; the transfer function approach may permit a more parsimonious representation of the lag structure than what was described there.

Since model estimation involves the specification both of the appropriate variables *and* the appropriate error structure, the filtering approach may provide the capability of specifying better error structures for some problems.

When two time series are strongly autocorrelated (e.g., show a strong trend), a significant correlation often exists between the two series. Jenkins (1979) argues that by prewhitening the series, one may conclude that most of the information about the dependent variable is captured in its own history and not by the independent variable. Thus R^2 tends to overstate the strength of the relationship because of an inappropriate error structure. To be satisfied that X explains Y, you must have a strong belief in the economic theory that is being used, and Jenkins argues that the current state of economic theory is not well enough developed to warrant such satisfaction.

TRANSFER FUNCTION MODEL BUILDING CHECKLIST

_____ Can the output variable be described conceptually in terms of a small number of related input variables?

_____ Have the various time series been plotted to look for unusual values and patterns (trend or seasonality)?

_____ Are transformations required to meet modeling assumptions?

_____ Have differences been taken to eliminate nonstationarity?

_____ Have univariate models been developed for the output series and each of the inputs?

_____ Have prewhitened input and output models for the output been used to obtain the residual cross-correlograms?

_____ Are adequate representations available for the transfer function and noise models?

_____ Are the parameter estimates, standard deviations, and residual plots available for interpretation?

_____ Has the adequacy of the model been checked?

_____ If inadequate, have one or more iterations been performed to reidentify the model for fitting and checking?

_____ If adequate, have forecasts and confidence limits been generated?

_____ Have decisions been made as to the source of forecast inputs, time origins, and lead times?

_____ Are the forecasts generated from the model reasonable?

SUMMARY

The case studies in this chapter demonstrated that

- Considerable insight can be gained into the nature of "interventions" in a model. In particular, strike effects were made explicit in a model for main telephone gain.
- Transfer function models can be developed with realistic interpretations. In particular, elasticity estimates were obtained for a toll revenue model.
- A complex model for main telephone gain in terms of two related variables could be built economically as a transfer function-noise model. Forecasts and associated confidence limits can also be derived.

USEFUL READING

BOX, G. E. P. and G. M. JENKINS (1976). *Time Series Analysis, Forecasting and Control, Revised Edition*. San Francisco, CA: Holden-Day.

BOX, G. E. P., and G. C. TIAO (1975). Intervention Analysis with Applications to Economic and Environmental Problems. *Journal of the American Statistical Association* 70, 70–79.

DAMSLETH, E., and B. H. SOLLIE (1980). Forecasting the Demand for New Telephone Main Stations in Norway by Using ARIMA Models. *In* O. D. Anderson, ed., *Forecasting Public Utilities*. Amsterdam, Netherlands: North-Holland Publishing Co.

GLASS, G. V., V. L. WILSON, and J. M. GOTTMAN (1975). *Design and Analysis of Time Series Experiments*. Boulder, CO: Colorado Associated University Press.

JENKINS, G. M. (1979). *Practical Experiences with Modelling and Forecasting Time Series*. Jersey, Channel Islands: GJ&P (Overseas) Ltd.

LIU, L. (1980). Analysis of Time Series with Calendar Effects. *Management Science* 26, 106–12.

WALL, K. D., A. J. PRESTON, J. W. BRAY, and M. H. PESTON (1975). Estimates of a Simple Control Model of the UK Economy. In G. A. Reaton, ed. *Modelling the Economy*. London, England: Heinemann.

Part 5

Principles of
Forecast Management

CHAPTER 25

Measuring Forecast Performance

There are two areas of forecast measurement the forecaster will find worthy of consideration.

- The first area is *forecast monitoring:* in this the manager hopes to identify the need for a forecast revision as soon as possible, and then to make the necessary revision.

- The second area is *results analysis:* the goal of this is to quantify the accuracy of prior forecasts as part of an overall measurement of the forecaster's effectiveness.

FORECAST MONITORING

Monitoring a forecast entails comparing recently published actual results against current forecasts and deciding if the forecast should be revised. Obviously, it is necessary to monitor the assumptions upon which the forecast was based as well as the numerical comparisons of forecasts and actuals.

Forecast monitoring can be thought of as a form of managerial control. Managerial control is also a process that measures current performance, based on the information available, and guides performance toward a predetermined goal. Forecasters are involved in a function whose primary output is wholly related to the future. The forecaster cannot change the future; yet the course the future takes determines the validity of a forecast; therefore a function of managerial control is to revise a forecast when it becomes evident that the original forecast or goal cannot be met. In doing so, the forecaster changes the predetermined goal so that it more closely approximates expected performance.

Developing an ability to anticipate change and to advise others in an organization of an impending change is a valuable asset for a forecaster, since these warnings, made at the appropriate time, can give a business using the forecast time to adjust its operations to changing conditions.

Preventing Surprise

A primary objective of forecast monitoring is to prevent a company that has requested a forecast from finding itself surprised by a turn of events: the firm should have sufficient time to evaluate alternative courses of action and not be forced to react in an unprepared way to unpredicted events. Another objective of monitoring is to accurately predict reversal in the direction of growth. This involves predicting turning points in the economy and in the demand for the firm's products. Common mistakes are to miss predicting a downturn in demand and to predict an upturn too soon.

At a more demanding level, objectives of monitoring are to predict a change in the rate of growth, to predict the level of growth, and to minimize any adverse impact that changing a forecast may have. The ability to accurately predict any speeding up or slowing down of growth enables a company to decide on the proper timing of company plans and programs. An accurate prediction of the level of growth—the forecast numbers themselves—allows the company to make sizing decisions about investment in facilities, about personnel and payroll, and to choose appropriate financing arrangements. Finally, it is necessay to avoid changing a forecast too frequently, since any change disrupts a company's plans to some extent. Obviously, the forecaster could revise the forecast every month, so that its final model and the actual data would be almost identical. However, this does not serve the needs of forecast users, since they require time to plan and implement programs in response to revised forecasts. The more carefully thought out and thoroughly researched the initial forecast is, the less will be the need to revise it.

Items To Be Monitored

The specific items that a forecaster monitors will naturally depend upon assigned areas of responsibility, but the items should clearly be selected because they relate to the purpose and objectives for which the forecast is being prepared.

The forecaster should consider monitoring composites or groups of items. Composites often serve as indicators of overall forecast quality and are frequently used as a base for making decisions about the forecast. They are resistant to individual deviations, which may merely be measurement aberrations and not of significance to a monitor. For example, a total forecast of total product line revenues might appear to need no adjustment, even though individual product revenue forecasts that are parts of the larger forecast may need to be adjusted.

Another monitoring concept is to compare the sum of the components to the whole. This helps to insure that there is a reasonable relationship between the more stable"top-down" or "macro" forecast and the more volatile "bottom-up" forecasts of many small components. For example, forecasts for products and services may be required for the various regions in which a company operates (including international operations). A separately prepared macro forecast of products and services for the sum of the regions is likely to provide a valuable check of the reasonableness of the sum of the individual forecasts made for component regions. The forecaster can use the macro forecast to be certain that both upward and downward revisions to the component parts are made to keep them in reasonable agreement with the total forecast. In managing inventory and production functions in a business, it is important to know if a forecast reveals a problem of distribution (some regions with high inventory, others low) or of production (adequate or inadequate capacity to meet total requirements).

Another useful concept is to monitor ratios or relationships among different items. A comparison of the percent of sales within a given geographic area to rates of total corporate sales is an example of this approach. Another example would be the ratio of residential main telephone gain to the establishment of new households. Electricity consumption per capita, the ratio of airline to total passenger-miles of traffic, and the percent crude oil of total energy consumption are additional examples of this concept.

Time relationships are another thing to be monitored: it may be appropriate to monitor changes or the percent of changes over time. A seasonally adjusted annual rate would be used for this; another example is the ratio of first-quarter sales to total annual sales. In addition to monitoring on a monthly basis, consider monitoring on a cumulative basis: compare the sum of the actuals since the beinning of the year with the sum of the forecasts. This has the advantage of smoothing out irregular, random, month-to-month variations.

In all cases it will be necessary to monitor external factors. These factors are bases for key assumptions made—for example—about business conditions or economic outlook. Corporate policy assumptions also need to be monitored. These might include credit and collection policies, or pricing policies (e.g., how price changes are decided, or whether a company encourages owning rather than leasing its products).

User Needs

One area of monitoring that is especially critical for the forecaster and managers of forecasting organizations is that of the forecast user's needs. It is possible that budgetary or organizational changes, introduction of a new product or discontinuation of an old product, or changes in management will cause changes in the user's forecast needs.

Since forecasting is a service function, forecasters and their managers need to monitor user needs to be certain that the service provided is consistent with evolving business needs. Questionnaires or periodic discussions with users will indicate if such changes have occurred.

A final concept worthy of consideration is the monitoring of similar forecast results for different regions or countries. This will help the forecaster to determine if a pattern is developing elsewhere that may impact the forecaster's area in the near future. Are there areas of the country or other countries that generally lead or lag your area? You may discover that your area is not the only area with weak or strong demand. A national pattern may be emerging that needs to be tracked. If forecasters concentrate only on their own geographic areas, they may overlook important events or trends in other areas that could help them to interpret their own results.

RESULTS ANALYSIS

The second aspect of forecast measurement referred to at the beginning of the chapter is results analysis: in this the goal is to develop meaningful ways of measuring the performance of forecasters.

In any managerial situation, improvement in organizational effectiveness is dependent upon measurement. The forecaster or managers of forecasting groups will find it desirable to establish a forecast measurement plan to provide indications of overall staff performance that can be reviewed with clients. A properly developed plan will indicate if performance is improving or worsening and will highlight trouble areas.

The measurement plan will provide managers a tool to assist in evaluating forecasts. When a measurement plan exists, forecasters know that they are likely to have to explain forecasts that have missed the mark. This will compel forecasters to structure and quantify their assumptions so that they will be able to document the reasons a forecast has missed the mark. Without structure and quantification of assumptions, there is no way for the forecaster to explain what went wrong.

More importantly, adequate documentation enables the forecaster to learn from past mistakes. From review of these after-the-fact reports, it can be determined if the assumptions were reasonable at the time they were made. Which assumptions turned out to be incorrect? Why? Were all sources of information reviewed? Were there any obvious breakdowns in communications? Was the forecast methodology appropriate for the particular problem? The answers to these questions become the information needed to evaluate the forecast.

In reviewing cases that did particularly well, it may be discovered that new methods or new sources of information were uncovered which were responsible for superior forecast performance. Perhaps the approach can be tried in areas where performance is not as good. The documentation of superior and substandard performance, which is a natural outgrowth of a forecast measurement plan, provides the needed inputs to determine areas where methods or data improvement are required. This documentation can also be used to support requests for additional people, data, or items as these are needed to improve performance.

The presence of a measurement plan will also be of value to the users of the forecasts. It will improve their understanding of the limitations that must be placed on the accuracy of the forecast they receive. For example, suppose the forecaster considers a ± 2 percent miss to be a good job for a given series, and the measurement plan takes this into account. A user would then be foolish to establish plans based on a ± ½ percent accuracy, which the user might otherwise think is reasonable accuracy. By providing users with forecasting accuracy objectives, you are in effect providing a range forecast that will enable the user to make plans that are sensitive to differing forecast levels. In addition, the credibility of the forecasting organization will be improved, because it is reporting on its performance.

SUMMARY

This chapter has provided insight into the important areas of forecast monitoring and results analysis, by focusing on concepts of forecast performance measurement. The forecaster is encouraged to consider monitoring

- Composites or groups of items.
- The individual component forecasts in relation to a sum-of-the-components (macro) forecast.
- Ratios of related items.
- Changes and percent changes over time.
- Forecast assumptions.
- User needs.

Ways of reporting on the accuracy of the forecasts should be considered. In addition to improving the quality of the initial forecast, the presence of a report on performance will enhance the credibility of the forecasting organization.

CHAPTER 26

Managing The
Forecasting Function

The success of a forecasting organization depends upon the extent to which traditional management philosophies and practices are applied to an unconventional business discipline. For the purposes of this discussion, the forecasting organization will be considered to be an in-company staff that is part of a larger company. To clarify the distinction between managers of forecasters and other managers within the business who are using the forecast the "other managers" will be referred to as "clients."

Specific managerial approaches for strengthening forecasting organizations can result from

- Setting goals for the organization.
- Establishing standards of performance.
- Measuring performance.
- Implementing new methods.
- Making optimum use of outside consulting opinion.

MANAGEMENT BY RESULTS

Managers are concerned with making decisions in the presence of *uncertainty* about the future. In many regards the future seems increasingly unstable: this is reflected in increasing prices, costs, and availability of goods; in inflation; in concern about environmental issues; and in demographic patterns. Consequently, bad decision making is becoming increasingly costly, not only in economic terms but also to society. The managers of the nation's social and economic systems should therefore be increasingly concerned with improved planning and forecasting as a *management* activity (Jenkins, 1979; Wheelwright and Makridakis, 1980).

The first step in improving the effectiveness of a forecasting staff is to develop a management plan that includes six key components:

- A statement of the purpose of the forecasting organization.
- A definition of major areas of responsibility.
- Long-term objectives for each area of the forecaster's responsibility.
- Indicators of performance.
- Short-term goals.
- Measurement of performance relative to the goals.

The first of these six components—the statement of purpose—is best kept concise. It can be a one-sentence statement. While this is simple in concept, forecasters and their managers can spend many hours wrestling with the purpose of their jobs. Since the forecaster is an advisor to management, a purpose might be "to provide the best possible advice to the client about the future demand for the company's products and services." For managers of independent forecasting companies (i.e., consulting firms), this statement may need to be expanded to include the need to generate new projects in addition to continuing (or follow-up) projects, so that an independent firm will remain financially viable.

The *areas of responsibility* may include revenue forecasting, expense forecasting, product or service forecasting; or other areas, such as forecast evaluation, measurement, monitoring, and presentation; or managerial responsibilities, such as forecaster appraisal and development.

The *objectives* for these areas of responsibility should be sufficiently general to have long-term significance and ought to indicate the striving for improvement that will make each an objective rather than simply a description. Some examples might be: "to improve the accuracy of . . . ," "to improve managerial and technical skills," "to improve the credibility of the forecasting organization," and "to insure the continuing relevance of . . . ."

Developing meaningful *indicators of performance* can be the source of much debate among staff members. Experience will cause you to reject some indicators and replace them with others more relevant. However, until this is done it will not be possible to state explicitly that the forecasting organization is achieving its objectives. The key to success is to make the areas of responsibility, indicators, and goals all job-relevant. An emphasis on joint manager-forecaster goal setting and communications will increase the probability that the organizational goals will be internalized by the members of the organization.

Emphasis should be placed on developing meaningful indicators of performance that can be directly applied to an organization engaged in forecasting in a business environment. Of course, modification of the recommendations can be made to more closely approximate the forecasting requirements of specific firms or organizations.

ESTABLISHING FORECASTING STANDARDS OF PERFORMANCE

How can a forecast manager tell a good forecast from a bad forecast at the time it is presented for approval? This is one of the most perplexing problems that forecast managers face in the normal course of events. Certainly, after the forecast time period has elapsed, anyone can look back and determine how closely the forecast predicted actual results; but this is after the fact—what in football is called "Monday morning quarterbacking." The manager wants to be confident that the forecast is reasonable at the time it is prepared.

What is needed is a *process* that, if followed, will increase the likelihood of good forecasting performance. In other words, it is necessary to establish standards of performance for forecasters that will increase the probability of improved forecast accuracy. A checklist that can be used by both forecasters and forecast managers to measure a specific forecast relative to some established standards is included at the end of this chapter, and its use is recommended.

Setting Down Basic Facts

The checklist is general in nature and covers the essential elements that must be a part of an effective forecasting process. It begins with the establishment of basic facts concerning past trends and forecasts. To be satisfied that these facts have been adequately researched, a forecast manager should expect to see tables and plots of historical data. The data should be adjusted to account for changes in geographic boundaries or other factors that would distort analyses and forecasts. If appropriate, the data should be seasonally adjusted to give a better representation of trend-cycle. Outliers or other unusual data values should be explained and replaced, if this is warranted. It is useful to indicate the National Bureau of Economic Research reference dates for peaks and troughs of business cycles. This provides the manager with an indication of the extent to which a client's data series are impacted by the national business cycles. Knowing this relationship will be helpful when the manager reviews the assumptions about the future state of the economy and assesses how these assumptions are reflected in the forecast.

Tables and plots of annual percentage changes provide an indication of the volatility of a series and are useful later in checking the reasonableness of the forecast compared to history. If possible, ratios should be developed between the forecast series and other stable data series that are based on company or regional performance. These ratios should be shown in tables or plots. Once again, these ratios provide reasonableness checks. If some major change is expected in the forecast period, these ratios should help identify the change.

Whenever possible, there should be at least a decade of data available for the manager's review. It may not be necessary to show this much history when presenting the forecast to the clients; but it will be necessary to have this much data available to analyze the impact of business cycles. If possible, data back to the 1957–58 recession should be available since, with the exception of 1973–75, that was the last major recession. However, in many forecasting circumstances, data this old may no longer be relevant.

There should also be available a record of forecasts and actual performance for at least the past three years. This will allow the manager to know how well the organization has done in the past and to gauge the possible reaction of the clients to change in the forecast. It will also be possible to determine from these data if any or all of the forecasters on the staff have a tendency to be too optimistic or too pessimistic over time.

Causes of Change

The next segment of the checklist deals with the causes of changes in past demand trends or levels. The first step is to identify the trend in the data. Regression analysis is an excellent tool for this. A regression against time, as a starting point, will provide a visual indication as to whether the trend is linear or nonlinear. There should be a plot of the series and its fitted trend on a scale of sufficient breadth to clearly identify deviations from trend. The reasons for the deviations should then be identified and explained in writing. These explanations need to be specific. Was there unusual construction activity? Was there a change in corporate policy or prices? Did the deviation correspond to a regional or national economic pattern? What was the source of the explanation—the forecaster, or someone else? Finally, how certain is the forecaster that the reason or explanation stated is correct? Documentation of history is an important step that can serve as reference material for all future forecasts and forecasters. It is particularly helpful to a new forecaster and improves productivity.

Causes of Differences

The next segment of the checklist is concerned with the reasons for the differences between previous forecasts and actual results. This form of results analysis is useful for uncovering problem areas, for identifying the need for new or improved methods, and for determining the quality of the prior forecasts. At this time, however, the manager is merely looking for a pattern of overforecasting or underforecasting. The key to identifying the reasons for forecast deviations is to have written records of basic assumptions, which should be reviewed. These assumptions should then be tested for specificity against the standards shown on the checklist.

Accompanying each assumption should be a rationale indicating why the assumption is necessary. The source of the assumption should be identified and the degree of confidence in the assumption should be stated. The source might be the forecaster, company economists, industry analysts, government publications, or newspaper clippings and journal articles. The forecaster may be absolutely certain that the assumption will prove correct. On the other hand, the forecaster may indicate that it was necessary to make the assumption, but that considerable doubt existed as to whether or not the future would be as assumed.

Factors Affecting Future Demand

The next segment of the checklist is concerned with the factors likely to affect future demand and, therefore, the forecast. Assumptions will have to be made about factors such as income, habit, price of a company's product, price of competing goods, and market potential. In addition, the manager should check to see that there is a logical time integration between historical actuals and the short- and long-term forecasts. Also, there should be a logical time integration between related forecast items, such as forecasts of economic conditions, revenues, and expenses.

At this time, the manager can ascertain that the forecasting methodologies used represent the best methods available at the time.

The purpose of the checklist is to establish standards for the forecasting organization: both the forecaster and the forecast manager can use the checklist in the preparation and subsequent review of the forecast. By establishing meaningful forecasting standards, forecast evaluation can be greatly simplified. The philosophy of forecast evaluation is one in which primary emphasis is placed on the process rather than the numbers. If a proper forecasting process has been meticulously followed by the forecaster, the end result will be as good a forecast as can be developed. If that result does not come about, then the manager needs to train the forecaster better or to select a different person to make the forecast.

Within limits, it is very difficult for a manager to "fine-tune" the numbers presented with any degree of confidence that the changes are appropriate. However, the manager can carefully review the forecast assumptions for reasonableness. The assumptions are the heart of the forecast, and considerable probing of these assumptions can satisfy the manager as to their appropriateness for the forecast period. The manager can also review the technical soundness of the analysis and be satisfied that no errors were made. Having performed these forecast evaluations, the manager can discriminate between a good forecast and a bad one at the time he or she is asked to approve it.

The continued development of checklists such as the one that follows will go a long way towards improving the quality of a forecast.

FORECAST MANAGEMENT CHECKLIST

Step 1: SETTING DOWN BASIC FACTS ABOUT PAST TRENDS AND FORECASTS

_____ Are historical tables and plots available?

_____ Are base-adjusted data available? (A "constant base" is what is wanted: for examples, have historical revenues been adjusted to today's price; have data been adjusted for mergers, acquisitions, etc.?)

_____ Are seasonally adjusted data available?

_____ Have outliers been explained? (As discussed in the treatment of ARIMA modeling, they may significantly affect the forecasts).

_____ Have NBER cyclical reference dates been overlaid?

_____ Are percentage changes shown in tables and plots?

_____ Have forecast-versus-actual comparisons been made for one or more forecast periods?

Step 2: DETERMINING CAUSES OF CHANGE IN PAST DEMAND TRENDS

_____ Is a trend identified?

_____ Is it linear or nonlinear?

_____ Are there plots of data and fitted trends?

_____ Is the scale of sufficient breadth to see deviations?

_____ Have the deviations been explained in writing?

_____ Are explanations about causes specific?

_____ Has the source of explanations been identified?

_____ Is the degree of certainty about the explanation noted?

Step 3: DETERMINING CAUSES OF DIFFERENCES BETWEEN PRE-VIOUS FORECASTS AND ACTUAL DATA

_____ Are differences explained?

_____ Are there any patterns to the explanations?

_____ Are there basic assumptions that can be reviewed?

Step 4: DETERMINING FACTORS LIKELY TO AFFECT FUTURE DEMAND

_____ Do assumptions relate to the future?

_____ Do assumptions indicate direction of impact?

_____ Do assumptions indicate the amount or rate of impact, the timing of the impact, and the duration of the impact on demand?

_____ Are there rationale statements for each assumption?

_____ Are the sources for a rationale statement identified?

Step 5: MAKING THE FORECAST FOR FUTURE PERIODS

_____ Time integration: are long-term and short-term forecasts and history all shown on one chart? Are the transitions reasonable?

_____ Item integration: are ratios of related items shown as well as their history through the long-term forecast?

_____ Functional integration: are related forecasts identified and relationships quantified? (For example, are revenue and product forecasts consistent?)

_____ Have multiple methods been used for key items and have results been compared?

_____ Has the impact on the user of the forecast been considered?

SUMMARY

A basic grounding in management is essential for anyone assuming control of a forecasting organization. The "management by objectives" or "management by results" framework is recommended for implementation in a forecasting organization. This chapter has provided ideas on approaches that can be helpful in implementing such a plan.

The availability of time-shared computers has brought many sophisticated forecasting methodologies out of the theoretical environment of the classroom into the business world. However, as is true in most areas of endeavor, business organizations must have managers with the foresight to try new approaches and management skills in order to set goals, establish standards, and measure performance before they can profit from the new knowledge. Without proper management, considerable resources can be expended with little commensurate payback. This invariably leads to frustration and abandonment of the effort.

The ideas presented here, and the checklists that summarize them, provide managers with a framework to increase the professionalism of their organization, improve forecast accuracy, and strengthen user acceptance of their forecast products.

USEFUL READING

JENKINS, G. M. (1979). *Practical Experiences with Modelling and Forecasting Time Series.* Jersey, Channel Islands: Gwilyn Jenkins and Partners Ltd.

LEVENBACH, H., and J. P. CLEARY (1981). *The Beginning Forecaster*. Belmont, CA: Lifetime Learning Publications.

WHEELWRIGHT, S. C. and S. MAKRIDAKIS (1980). *Forecasting Methods for Management*. New York, NY: John Wiley and Sons.

Appendixes
Bibliography
Index

Appendix A

Table 1   Standardized Normal Distribution

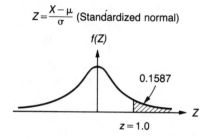

$$Z = \frac{X - \mu}{\sigma} \text{ (Standardized normal)}$$

z	.00	.01	.02	.03	.04	.05	.06	.07	.08	.09
0.0	.5000	.4960	.4920	.4880	.4840	.4801	.4761	.4721	.4681	.4641
0.1	.4602	.4562	.4522	.4483	.4443	.4404	.4364	.4325	.4286	.4247
0.2	.4207	.4168	.4129	.4090	.4052	.4013	.3974	.3936	.3897	.3859
0.3	.3821	.3783	.3745	.3707	.3669	.3632	.3594	.3557	.3520	.3483
0.4	.3446	.3409	.3372	.3336	.3300	.3264	.3228	.3192	.3156	.3121
0.5	.3085	.3050	.3015	.2981	.2946	.2912	.2877	.2843	.2810	.2776
0.6	.2743	.2709	.2676	.2643	.2611	.2578	.2546	.2514	.2483	.2451
0.7	.2420	.2389	.2358	.2327	.2296	.2266	.2236	.2206	.2177	.2148
0.8	.2119	.2090	.2061	.2033	.2005	.1977	.1949	.1922	.1894	.1867
0.9	.1841	.1814	.1788	.1762	.1736	.1711	.1685	.1660	.1635	.1611
1.0	.1587	.1562	.1539	.1515	.1492	.1469	.1446	.1423	.1401	.1379
1.1	.1357	.1335	.1314	.1292	.1271	.1251	.1230	.1210	.1190	.1170
1.2	.1151	.1131	.1112	.1093	.1075	.1056	.1038	.1020	.1003	.0985
1.3	.0968	.0951	.0934	.0918	.0901	.0885	.0869	.0853	.0838	.0823
1.4	.0808	.0793	.0778	.0764	.0749	.0735	.0721	.0708	.0694	.0681
1.5	.0668	.0655	.0643	.0630	.0618	.0606	.0594	.0582	.0571	.0559
1.6	.0548	.0537	.0526	.0516	.0505	.0495	.0485	.0475	.0465	.0455
1.7	.0446	.0436	.0427	.0418	.0409	.0401	.0392	.0384	.0375	.0367
1.8	.0359	.0351	.0344	.0336	.0329	.0322	.0314	.0307	.0301	.0294
1.9	.0287	.0281	.0274	.0268	.0262	.0256	.0250	.0244	.0239	.0233
2.0	.0228	.0222	.0217	.0212	.0207	.0202	.0197	.0192	.0188	.0183
2.1	.0179	.0174	.0170	.0166	.0162	.0158	.0154	.0150	.0146	.0143
2.2	.0139	.0136	.0132	.0129	.0125	.0122	.0119	.0116	.0113	.0110
2.3	.0107	.0104	.0102	.0099	.0096	.0094	.0091	.0089	.0087	.0084
2.4	.0082	.0080	.0078	.0075	.0073	.0071	.0069	.0068	.0066	.0064
2.5	.0062	.0060	.0059	.0057	.0055	.0054	.0052	.0051	.0049	.0048
2.6	.0047	.0045	.0044	.0043	.0041	.0040	.0039	.0038	.0037	.0036
2.7	.0035	.0034	.0033	.0032	.0031	.0030	.0029	.0028	.0027	.0026
2.8	.0026	.0025	.0024	.0023	.0023	.0022	.0021	.0021	.0020	.0019
2.9	.0019	.0018	.0018	.0017	.0016	.0016	.0015	.0015	.0014	.0014
3.0	.0013	.0013	.0013	.0012	.0012	.0011	.0011	.0011	.0010	.0010

Source: Based on *Biometrika Tables for Statisticians*, Vol. 1, 3rd ed. (1966), with the permission of the *Biometrika* trustees.

Note: The table plots the cumulative probability $Z > z$.

Table 2 Percentiles of the *t*-Distribution

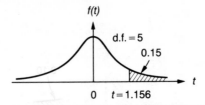

Degrees of freedom	Probability of a value at least as large as the table entry					
	0.15	0.1	0.05	0.025	0.01	0.005
1	1.963	3.078	6.314	12.706	31.821	63.657
2	1.386	1.886	2.920	4.303	6.965	9.925
3	1.250	1.638	2.353	3.182	4.541	5.841
4	1.190	1.533	2.132	2.776	3.747	4.604
5	1.156	1.476	2.015	2.571	3.365	4.032
6	1.134	1.440	1.943	2.447	3.143	3.707
7	1.119	1.415	1.895	2.365	2.998	3.499
8	1.108	1.397	1.860	2.306	2.896	3.355
9	1.100	1.383	1.833	2.262	2.821	3.250
10	1.093	1.372	1.812	2.228	2.764	3.169
11	1.088	1.363	1.796	2.201	2.718	3.106
12	1.083	1.356	1.782	2.179	2.681	3.055
13	1.079	1.350	1.771	2.160	2.650	3.012
14	1.076	1.345	1.761	2.145	2.624	2.977
15	1.074	1.341	1.753	2.131	2.602	2.947
16	1.071	1.337	1.746	2.120	2.583	2.921
17	1.069	1.333	1.740	2.110	2.567	2.898
18	1.067	1.330	1.734	2.101	2.552	2.878
19	1.066	1.328	1.729	2.093	2.539	2.861
20	1.064	1.325	1.725	2.086	2.528	2.845
21	1.063	1.323	1.721	2.080	2.518	2.831
22	1.061	1.321	1.717	2.074	2.508	2.819
23	1.060	1.319	1.714	2.069	2.500	2.807
24	1.059	1.318	1.711	2.064	2.492	2.797
25	1.058	1.316	1.708	2.060	2.485	2.787
26	1.058	1.315	1.706	2.056	2.479	2.779
27	1.057	1.314	1.703	2.052	2.473	2.771
28	1.056	1.313	1.701	2.048	2.467	2.763
29	1.055	1.311	1.699	2.045	2.462	2.756
30	1.055	1.310	1.697	2.042	2.457	2.750
(Normal) ∞	1.036	1.282	1.645	1.960	2.326	2.576

Source: Abridged from Table IV in Sir Ronald A. Fisher, *Statistical Methods for Research Workers*, 14th ed. (copyright © 1970 by University of Adelaide, a Division of Macmillan Publishing Co., Inc.) with the permission of the publisher and the late Sir Ronald Fisher's Literary Executor.

Table 3   Percentiles of the Chi-Squared Distribution

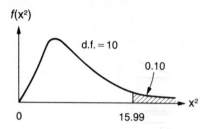

Degrees of freedom	Probability of a value at least as large as the table entry								
	0.90	0.75	0.50	0.25	0.10	0.05	0.025	0.01	0.005
1	0.0158	0.102	0.455	1.323	2.71	3.84	5.02	6.63	7.88
2	0.211	0.575	1.386	2.77	4.61	5.99	7.38	9.21	10.60
3	0.584	1.213	2.37	4.11	6.25	7.81	9.35	11.34	12.84
4	1.064	1.923	3.36	5.39	7.78	9.49	11.14	13.28	14.86
5	1.610	2.67	4.35	6.63	9.24	11.07	12.83	15.09	16.75
6	2.20	3.45	5.35	7.84	10.64	12.59	14.45	16.81	18.55
7	2.83	4.25	6.35	9.04	12.02	14.07	16.01	18.48	20.3
8	3.49	5.07	7.34	10.22	13.36	15.51	17.53	20.1	22.0
9	4.17	5.90	8.34	11.39	14.68	16.92	19.02	21.7	23.6
10	4.87	6.74	9.34	12.55	(15.99)	18.31	20.5	23.2	25.2
11	5.58	7.58	10.34	13.70	17.28	19.68	21.9	24.7	26.8
12	6.30	8.44	11.34	14.85	18.55	21.0	23.3	26.2	28.3
13	7.04	9.30	12.34	15.98	19.81	22.4	24.7	27.7	29.8
14	7.79	10.17	13.34	17.12	12.1	23.7	26.1	29.1	31.3
15	8.55	11.04	14.34	18.25	22.3	25.0	27.5	30.6	32.8
16	9.31	11.91	15.34	19.37	23.5	26.3	28.8	32.0	34.3
17	10.09	12.79	16.34	20.5	24.8	27.6	30.2	33.4	35.7
18	10.86	13.68	17.34	21.6	26.0	28.9	31.5	34.8	37.2
19	11.65	14.56	18.34	22.7	27.2	30.1	32.9	36.2	38.6
20	12.44	15.45	19.34	23.8	28.4	31.4	34.2	37.6	40.0

Source: Based on *Biometrika Tables for Statisticians*, Vol. 1, 3rd ed. (1966), with the permission of the *Biometrika* trustees.

Table 4   *F*-Distribution, 5 Percent Significance

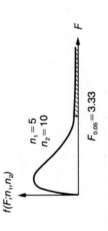

$f(F;n_1,n_2)$

$n_1 = 5$
$n_2 = 10$

$F_{0.05} = 3.33$

Degrees of freedom for denominator

	1	2	3	4	5	6	7	8	9	10	12	15	20	24	30	40	60	120	∞
1	161	200	216	225	230	234	237	239	241	242	244	246	248	249	250	251	252	253	254
2	18.5	19.0	19.2	19.2	19.3	19.3	19.4	19.4	19.4	19.4	19.4	19.4	19.5	19.5	19.5	19.5	19.5	19.5	19.5
3	10.1	9.55	9.28	9.12	9.01	8.94	8.89	8.85	8.81	8.79	8.74	8.70	8.66	8.64	8.62	8.59	8.57	8.55	8.53
4	7.71	6.94	6.59	6.39	6.26	6.16	6.09	6.04	6.00	5.96	5.91	5.86	5.80	5.77	5.75	5.72	5.69	5.66	5.63
5	6.61	5.79	5.41	5.19	5.05	4.95	4.88	4.82	4.77	4.74	4.68	4.62	4.56	4.53	4.50	4.46	4.43	4.40	4.37
6	5.99	5.14	4.76	4.53	4.39	4.28	4.21	4.15	4.10	4.06	4.00	3.94	3.87	3.84	3.81	3.77	3.74	3.70	3.67
7	5.59	4.74	4.35	4.12	3.97	3.87	3.79	3.73	3.68	3.64	3.57	3.51	3.44	3.41	3.38	3.34	3.30	3.27	3.23
8	5.32	4.46	4.07	3.84	3.69	3.58	3.50	3.44	3.39	3.35	3.28	3.22	3.15	3.12	3.08	3.04	3.01	2.97	2.93
9	5.12	4.26	3.86	3.63	3.48	3.37	3.29	3.23	3.18	3.14	3.07	3.01	2.94	2.90	2.86	2.83	2.79	2.75	2.71
10	4.96	4.10	3.71	3.48	3.33	3.22	3.14	3.07	3.02	2.98	2.91	2.85	2.77	2.74	2.70	2.66	2.62	2.58	2.54
11	4.84	3.98	3.59	3.36	3.20	3.09	3.01	2.95	2.90	2.85	2.79	2.72	2.65	2.61	2.57	2.53	2.49	2.45	2.40
12	4.75	3.89	3.49	3.26	3.11	3.00	2.91	2.85	2.80	2.75	2.69	2.62	2.54	2.51	2.47	2.43	2.38	2.34	2.30
13	4.67	3.81	3.41	3.18	3.03	2.92	2.83	2.77	2.71	2.67	2.60	2.53	2.46	2.42	2.38	2.34	2.30	2.25	2.21
14	4.60	3.74	3.34	3.11	2.96	2.85	2.76	2.70	2.65	2.60	2.53	2.46	2.39	2.35	2.31	2.27	2.22	2.18	2.13
15	4.54	3.68	3.29	3.06	2.90	2.79	2.71	2.64	2.59	2.54	2.48	2.40	2.33	2.29	2.25	2.20	2.16	2.11	2.07

Table 4 (continued)

						Degrees of freedom for denominator												
	1	2	3	4	5	6	7	8	9	10	12	15	20	24	30	40	60	120	∞
16	4.49	3.63	3.24	3.01	2.85	2.74	2.66	2.59	2.54	2.49	2.42	2.35	2.28	2.24	2.19	2.15	2.11	2.06	2.01
17	4.45	3.59	3.20	2.96	2.81	2.70	2.61	2.55	2.48	2.45	2.38	2.31	2.23	2.19	2.15	2.10	2.06	2.01	1.96
18	4.41	3.55	3.16	2.93	2.77	2.66	2.58	2.51	2.46	2.41	2.34	2.27	2.19	2.15	2.11	2.06	2.02	1.97	1.92
19	4.38	3.52	3.13	2.90	2.74	2.63	2.54	2.48	2.42	2.39	2.31	2.23	2.16	2.11	2.07	2.03	1.98	1.93	1.88
20	4.35	3.49	3.10	2.87	2.71	2.60	2.51	2.45	2.39	2.35	2.28	2.20	2.12	2.08	2.04	1.99	1.95	1.90	1.84
21	4.32	3.47	3.07	2.84	2.68	2.57	2.49	2.42	2.37	2.32	2.25	2.18	2.10	2.05	2.01	1.96	1.92	1.87	1.81
22	4.30	3.44	3.05	2.82	2.66	2.55	2.46	2.40	2.34	2.30	2.23	2.15	2.07	2.03	1.98	1.94	1.89	1.84	1.78
23	4.28	3.42	3.03	2.80	2.64	2.53	2.44	2.37	2.32	2.27	2.20	2.13	2.05	2.01	1.96	1.91	1.86	1.81	1.76
24	4.26	3.40	3.01	2.78	2.62	2.51	2.42	2.36	2.30	2.25	2.18	2.11	2.03	1.98	1.94	1.89	1.84	1.79	1.73
25	4.24	3.39	2.99	2.76	2.60	2.49	2.40	2.34	2.28	2.24	2.16	2.09	2.01	1.96	1.92	1.87	1.82	1.77	1.71
30	4.17	3.32	2.92	2.69	2.53	2.42	2.33	2.27	2.21	2.16	2.09	2.01	1.93	1.89	1.84	1.79	1.74	1.68	1.62
40	4.08	3.23	2.84	2.61	2.45	2.34	2.25	2.18	2.12	2.08	2.00	1.92	1.84	1.79	1.74	1.69	1.64	1.58	1.51
60	4.00	3.15	2.76	2.53	2.37	2.25	2.17	2.10	2.04	1.99	1.92	1.84	1.75	1.70	1.65	1.59	1.53	1.47	1.39
120	3.92	3.07	2.68	2.45	2.29	2.18	2.09	2.02	1.96	1.91	1.83	1.75	1.66	1.61	1.55	1.50	1.43	1.35	1.25
∞	3.84	3.00	2.60	2.37	2.21	2.10	2.01	1.94	1.88	1.83	1.75	1.67	1.57	1.52	1.46	1.39	1.32	1.22	1.00

Source: Reproduced with the permission of the Biometrika Trustees from M. Merrington and C. M. Thompson, "Tables of percentage points of the inverted beta (F) distribution," *Biometrika* 33(1943), 73.

Table 5 The Durbin-Watson Test Statistic d: 5 Percent
Significance of d_l and d_μ

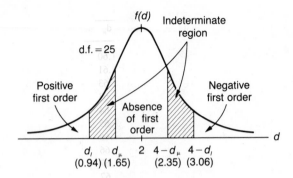

Degree of freedom	$k = 1$		$k = 2$		$k = 3$		$k = 4$		$k = 5$	
n	d_l	d_μ	d_l	d_μ	d_l	d_μ	d_l	d_μ	d_l	d_μ
15	0.95	1.23	0.83	1.40	0.71	1.61	0.59	1.84	0.48	2.09
16	0.98	1.24	0.86	1.40	0.75	1.59	0.64	1.80	0.53	2.03
17	1.01	1.25	0.90	1.40	0.79	1.58	0.68	1.77	0.57	1.98
18	1.03	1.26	0.93	1.40	0.82	1.56	0.72	1.74	0.62	1.93
19	1.06	1.28	0.96	1.41	0.86	1.55	0.76	1.73	0.66	1.90
20	1.08	1.28	0.99	1.41	0.89	1.55	0.79	1.72	0.70	1.87
21	1.10	1.30	1.01	1.41	0.92	1.54	0.83	1.69	0.73	1.84
22	1.12	1.31	1.04	1.42	0.95	1.54	0.86	1.68	0.77	1.82
23	1.14	1.32	1.06	1.42	0.97	1.54	0.89	1.67	0.80	1.80
24	1.16	1.33	1.08	1.43	1.00	1.54	0.91	1.66	0.83	1.79
25	1.18	1.34	1.10	1.43	1.02	1.54	(0.94)	(1.65)	0.86	1.77
26	1.19	1.35	1.12	1.44	1.04	1.54	0.96	1.65	0.88	1.76
27	1.21	1.36	1.13	1.44	1.06	1.54	0.99	1.64	0.91	1.75
28	1.22	1.37	1.15	1.45	1.08	1.54	1.01	1.64	0.93	1.74
29	1.24	1.38	1.17	1.45	1.10	1.54	1.03	1.63	0.96	1.73
30	1.25	1.38	1.18	1.46	1.12	1.54	1.05	1.63	0.98	1.73
31	1.26	1.39	1.20	1.47	1.13	1.55	1.07	1.63	1.00	1.72
32	1.27	1.40	1.21	1.47	1.15	1.55	1.08	1.63	1.02	1.71
33	1.28	1.41	1.22	1.48	1.16	1.55	1.10	1.63	1.04	1.71
34	1.29	1.41	1.24	1.48	1.17	1.55	1.12	1.63	1.06	1.70
35	1.30	1.42	1.25	1.48	1.19	1.55	1.13	1.63	1.07	1.70
36	1.31	1.43	1.26	1.49	1.20	1.56	1.15	1.63	1.09	1.70
37	1.32	1.43	1.27	1.49	1.21	1.56	1.16	1.62	1.10	1.70
38	1.33	1.44	1.28	1.50	1.23	1.56	1.17	1.62	1.12	1.70
39	1.34	1.44	1.29	1.50	1.24	1.56	1.19	1.63	1.13	1.69
40	1.35	1.45	1.30	1.51	1.25	1.57	1.20	1.63	1.15	1.69
45	1.39	1.48	1.34	1.53	1.30	1.58	1.25	1.63	1.21	1.69
50	1.42	1.50	1.38	1.54	1.34	1.59	1.30	1.64	1.26	1.69

Table 5   *(continued)*

Degree of freedom, n	$k = 1$		$k = 2$		$k = 3$		$k = 4$		$k = 5$	
	d_l	d_μ	d_l	d_μ	d_l	d_μ	d_l	d_μ	d_l	d_μ
55	1.45	1.52	1.41	1.56	1.37	1.60	1.33	1.64	1.30	1.69
60	1.47	1.54	1.44	1.57	1.40	1.61	1.37	1.65	1.33	1.69
65	1.49	1.55	1.46	1.59	1.43	1.63	1.40	1.66	1.36	1.69
70	1.51	1.57	1.48	1.60	1.45	1.63	1.42	1.66	1.39	1.70
75	1.53	1.58	1.50	1.61	1.47	1.64	1.45	1.67	1.42	1.70
80	1.54	1.59	1.52	1.63	1.49	1.65	1.47	1.67	1.44	1.70
85	1.56	1.60	1.53	1.63	1.51	1.66	1.49	1.68	1.46	1.71
90	1.57	1.61	1.55	1.64	1.53	1.66	1.50	1.69	1.48	1.71
95	1.58	1.62	1.56	1.65	1.54	1.67	1.52	1.69	1.50	1.71
100	1.59	1.63	1.57	1.65	1.55	1.67	1.53	1.70	1.51	1.72

Source: Reprinted with permission from J. Durbin and G. S. Watson, "Testing for Serial Correlation in Least Squares Regression: II" *Biometrika* 38(1951), 159–77.

Appendix B

Time series data used for examples in the text

	Toll revenue volumes for the telecommunications example (REV) for a given region					
	1969	1970	1971	1972	1973	1974
1	21821.	25157.	27163.	30653.	34306.	35120.
2	25490.	29910.	31927.	35349.	38617.	41843.
3	25565.	29082.	30484.	34388.	37187.	39401.
4	27819.	31526.	33327.	35342.	40623.	43236.
5	26427.	29165.	31325.	35585.	38490.	41350.
6	26259.	29508.	31036.	34040.	36847.	40786.
7	24614.	25886.	28622.	32044.	35398.	38596.
8	25792.	27840.	30660.	35063.	37774.	40991.
9	26970.	29082.	31147.	34365.	37335.	40425.
10	28605.	29900.	31572.	35634.	39074.	41656.
11	28987.	30651.	34576.	37389.	39642.	42991.
12	27023.	28678.	31502.	35060.	37362.	38881.
SUM	315372.	346385.	373341.	414912.	452655.	485276.
	1975	1976	1977	1978	1979	
1	36133.	37656.	41423.	45238.	50687.	
2	44003.	46263.	49410.	52583.	61547.	
3	41120.	44386.	46222.	50869.	57783.	
4	45671.	47650.	50690.	56078.	63420.	
5	43914.	45748.	48829.	54799.	60589.	
6	43167.	44886.	48112.	53109.	59672.	
7	40303.	42753.	45271.	50974.	57707.	
8	42452.	44843.	48678.	53265.	59672.	
9	42320.	45203.	49029.	54213.	59973.	
10	43825.	45995.	50191.	56892.	63458.	
11	44955.	47341.	51961.	58695.	65100.	
12	41121.	43658.	48199.	54142.	60500.	
SUM	508984.	536382.	578015.	640857.	720108.	

Monthly message volumes for the telecommunications example
(MSG) for a given region

	1969	1970	1971	1972	1973	1974
1	7600.	8391.	8775.	9851.	11003.	11638.
2	7630.	8244.	9079.	10268.	10988.	11575.
3	8131.	9388.	10017.	10893.	11874.	12456.
4	8104.	8946.	9719.	10285.	11609.	12437.
5	8207.	9102.	9638.	10858.	11986.	12807.
6	8577.	9594.	10343.	11408.	12067.	12651.
7	8150.	8946.	9404.	10159.	11242.	12101.
8	8583.	9333.	10175.	11321.	12393.	12855.
9	8662.	9491.	10205.	10993.	11845.	12685.
10	8862.	9340.	9916.	11293.	12311.	12973.
11	8175.	8961.	10003.	10946.	11885.	11955.
12	8444.	9084.	9583.	10510.	11307.	11739.
SUM	99125.	108820.	116857.	128785.	140510.	147872.

	1975	1976	1977	1978	1979
1	12330.	12572.	13621.	15299.	17092.
2	12029.	12857.	13681.	15195.	16894.
3	12972.	13882.	15172.	16782.	18210.
4	12759.	13531.	14285.	15552.	17746.
5	12810.	13375.	14561.	16364.	18176. -
6	13286.	14110.	15277.	16939.	18100.
7	12488.	12869.	13828.	15115.	16853.
8	13054.	13897.	15402.	17052.	18498.
9	13162.	13997.	15257.	16777.	
10	13266.	13690.	15032.	16928.	
11	12044.	13474.	14700.	16498.	
12	12377.	13171.	14190.	15539.	
SUM	152577.	161425.	175006.	194040.	141569.

	Number of business telephones for the telecommunications example (BMT) for a given region					
	1970	1971	1972	1973	1974	1975
1	491,195	502,336	504,071	511,665	518,522	517,100
2	493,068	503,013	503,785	512,555	518,797	516,520
3	495,482	504,042	505,736	513,653	519,290	516,136
4	498,042	505,292	507,537	514,959	520,869	515,213
5	499,660	504,720	507,267	514,576	519,606	513,040
6	499,004	504,378	506,704	515,345	520,006	512,909
7	499,602	503,571	507,436	515,712	519,274	512,306
8	500,065	503,158	508,378	517,240	520,072	512,918
9	502,858	505,757	510,861	519,132	521,234	514,679
10	502,468	504,828	511,251	519,159	520,631	514,246
11	502,589	503,905	510,768	518,610	519,474	513,864
12	502,504	503,434	510,505	518,296	517,935	512,591
SUM	5,986,540	6,048,433	6,094,300	6,190,904	6,235,709	6,171,521

	1976	1977	1978	1979
1	512,615	512,286	515,163	520,438
2	512,553	512,363	515,423	521,051
3	512,702	512,626	516,766	522,419
4	513,165	513,296	517,439	523,345
5	510,701	511,400	515,357	521,849
6	510,229	511,621	515,885	522,547
7	510,610	511,656	516,345	522,333
8	510,771	512,889	517,708	523,943
9	512,981	514,907	520,203	526,622
10	513,713	515,563	521,043	526,921
11	512,950	515,422	520,522	526,194
12	512,042	514,829	519,938	525,297
SUM	6,145,032	6,158,857	6,211,791	6,282,958

	Nonfarm employment for the telecommunications example (NFRM) for a given region					
	1970	1971	1972	1973	1974	1975
1	7063	6926	6866	6977	6985	6792
2	7092	6929	6871	6995	6974	6763
3	7157	6982	6960	7059	7024	6782
4	7202	7007	7000	7102	7064	6801
5	7221	7059	7062	7149	7130	6865
6	7282	7112	7122	7224	7196	6903
7	7191	7050	7020	7154	7135	6866
8	7189	7022	7083	7191	7149	6900
9	7140	6973	7045	7148	7080	6806
10	7106	6982	7120	7181	7087	6818
11	7102	7000	7149	7204	7080	6827
12	7116	7024	7164	7200	7022	6839
SUM	85861	84066	84462	85584	84926	81962
	1976	1977	1978	1979		
1	6666	6645	6804	6993		
2	6665	6662	6816	7011		
3	6715	6718	6903	7065		
4	6758	6781	6984	7122		
5	6781	6847	7055	7196		
6	6837	6920	7137	7269		
7	6883	6916	7086	7220		
8	6868	6948	7133	7235		
9	6806	6920	7102	7202		
10	6816	6957	7142	7243		
11	6834	6980	7188	7267		
12	6845	6998	7192	7272		
SUM	81474	82292	84542	86095		

Nonfarm less manufacturing employment for the telecommunications example
NFMA, for a given region

	1970	1971	1972	1973	1974	1975
1	5275	5282	5295	5393	5413	5365
2	5279	5279	5285	5391	5393	5344
3	5336	5326	5361	5443	5436	5365
4	5403	5365	5406	5495	5479	5393
5	5439	5418	5465	5534	5540	5457
6	5489	5465	5507	5589	5588	5483
7	5463	5457	5469	5558	5570	5474
8	5429	5397	5472	5560	5558	5478
9	5380	5335	5420	5506	5486	5362
10	5400	5350	5491	5541	5511	5374
11	5411	5368	5516	5562	5525	5387
12	5427	5421	5548	5586	5531	5417
SUM	64731	64463	65235	66157	66031	64897

	1976	1977	1978	1979
1	5267	5231	5365	5527
2	5252	5237	5370	5528
3	5286	5276	5437	5575
4	5325	5332	5514	5631
5	5340	5389	5578	5696
6	5382	5442	5636	5748
7	5458	5472	5624	5727
8	5418	5475	5636	5730
9	5342	5436	5603	5686
10	5358	5468	5641	5725
11	5375	5490	5673	5758
12	5406	5527	5693	5777
SUM	64207	64776	66768	68109

BIBLIOGRAPHY

[numbers in brackets refer to page number in this book]

ALLEN, R. E. D. (1975). *Index Numbers in Theory and Practice*. Chicago, IL: Aldine Publishing Co.

ALMON, S. (1965). The Distributed Lag Between Capital Appropriations and Expenditures. *Econometrica* 33, 178–96. [165]

ANDERSON, O. D. (1976). *Time Series Analysis and Forecasting, The Box-Jenkins Approach*. London, England: Butterworth. [220, 251, 262]

ANDERSON, R. L. (1942). Distribution of the Serial Correlation Coefficient. *Annals of Mathematical Statistics* 13, 1–13. [251]

ANDREWS, D. F., P. J. BICKEL, F. R. HAMPEL, P. J. HUBER, W. H. ROGERS, and J. W. TUKEY (1972). *Robust Estimates of Location: Survey and Advances*. Princeton, NJ: Princeton University Press.

ARMSTRONG, J. S. (1978). Forecasting with Econometric Methods: Folklore versus Fact. *Journal of Business* 51, 549–64. [12, 154, 197]

ARMSTRONG, J. S. (1978). *Long-Range Forecasting: From Crystal Ball to Computer*. New York, NY: John Wiley and Sons. [78]
This book concentrates on methods for long-range forecasting that are applicable to all areas of social, behavioral, and management sciences.

ARMSTRONG, J. S., and M. C. GROHMAN (1972). A Comparative Study of Methods for Long-Range Forecasting. *Management Science* 19, 211–21.

ARTLE, R., and C. AVEROUS (1973). The Telephone System as a Public Good: Static and Dynamic Aspects. *The Bell Journal of Economics and Management Science* 4, 89–100.

BALESTRA, P., AND M. NERLOVE (1966). Pooling Cross Section and Time Series Data in the Estimation of a Dynamic Model: The Demand for Natural Gas. *Econometrica* 31, 585–612. [201]

BARTLETT, M. S. (1935). Some Aspects of the Time-Correlation Problem in Regard to Tests of Significance. *Journal of the Royal Statistical Society* B 98, 536–43. [326]

BARTLETT, M. S. (1946). On the Theoretical Specifications of the Sampling Properties of Autocorrelated Time Series. *Journal of the Royal Statistical Society* B 8, 27–41. [250]

BASS, F. M., and R. R. WITTINK (1975). Pooling Issues and Methods in Regression Analysis with Examples in Marketing Research. *Journal of Marketing Research* 12, 414–25. [201]

BAUMOL, W. J. (1972). *Economic Theory and Operations Analysis.* Englewood Cliffs, NJ: Prentice-Hall.

BECKER, R. A. and J. M. CHAMBERS (1977). GR-Z: A System of Graphical Subroutines for Data Analysis. *Proceedings of Computer Science and Statistics, Tenth Annual Symposium on the Interface.* National Bureau of Standards Special Publication 503, 409–15.

BELSLEY, D. A., E. KUH, and R. E. WELSCH (1980). *Regression Diagnostics: Identifying Influential Data and Sources of Collinearity.* New York, NY: John Wiley and Sons. [49, 181, 184]

Presents detailed discussions of diagnostic techniques useful in regression analysis, to identify influential or unusual subsets of data points and outliers, and to identify the type and extent of multicollinearity among the regression variables.

BHATTACHARYYA, M. N. (1974). Forecasting the Demand for Telephones in Australia. *Applied Statistics* 23, 1–10. [319]

Estimates an ARIMA model over the period from July 1962 through June 1971, in which the dependent variable is the 12-month percentage change in the number of new telephones installed and the independent variable is the 12-month percentage change in the connection charge plus annual rental.

BLOOMFIELD, P. (1976). *Fourier Analysis of Time Series—An Introduction.* New York, NY: John Wiley and Sons.

Treats time series forecasting methodologies of increasing complexity, including harmonic regression, the fast Fourier transform, complex demodulation, and spectrum analysis.

BOWERMAN, B. L., and R. T. O'CONNELL (1979). *Time Series and Forecasting.* North Scituate, MA: Duxbury Press. [297, 300]

BOX, G. E. P. (1966). Use and Abuse of Regression. *Technometrics* 8, 625–29.

BOX, G. E. P., and D. R. COX (1964). An Analysis of Transformations. *Journal of the Royal Statistical Society* B 26, 211–52. [35]

BOX, G. E. P., and G. M. JENKINS (1976). *Time Series Analysis—Forecasting and Control, Revised Edition.* San Francisco, CA: Holden-Day. [38, 220, 232, 236, 239, 251, 262, 263, 276, 278, 297, 323, 326, 330]

BOX, G. E. P., G. M. JENKINS, and D. W. BACON (1967). Models for Forecasting Seasonal and Nonseasonal Time Series. *In* B. Harris, ed. *Advanced Seminar on Spectral Analysis.* New York, NY: John Wiley and Sons. [319]

BOX, G. E. P., and D. A. PIERCE (1970). Distribution of Residual Autocorrelations in Autoregressive Integrated Moving-Average Time-Series Models. *Journal of the American Statistical Association* 65, 1509–26. [238]

BOX, G. E. P., and G. C. TIAO (1975). Intervention Analysis with Applications to Economic and Environmental Problems. *Journal of the American Statistical Association* 70, 70–79. [348]

BRELSFORD, W. M., and D. A. RELLES (1981). *STATLIB—A Statistical Computing Library*. Englewood Cliffs, NJ: Prentice-Hall.

BRILLINGER, D. R. (1981). *Time Series—Data Analysis and Theory*. Expanded edition. San Francisco, CA: Holden-Day.

Deals with theoretical aspects of time series forecasting useful for graduate level courses and as reference source for researchers.

BROWN, R. G. (1959). *Statistical Forecasting for Inventory Control*. New York, NY: McGraw-Hill.

BROWN, R. G. (1963). *Smoothing, Forecasting, and Prediction of Discrete Time Series*. Englewood Cliffs, NJ: Prentice Hall.

BRUBACHER, S., and G. T. WILSON (1976). Interpolating Time Series with Applications to the Estimation of Holiday Effects on Electricity Demand. *Applied Statistics* 25, 107–16. [319]

BUTLER, W. F., R. A. KAVESH, and R. B. PLATT, eds. (1974). *Methods and Techniques of Business Forecasting*. Englewood Cliffs, NJ: Prentice-Hall. [17]

CHAMBERS, J. C., S. K. MULLICK, and D. D. SMITH (1971). How to Choose the Right Forecasting Technique. *Harvard Business Review* 49, 45–74.

CHAMBERS, J. C., S. K. MULLICK, and D. D. SMITH (1974). *An Executive's Guide to Forecasting*. New York, NY: John Wiley and Sons. [4]

A nontechnical description of forecasting aimed at practitioners who are concerned with how different forecasting techniques can be used.

CHATFIELD, C. (1980). *The Analysis of Time Series: An Introduction, 2nd Ed*. New York, NY: Chapman and Hall.

An introductory book intended to bridge the gap between theory and practice in time series analysis.

CHATFIELD, C. (1978). Adaptive Filtering: A Critical Assessment. *Journal of Operational Research Society* 29, 891–96.

CHATFIELD, C. (1978). The Holt-Winters Forecasting Procedure. Applied Statistics 27, 264–79.

CHATFIELD, C., and D. L. PROTHERO (1973). Box-Jenkins Seasonal Forecasting: Problems in a Case Study. *Journal of the Royal Statistical Society* A 136, 295–336. [319]

CHATTERJEE, S., and B. PRICE (1977). *Regression Analysis by Example*. New York, NY: John Wiley and Sons. [49, 50]

Emphasizes informal data analysis techniques for exploring interrelationships among a given set of variables for developing regression equations.

CHOW, G. C. (1960). Tests of Equality between Sets of Coefficients in Two Linear Regressions. *Econometrica* 28, 591–605. [180]

CHRIST, C. F. (1975). Judging the Performance of Econometric Models of the U.S. Economy. *International Economic Review* 16, 54–74.

CLEVELAND, W. S., and S. J. DEVLIN (1980). Calendar Effects in Monthly Time Series; Detection by Spectrum. *Journal of the American Statistical Association* 75, 487–96. [38]

CLEVELAND, W. S., D. M. DUNN, and I. J. TERPENNING (1979). SABL— A Resistant Seasonal Adjustment Procedure with Graphical Methods for Interpretation and Diagnosis. *In* A. Zellner, ed., *Seasonal Analysis of Economic Time Series*, Washington, DC: U.S. Government Printing Office. [37]

CLEVELAND, W. S., and G. C. TIAO (1976). Decomposition of Seasonal Time Series: A Model for the Census X-11 Program. *Journal of the American Statistical Association* 71, 581–87. [37]

COCHRANE, D., and G. N. ORCUTT (1949). Application of Least Squares to Relationships Containing Autocorrelated Error Terms. *Journal of the American Statistical Association* 44, 32–61. [174]

COOPER, R. L. (1972). The Predictive Performance of Quarterly Econometric Models of the United States. *In* B. Hickman, ed., *Econometric Models of Cyclical Behavior*, 2. Conference on Research in Income and Wealth 36, New York, NY: Columbia University Press.

DAMSLETH, E., and B. H. SOLLIE (1980). Forecasting the Demand for New Telephone Main Stations in Norway by Using ARIMA Models. *In* O. D. Anderson, ed., *Forecasting Public Utilities*. Amsterdam, Netherlands: North-Holland Publishing Company. [348]

DANIEL, C., and F. S. WOOD (1977). *Fitting Equations to Data*. New York, NY: John Wiley and Sons.

DANIELS, L. M. (1980). *Business Forecasting for the 1980's—and Beyond*. Baker Library, Reference List, No. 31, Boston, MA: Harvard Business School.

This bibliography concentrates on forecasting books, services, and a few articles published since 1972. Most of the books are briefly annotated.

DAVIS, B. E., G. J. CACCAPPOLO, and M. A. CHAUDRY (1973). An Econometric Planning Model for American Telephone and Telegraph Company. *The Bell Journal of Economics and Management Science* 4, 29–56.

DEVLIN, S., R. GNANADESIKAN, and J. R. KETTENRING (1975). Robust Estimation and Outlier Detection with Correlation Coefficients. *Biometrika* 62, 531–45. [48]

DHRYMES, P. J. (1971). *Distributed Lags; Problems of Formulation and Estimation*, San Francisco, CA: Holden-Day. [162]

DOBELL, A. R., L. D. TAYLOR, L. WAVERMAN, T. H. LIU, and M. D. G. COPELAND (1972). Telephone Communications in Canada: Demand Production, and Investment Decisions. *The Bell Journal of Economics and Management Science* 3, 175–219.

DRAPER, N. R., and H. SMITH (1981). *Applied Regression Analysis, 2nd ed.* New York, NY: John Wiley and Sons. [43, 49, 50, 78, 184]

DUNN, D. M., W. H. WILLIAMS, and W. A. SPIVEY (1971). Analysis and Prediction of Telephone Demand in Local Geographical Areas. *The Bell Journal of Economics and Management Science* 2, 561–76.

Paper focuses on forecasting inward and outward movement of business and residence main stations in three Michigan cities over the period 1954–1968 by using monthly models including simple trend-seasonal-irregular time series decomposition and adaptive exponential smoothing with and without exogenous variables.

DURBIN, J. (1970). Testing for Serial Correlation in Least Squares Regression When Some of the Regressors are Lagged Dependent Variables. *Econometrica* 38, 410–21.

DURBIN, J., and G. S. WATSON (1950). Testing for Serial Correlation in Least Squares Regression: I. *Biometrika* 37, 409–28. [168]

DURBIN, J., and G. S. WATSON (1951). Testing for Serial Correlation in Least Squares Regression: II. *Biometrika* 38, 159–78. [168]

DURBIN, J., and G. S. WATSON (1971). Testing for Serial Correlation in Least Squares Regression III. *Biometrika* 58, 1–19.

ECKSTEIN, O., E. W. GREEN, and A. SINAI (1974). The Data Resources Model: Uses, Structure, and Analysis of the U.S. Economy. *International Economic Review* 15, 595–615.

ENRICK, N. L. (1979) *Market and Sales Forecasting—A Quantitative Approach.* Huntington, NY: Krieger Publishing Co.

EVERITT, B. S. (1978). *Graphical Techniques For Multivariate Data.* New York, NY: North-Holland Publishing Co.

FAIR, R. C. (1970). *A Short-Run Forecasting Model of the U.S. Economy.* Lexington, MA: D. C. Heath.

FARRAR, D. E., and R. R. GLAUBER (1967). Multicollinearity in Regression Analysis: The Problem Revisited. *Review of Economics and Statistics* 49, 92–107. [181]

FILDES, R. (1979). Quantitative Forecasting—The State of the Art: Extrapolative Models. *Journal of the Operational Research Society* 30, 691–710.

FILDES, R., and D. WOODS, eds. (1978). *Forecasting and Planning.* Farnborough, England: Saxon House.

A collection of readings dealing with the application of advanced forecasting techniques in practice.

FISHER, F. M. (1966). *The Identification Problem.* New York, NY: McGraw-Hill. [154, 162, 191, 197]

FULLER, W. A. (1976). *Introduction to Statistical Time Series.* New York, NY: John Wiley and Sons.

GEURTS, M. D., and I. B. IBRAHIM (1975). Comparing the Box-Jenkins Approach with the Exponentially Smoothed Forecasting Model Application to Hawaii Tourists. *Journal of Marketing Research,* 182–88.

GILCHRIST, W. (1976). *Statistical Forecasting.* New York, NY: John Wiley and Sons.

The author describes a number of forecasting methods including, linear trend models, growth curves, adaptive methods, and other extensions.

GLASS, G. V., V. L. WILLSON, and J. M. GOTTMAN (1975). *Design and Analysis of Time Series Experiments.* Boulder, CO: Colorado Associated University Press. [348]

GNANADESIKAN, R. (1977). *Methods for Statistical Data Analysis of Multivariate Observations.* New York, NY: John Wiley and Sons.

GOLDFELD, S. M., and R. E. QUANDT (1965). Some Tests for Homoscedasticity. *Journal of the American Statistical Association* 60, 539–47. [178]

GOODMAN, M. L. (1974). A New Look at Higher-order Exponential Smoothing for Forecasting. *Operations Research* 22, 880–88.

GRAFF, P. (1977). *Die Wirtshaftsprognose.* Tubingen, West Germany: J. C. B. Mohr.
Describes the analysis and utilization of long-term trends for forecasting. In German.

GRANGER, C. W. J. (1980). *Forecasting in Business and Economics.* New York, NY: Academic Press. [78, 251, 256]

GRANGER, C. W. J., and O. MORGENSTERN (1970). *Predictability of Stock Market Prices.* Lexington, MA: D. C. Heath.

GRANGER, C. W. J., and P. NEWBOLD (1977). *Forecasting Economic Time Series.* New York, NY: Academic Press. [6, 154, 197, 220, 251, 300]
Treats the analysis of economic data from the perspective of time series (ARIMA) methods and classical econometrics.

GRANGER, C. W. J., and P. NEWBOLD (1976). Forecasting Transformed Series. *Journal of the Royal Statistical Society* B 38, 189–203. [224]

GREGG, J. V., C. H. HASSELL, and J. T. RICHARDSON (1964). *Mathematical Trend Curves: An Aid to Forecasting.* ICI Monograph No. 1. Edinburgh, Scotland: Oliver and Boyd. [78]

GROFF, G. K. (1973). Empirical Comparison of Models for Short-range Forecasting. *Management Science* 20, 22–31.

GROSS, C. W., and R. T. PETERSON (1976). *Business Forecasting.* Boston, MA: Houghton Mifflin Company.
An introduction to business forecasting, with emphasis on forecasting at the firm and industry level, as opposed to aggregate economic forecasting.

HAITOVSKY, Y., and G. I. TREYZ (1972). Forecasts With Quarterly Macroeconometric Models, Equation Adjustments, and Benchmark Predictions: The U.S. Experience. *Review of Economic Statistics* 54, 317–25.

HAITOVSKY, Y., G. I. TREYZ, and V. SU (1974). *Forecasts with Quarterly Macroeconometric Models.* New York, NY: National Bureau of Economic Research.
Examines macroeconomic forecasting models in order to analyze the magnitude and source of errors in the resulting forecast.

HANNAN, E. J. (1970). *Multiple Time Series.* New York, NY: John Wiley and Sons.

HARRISON, P. J. (1965). Short-term Forecasting. *Applied Statistics* 14, 102–39. [302]

HARRISON, P. J., and C. F. STEVENS (1971). A Bayesian Approach to Short-term Forecasting. *Operational Research Quarterly* 22, 341–62.

HARRISON, P. J., and O. L. DAVIES (1964). The Use of Cumulative Sum (Cusum) Techniques for the Control of Routine Forecasts of Product Demand. *Operations Research* 12, 325–33.

HARRISON, P. J., and C. F. STEVENS (1976). Bayesian Forecasting (with discussion). *Journal of the Royal Statistical Society* B 38, 205–47.

HELMER, R. A., and J. K. JOHANSSON (1977). An Exposition of the Box-Jenkins Transfer Function Analysis with Application to the Advertising-Sales Relationship. *Journal of Marketing Research* 14, 227–39.

HICKMAN, B. G. (1975). *Econometric Models of Cyclical Behavior*. New York, NY: Columbia University Press. [6]

HILDRETH, G., and J. Y. LU (1960). *Demand Relations with Autocorrelated Disturbances*. Agricultural Experiment Station, Technical Bulletin 276. Lansing, MI: Michigan State University. [174]

HIRSCH, A. A. (1973). The B. E. A. Quarterly Model As a Forecasting Instrument. *Survey of Current Business* 53, 24–38.

HOCKING, R. R. (1976). The Analysis and Selection of Variables in Linear Regression. *Biometrics* 32, 1–49.
Problems of subset selection and variable analysis in linear regression are reviewed. The discussion covers the underlying theory, computational techniques, and selection criteria.

HOERL, A. E., and R. W. KENNARD (1970). Ridge Regression: Biased Estimation for Nonorthogonal Problems. *Technometrics* 12, 55–67.
An historical paper on ridge regression.

HOLT, C. C. (1957). *Forecasting Trends and Seasonals by Exponentially Weighted Moving Averages*. O.N.R. Memorandum No. 52. Pittsburgh, PA: Carnegie Institute of Technology. [295]

HOTELLING, H. (1927). Differential Equations Subject to Error and Population Estimates. *Journal of the American Statistical Association*, 22, 283–314. [81]

INTRILIGATOR, M. D. (1978). *Econometric Models, Techniques, and Applications*. Englewood Cliffs, NJ: Prentice-Hall.

JENKINS, G. M. (1979). *Practical Experiences with Modeling and Forecasting Time Series*. Jersey, Channel Islands: GJ&P (Overseas) Ltd. [12, 323, 330, 348, 349, 361]

JENKINS, G. M., and D. G. WATTS (1968). *Spectral Analysis and its Applications*. San Francisco, CA: Holden-Day. [238]

JENSEN, D. (1979). General Telephone of the Northwest: A Forecasting Case Study. *Journal of Contemporary Business* 8, 19–34. [319]

JOHNSTON, J. (1972). *Econometric Methods*, 2nd edition, New York, NY: McGraw-Hill. [154, 168, 169, 181, 184, 186, 191, 204]

KENDALL, M. (1975). *Multivariate Analysis, second edition*. London, England: Charles Griffin and Company, Ltd.

Describes the full range of multivariate statistical techniques, including regression analysis, with an orientation to applications.

KENDALL, M. (1976) *Time Series*, 2nd ed. London, England: Charles Griffin and Company.

KLEIN, L. R. (1971). *An Essay on the Theory of Economic Prediction*. Chicago, IL: Markham.

KLEIN, L. R. (1971). Forecasting and Policy Evaluation Using Large Scale Econometric Models: The State of the Art. *In* M. D. Intriligator, ed., *Frontiers of Quantitative Economics*. Amsterdam, Netherlands: North-Holland Publishing Company.

KMENTA, J. (1971). *Elements of Econometrics*. New York, NY: MacMillan Publishing Co. [204, 213]

KLEIN, L. R., and R. M. YOUNG (1980). *An Introduction to Econometric Forecasting and Forecasting Models*. Lexington, MA: Lexington Books, D. C. Heath and Company.

Description of the techniques and experiences from sixteen years of continuous forecasting and modeling experience of the Wharton forecasting group.

KOLMOGOROV, A. N. (1941). Interpolation and Extrapolation of Stationary Random Sequences. Bulletin Moscow University. *URRS. Ser. Math. 5*.

KOYCK, L. M. (1954). *Distributed Lags and Investment Analysis*. Amsterdam, Netherlands: North-Holland Publishing Co. [162]

LARSEN, W. A., and S. J. McCLEARY (1972). The Use of Partial Residual Plots in Regression Analysis. *Technometrics* 14, 781–90. [62]

LEUTHOLD, R. M., A. J. MACCORMICK, A. SCHMITZ, and D. G. WATTS (1970). Forecasting Daily Hog Prices and Quantities: A Study of Alternative Forecasting Techniques. *Journal of the American Statistical Association* 65, 90–107.

LEVENBACH, H. (1980). A Comparative Study of Time Series Models for Forecasting Telephone Demand. *In* O. D. Anderson, ed., *Forecasting Public Utilities*. Amsterdam, Netherlands: North-Holland Publishing Co. [255, 309]

LEVENBACH, H., and J. P. CLEARY (1981). *The Beginning Forecaster*. Belmont, CA: Lifetime Learning Publications. [4, 26, 37, 55, 72, 160, 178, 181, 274, 297]

LEVENBACH, H., and B. E. REUTER (1976). Forecasting Trending Time Series with Relative Growth Rate Models. *Technometrics* 18, 261–72. [82, 83]

LEWIS, C. E. (1975). *Demand Analysis and Inventory Control*. Lexington, MA: Saxon House/Lexington Books.

Deals with short-term forecasting methodologies suitable for production planning and control systems as well as the interface between the forecasting system and the production planning and inventory control system.

LINSTONE, H. A., and D. SAHAL, eds. (1976). *Technological Substitution—Forecasting Techniques and Applications*. New York, NY: American Elsevier.

LINSTONE, H. A., and M. TUROFF, eds. (1975). *The Delphi Method*. Reading, MA: Addison-Wesley.

LITTLECHILD, S. C. (1975). Two-part Tariffs and Consumption Externalities. *The Bell Journal of Economics* 6, 661–670.

LIU, L. (1980). Analysis of Time Series with Calendar Effects. *Management Science* 26, 106–12. [348]

LJUNG, G. M., and G. E. P. BOX (1976). *A Modification of the Overall χ^2 Test for Lack of Fit in Time Series Models*. Technical Report 477, Department of Statistics. Madison, WI: University of Wisconsin.

LJUNG, G. M., and G. E. P. BOX (1978). On a Measure of Fit in Time Series Models. *Biometrika* 65, 297–303. [238]

MABERT, V. A. (1975). *An Introduction to Short-term Forecasting Using the Box-Jenkins Methodology*. Production Planning and Control Monograph Series No. 2. Atlanta, GA: American Institute of Industrial Engineers. [220, 251]

MADDALA, G. S. (1971). The Use of Variance Components Models in Pooling Cross Section and Time Series Data. *Econometrica* 39, 341–58. [201]

MADDALA, G. S. (1977). *Econometrics*. New York, NY: McGraw-Hill. [204]

MAKRIDAKIS, S., and M. HIBON (1979). The Accuracy of Forecasting: An Empirical Investigation (with Discussion). *Journal of the Royal Statistical Association* A 142, 97–145.

MAKRIDAKIS, S., and S. C. WHEELWRIGHT (1977). Adaptive Filtering: An Integrated Autoregressive/Moving Average Filter for Time Series Forecasting. *Operational Research Quarterly* 28, 425–37.

MAKRIDAKIS, S., and S. C. WHEELWRIGHT (1977). Forecasting: Issues and Challenges for Marketing Management, *Journal of Marketing*, 24–38.

MAKRIDAKIS, S., and S. C. WHEELWRIGHT (1978). *Interactive Forecasting—Univariate and Multivariate Methods*. San Francisco, CA: Holden-Day.

A basic description of a wide range of both univariate and multivariate time series and regression forecasting methodologies. Has practical use as a user's manual for the SIBYL/RUNNER interactive forecasting computer programs.

MAKRIDAKIS, S., and S. C. WHEELWRIGHT (1978). *Forecasting—Methods and Applications*. Santa Barbara, CA: Wiley-Hamilton. [38, 220, 251, 300]

A comprehensive text describing the fundamentals of a wide range of forecasting methodologies, including numerous examples and sample problems.

MAKRIDAKIS, S., and S. C. WHEELWRIGHT, eds. (1979). *Forecasting*. New York, NY: North-Holland Publishing Co.

A special issue of *Management Science* dealing with current forecasting methodologies for both practitioners and researchers.

MARTINO, J. P. (1972). *Technological Forecasting for Decision making*. New York, NY: American Elsevier.

MARTINO, J. P. (1972). *An Introduction to Technological Forecasting*. London, England: Gordon and Breach Science Publishers.

MASS, N. J. (1975). *Economic Cycles: An Analysis of Underlying Causes*. Cambridge, MA: Write-Allen Press.

Describes a series of system dynamics models (of the Forrester-type) used to explore the basic factors underlying short-term and long-term cyclical movements in the economy.

McCARTHY, M. D. (1972). *The Wharton Quarterly Econometric Forecasting Model (Mark III)*. Philadelphia, PA: University of Pennsylvania.

McLAUGHLIN, R. L. (1975). A New Five-Phase Economic Forecasting System. *The Journal of Business Economics,* 49–60.

McLAUGHLIN, R. L. (1979). Organizational Forecasting: Its Achievements and Limitations. *In* S. Makridakis and S. C. Wheelwright, eds., *Forecasting,* New York, NY: North-Holland Publishing Co.

McNEES, S. K. (1975). An Evaluation of Economic Forecasts. *New England Economic Review 4,* 3–39.

MEHRA, R. K. (1979). Kalman Filters and Their Applications to Forecasting. *In* S. Makridakis and S. C. Wheelwright, eds., *Forecasting*. New York, NY: North-Holland Publishing Co.

A survey of Kalman filtering techniques and their applications to forecasting. Includes a case study on forecasting of stock earnings for 49 U.S. companies, as well as numerous references.

MILNE, T. E. (1975). *Business Forecasting—A Managerial Approach*. London, England: Longman Group Ltd.

Aimed at the nontechnical or managerial reader interested in forecasting and seeking to gain a basic appreciation for the methodologies available and their application in practice.

MONTGOMERY, D. C., and L. A. JOHNSON (1976). *Forecasting and Time Series Analysis*. New York, NY: McGraw-Hill. [297, 300]

Covers the full range of short-term forecasting methods, including exponential smoothing methods, the Box-Jenkins approach, and Bayesian forecasting. Computer programs for several exponential smoothing methods are included.

MOSTELLER, F., and J. W. TUKEY (1977). *Data Analysis and Regression*. Reading, MA: Addison-Wesley. [50, 71, 72]

MUNDLAK, Y. (1978). On the Pooling of Time Series and Cross Section Data. *Econometrica* 46, 69–85. [201]

NAYLOR, T. H., T. G. SEAKS, and D. W. WICHERN (1972). Box-Jenkins Methods: An Alternative Approach to Econometric Models. *International Statistical Review 40,* 123–37.

NELSON, C. R. (1972). The Prediction Performance of the FRB-MIT-PENN Model of the U.S. Economy. *American Economic Review 62,* 902–17.

NELSON, C. R. (1973). *Applied Time Series Analysis for Managerial Forecasting*. San Francisco, CA: Holden-Day. [220, 251, 262]

NEWBOLD, P. (1979). Time-Series Model Building and Forecasting: A Survey. *In* S. Makridakis and S. C. Wheelwright, eds., *Forecasting*. New York, NY: North-Holland Publishing Company.
Reviews Box Jenkins methodology, surveys important developments since 1970, and indicates the range of applications of the procedures in practical forecasting problems.

NEWBOLD, P., and C. W. J. GRANGER (1974). Experience with Forecasting Univariate Time Series and the Combination of Forecasts. *Journal of the Royal Statistical Society* A 137, 131–46.

NEWBOLD, P., and G. V. REED (1979). The Implication For Economic Forecasting of Time Series Model Building Methods. *In* O. D. Anderson, ed., *Forecasting*, New York, NY: North-Holland Publishing Company.

OLIVER, F. R. (1964). Methods for Estimating the Logistic Growth Function. *Applied Statistics* 13, 57–66. [80]

PACK, D. J. (1977). Revealing Time Series Interrelationships. *Decision Sciences* 8, 377–402. [323]

PARKER, G. G. C., and E. L. SEGURA (1971). How to Get a Better Forecast, *Harvard Business Review,* 99–109.

PARSONS, L. J., and R. SCHULTZ (1976). *Marketing Models and Econometric Research.* New York, NY: North-Holland Publishing Co. [126]

PHLIPS, L. (1974). *Applied Consumption Analysis.* New York, NY: American Elsevier. [101, 126, 152]

PIERCE, D. A. (1980). A Survey of Recent Developments in Seasonal Adjustment. *The American Statistician* 34, 125–34. [38]

PINDYCK, R. S., and D. L. RUBINFELD (1976). *Econometric Models and Economic Forecasts.* New York, NY: McGraw-Hill. [120, 126, 154, 162, 190, 251]
Treats single- and multiple-equation models and time series models in detail.

POPKIN, J., ed. (1977). *Analysis of Inflation: 1965–1974.* Cambridge, MA: Ballinger Publishing Company for NBER.

PRIESTLEY, M. B. (1971). Fitting Relationships between Time Series. *Bulletin of the International Statistical Institute* 44, 295–324. [309]

PROTHERO, D. L., and K. F. WALLIS (1976). Modelling Macroeconomic Time Series (with Discussion). *Journal of the Royal Statistical Society* A 139, 468–86.

QUENOUILLE, M. H. (1949). Approximate Tests of Correlation in Time Series. *Journal of the Royal Statistical Society* B 11, 68–84. [251]

QUENOUILLE, M. H. (1957). *The Analysis of Multiple Time Series.* London, England: Griffin.

RAO, C. R. (1973). *Linear Statistical Inference and Its Applications.* New York, NY: John Wiley and Sons. [43]

RAO, P., and R. L. MILLER (1977). *Applied Econometrics.* Belmont, CA: Wadsworth Publishing Co.

RENTON, G. A., ed. (1975). *Modelling the Economy*. London, England: Heinemann Educational Books.

ROHLFS, J. (1974). A Theory of Interdependent Demand for a Communications Service. *The Bell Journal of Economics and Management Science*. 5, 16–37.

SAMUELSON, P. (1978). *Economics,* 9th ed. New York, NY: McGraw-Hill. [91, 101]

SEARLE, S. R. (1977). *Linear Models*. New York, NY: John Wiley and Sons. [43]

SEBER, G. F. (1977). *Linear Regression Analysis*. New York, NY: John Wiley and Sons. [43]

SHISKIN, J., A. H. YOUNG, and J. C. MUSGRAVE (1967). *The X-11 Variant of the Census Method-II Seasonal Adjustment Program:* Technical Paper No. 15, U.S. Department of Commerce, Bureau of the Census, Washington, DC: U.S. Government Printing Office. [37, 38]

SLUTSKY, E. (1937). The Summation of Random Causes as the Source of Cyclic Processes. *Econometrica* 5, 105–46. [220]

SOBEK, R. S. (1973). A Manager's Primer on Forecasting. *Harvard Business Review* 5, 1–9.

SQUIRE, L. (1973). Some Aspects of Optimal Pricing for Telecommunications. *The Bell Journal of Economics and Management Science* 4, 515–25.

STEKLER, H. O. (1970). *Economic Forecasting*. New York, NY: Praeger.

STRALKOWSKI, C. M., R. E. DEVOR, AND S. M. WU (1974). Charts for the Interpretation and Estimation of the Second-Order Moving Average and Mixed First-Order Autoregressive-Moving Average Models. *Technometrics* 16.

SULLIVAN, W. G., and W. W. CLAYCOMBE (1977). *Fundamentals of Forecasting*. Reston, VI: Reston Publishing Company. [300]
Enables the nonstatistical reader to apply popular forecasting techniques. Includes listings of computer programs that implement some of the simpler methodologies covered in the text.

SWAMY, P. A. V. B. (1970). Efficient Inference in a Random Coefficient Regression Model. *Econometrica* 38, 311–23. [201]

TAYLOR, L. D. (1975). The Demand for Electricity: A Survey. *The Bell Journal of Economics* 6, 74–110.

TAYLOR, L. D. (1980). *Telecommunications Demand: A Survey and Critique*. Cambridge, MA: Ballinger Press. [112]

THEIL, H. (1958). *Economic Forecasts and Policy*. Amsterdam, Netherlands: North-Holland Publishing Company.

THEIL, H. (1966). *Applied Economic Forecasting*. Amsterdam, Netherlands: North-Holland Publishing Company.

THOMAS, J. J., and K. F. WALLIS (1971). Seasonal Variation in Regression Analysis. *Journal of the Royal Statistical Society* A 134, 57–72. [302]

THOMOPOULOS, N. T. (1980). *Applied Forecasting Methods*. Englewood Cliffs, NJ: Prentice-Hall.

An introductory book on forecasting (primarily time series analysis) for practitioners and students in business administration, management science, and industrial engineering.

THOMPSON, H. E., and G. C. TIAO (1971). An Analysis of Telephone Data: A Case Study of Forecasting Seasonal Time Series. *The Bell Journal of Economics and Management Science* 2, 515–41. [319]

Develops ARIMA models for forecasting monthly telephone inward and outward movements in Wisconsin for the period from January 1951 through October 1966.

TOMASEK, O. (1972). Statistical Forecasting of Telephone Time Series. *Telecommunications Journal* 39, 1–7. [319]

Develops ARIMA models for forecasting monthly inward movement of telephone main stations in Montreal over the period 1961–70.

VON RABENAU, B., and K. STAHL (1974). Dynamic Aspects of Public Goods: A Further Analysis of the Telephone System. *The Bell Journal of Economics and Management Science* 5, 651–69.

WALL, K. D., A. J. PRESTON, J. W. BRAY, and M. H. PESTON (1975). Estimates of a Simple Control Model of the UK Economy. *In* G. A. Reaton, ed., *Modelling the Economy*. London, England: Heinemann. [348]

WALSH, J. E. (1962). *Handbook of Nonparametric Statistics*. Princeton, NJ: Van Nostrand.

WECKER, W. E. (1979). Predicting the Turning Points of a Time Series, *Journal of Business* 52, 35–50. [274]

WELLENIUS, B. (1970). A Method for Forecasting the Demand for Urban Residential Telephone Connections. *Telecommunication Journal* 37.

Develops a model for forecasting the residential demand for telephones in Santiago, Chile as a function of income.

WHEELWRIGHT, S. C., and D. G. CLARKE (1976). Corporate Forecasting: Promise and Reality, *Harvard Business Review* 54, 40–64.

WHEELWRIGHT, S. C., and S. MAKRIDAKIS (1980). *Forecasting Methods for Management, 3rd ed*. New York, NY: John Wiley and Sons. [7, 361]

For the practitioner who seeks to better understand the wide range of forecasting methods and their advantages and disadvantages.

WHITTLE, P. (1963). *Prediction and Regulation by Linear Least-Squares Methods*. London, England: English Universities Press.

WINTERS, P. R. (1960). Forecasting Sales by Exponentially Weighted Moving Averages. *Management Science* 6, 324–42. [295]

WOLD, H. (1954). *A Study in the Analysis of Stationary Time Series*, (First edition, 1938). Uppsala, Sweden: Almquist and Wiksell. [220]

WONNACOTT, R. J., and T. H. WONNACOTT (1979). *Econometrics, second edition*. New York, NY: John Wiley and Sons. [179, 192]

WONNACOTT, T. H., and R. J. WONNACOTT (1969). *Introductory Statistics*. New York, NY: John Wiley and Sons.

WOODS, D., and R. FILDES (1976). *Forecasting for Business, Methods and Applications*. New York, NY: Longman.

Aims to provide the practicing manager or student of management with the tools and the framework necessary to produce a good forecast without having to become a statistician or econometrician.

YULE, G. U. (1926). Why Do We Sometimes Get Nonsense Correlations Between Time Series? A Study In Sampling and The Nature of Time Series. *Journal of the Royal Statistical Society* B 89, 1–64. [220]

YULE, G. U. (1927). On a Method of Investigating Periodicities in Disturbed Series, with Special Reference to Wolfer's Sunspot Numbers. *Philosophical Transactions* A 226, 267–98.

ZELLNER, A. (1962). An Efficient Method of Estimating Seemingly Unrelated Regressions and Tests for Aggregation Bias. *Journal of the American Statistical Association* 57, 348–68. [201]

ZELLNER, A., and F. PALM (1974). Time Series Analysis and Simultaneous Equation Econometric Models. *Journal of Econometrics* 2, 17–54.

Index

DATE DUE

DEC APR 1 7 1990		
DEC 2 6 1983		
FEB 1 6 1984		
MAR 1 2 1985		
9. 9.'85		
APR 2 8 1986		
NOV 0 4 1986		
19'87		
MAR 2 7 1988		

DEMCO